# Innovations of Green Process Engineering for Sustainable Energy and Environment 2017



Topical Conference at the 2017 AIChE Annual Meeting

Minneapolis, Minnesota, USA
29 October – 3 November 2017

ISBN: 978-1-5108-5781-0

# Innovations of Green Process Engineering for Sustainable Energy and Environment 2017

Topical Conference at the 2017 AIChE Annual Meeting

Minneapolis, Minnesota, USA
29 October – 3 November 2017

AIChE
The Global Home of Chemical Engineers

**Printed from e-media with permission by:**

Curran Associates, Inc.
57 Morehouse Lane
Red Hook, NY 12571

**Some format issues inherent in the e-media version may also appear in this print version.**



Printed by Curran Associates, Inc. (2018)



AIChE
120 Wall Street, FL 23
New York, NY 10005-4020

Phone: (800) 242-4363
Fax: (203) 775-5177

www.aiche.org

**Additional copies of this publication are available from:**

Curran Associates, Inc.
57 Morehouse Lane
Red Hook, NY 12571 USA
Phone: 845-758-0400
Fax: 845-758-2633
Email: curran@proceedings.com
Web: www.proceedings.com

# TABLE OF CONTENTS

**Some abstracts in this book refer to images that were not made available by the publisher**

# 491710 Nitrogen Recovery and Biocrude Formation from Hydrothermal Liquefaction of Protein

**Abstract:**

*Nitrogen Recovery and Biocrude Formation from the Hydrothermal Liquefaction of Protein*

James D. Sheehan and Phillip E. Savage

Hydrothermal liquefaction is a thermochemical process that uses water at elevated temperatures (200-370°C) and pressures (10-22 MPa) to transform wet biomass into an energy dense "biocrude" oil and aqueous-phase nutrient coproducts.[1–3] The biocrude can be upgraded and co-refined with petroleum crudes into liquid fuels while the N- and P-containing nutrient coproducts can be recovered and recycled to cultivate additional biomass (e.g., microalgae). N is an essential element for cultivating microalgae and its demand can be met in part by recycling recovered N-containing nutrients from HTL, which are mostly derived from microalgal proteins. Proteins are major biochemical constituents of microalgae and contain approximately 90% of microalgal N.[4]The N atoms in the biomass proteins are ideally recovered as nutrients, however N-containing compounds can also partition into biocrude, which lowers the fuel quality and reduces the heating value of biocrude. Thus, the sustainability of HTL can be highly dependent on proteins and their transformation into valuable nutrient coproducts or detrimental N-containing compounds that partition into biocrude.

We performed HTL of bovine serum albumin (BSA), a well-defined model protein, and observed its transformation into valuable nutrient coproducts (e.g. amino acids, ammonia) and biocrude. We used temperatures ranging from 200 °C to 400 °C and holding times between 30 s to 1 hour. BSA readily decomposed during HTL into smaller peptides. The presence of peptides with molecular weights exceeding 5 kDa was no longer observable at temperatures at 250°C or higher. At 200°C, BSA aggregated into aqueous- and organic solvent-insoluble solids, in yields up to 80 wt%, that eventually decomposed into peptides which persevered in the hydrothermal media. The yields of amino acids were less than 6 wt% under all conditions, however, at temperatures of 350°C or higher the yields of primary and secondary amines in the aqueous-soluble products were upwards of 50 wt% and over 40% of N was recovered as ammonia, a preferred N-source for cultivating microalgae. The highest yields of biocrude were between 17 and 24 wt% and were achieved at temperatures ranging 300 to 400°C and holding times of 1 to 10 min. Mild (200°C) and severe (400°C, >10 min) conditions resulted with biocrude yields less than 5 wt%. This presentation will also include a discussion of the reaction pathways, kinetics, and mechanisms and thereby breaks new ground by connecting operationally defined lumped products (e.g., biocrude) with molecular-level chemistry for HTL of proteins.

References

(1) Peterson, A. A.; Vogel, F.; Lachance, R. P.; Fröling, M.; Antal, Jr., M. J.; Tester, J. W. Thermochemical Biofuel Production in Hydrothermal Media: A Review of Sub- and Supercritical Water Technologies. *Energy Environ. Sci.* **2008**, *1*, 32–65.

(2) Toor, S. S.; Rosendahl, L.; Rudolf, A. Hydrothermal Liquefaction of Biomass: A Review of Subcritical Water Technologies. *Energy* **2011**, *36*, 2328–2342.

(3) López Barreiro, D.; Prins, W.; Ronsse, F.; Brilman, W. Hydrothermal Liquefaction (HTL) of Microalgae for Biofuel Production: State of the Art Review and Future Prospects. *Biomass and Bioenergy* **2013**, *53*, 113–127.

(4) Becker, E. *Microalgae Biotechnology and Microbiology*; Cambridge University Press: New York, 1994.

# 491087 Bio-Succinic Acid Production from Tartaric Acid

**Abstract:**

Fossil fuels are currently the primary source of energy and chemicals because of their high availability and energy content.[1] However, the depletion of fossil fuels and environmental problems such as greenhouse gas emissions have triggered the development of biomass-based technologies, as biomass is the only source of renewable carbon.[2] Those new technologies also bring novel value added chemicals by utilizing the functional groups which are already present in biomass molecules.[3] Succinic acid ($C_4H_6O_4$), a four-carbon dicarboxylic acid, is recognized by the Department of Energy (DOE) as one of the top biomass-derived platform chemicals.[4] Its main applications include replacement of maleic anhydride as the C4 building block and synthesis of biodegradable and biobased polymers (Scheme 1).[4] The traditional method for succinic acid production relies on fossil fuels. Fermentation has been proposed as a renewable alternative, yet low product concentration and high concentration of organic acid are still drawbacks.[5,6]

Scheme 1. Succinic acid synthesis and applications

We developed a novel one-step catalytic route to produce bio-succinic acid using biomass-derived tartaric acid (Scheme 1), in high yields. Tartaric acid, which is naturally produced in grapes, can be recovered from winery waste streams as a byproduct of wine industry.[7] A liquid-phase system under hydrogen atmosphere comprised of molybdenum oxide catalyst supported on carbon black ($MoO_x$/BC) is effective in catalyzing C-O bond cleavage of tartaric acid. The role of solvent was investigated to gain a better understanding of the reaction system. The $MoO_x$/BC performance under different pre-reduction temperatures combined with characterization data such as temperature programmed reduction (TPR), X-ray powder diffraction (XRD), X-ray photoelectron spectroscopy (XPS) reveals that the high catalyst activity is correlated with the formation of specific Mo oxidation states. Finally, recyclability studies and structural characterization of the catalyst after reaction indicate that $MoO_x$/BC remains active upon reuse.

References

(1) Beerthuis, R.; Rothenberg, G.; Shiju, N. R. *Green Chem.* **2014**, *17*, 1341–1361.

(2) Ruppert, A. M.; Weinberg, K.; Palkovits, R. *Angew. Chemie - Int. Ed.* **2012**, *51* (11), 2564–2601.

(3) Corma, A.; Iborra, S.; Velty, A. *Chem. Rev.* **2007**, *107* (6), 2411–2502.

(4) Carlson, A.; Coggio, B.; Lau, K.; Mercogliano, C.; Millis, J. *Chemicals and Fuels from Bio-Based Building Blocks*, **2016**, 173–190.

(5) Cheng, K. K.; Zhao, X. B.; Zeng, J.; Wu, R. C.; Xu, Y. Z.; Liu, D. H.; Zhang, J. A. *Appl. Microbiol. Biotechnol.* **2012**, *95* (4), 841–850.

(6) Bechthold, I.; Bretz, K.; Kabasci, S.; Kopitzky, R.; Springer, A. *Chem. Eng. Technol.* **2008**, *31* (5), 647–654.

(7) Oliveira, M.; Duarte, E. **2016**, *10* (1), 168–176.

# 495881 1,6-Hexanediol Synthesis from Cellulose


**Abstract:**

1,6-hexanediol (HDO) is a high value commodity chemicals ( $4,400/ton) used as a polyester or polyurethane component traditionally produced from adipic acid. In this presentation we discuss a three step catalytic route to produce HDO from cellulose[1] that includes: i) hydrolysis/dehydration of cellulose into LGO and HMF with sulfuric acid, ii) hydrogenation/hydrogenolysis of LGO and HMF into THFDM over Pd/Si-$Al_2O_3$ [2] and iii) hydrogenolysis of THFDM into HDO over a bifunctional Pt-acid catalyst. The overall HDO yield is 50%. In this presentation we will discuss the catalytic chemistry of each of these steps and elucidate the relationship between the catalyst structure and performance. We use cellulose, white birch and corn stover as feedstocks.

The THFDM is converted into 1,2,6 hexanetriol and then HDO in the final step through a dehydration-hydrogenation mechanism on a bifunctional Pt-based catalysts in over 90% HDO selectivity (70% HDO yield). We will show that the active catalytic acid phase for the dehydration, a BrÃ¸nsted acid site, is generated on the oxophilic support through hydrogen spillover from the Pt.

Â


**Reference:**


1 J. He, K. Huang, K. J. Barnett, S. Krishna, D. M. Alonso, Z. J. Brentzel, S. P. Burt, T. Walker, W. F. Banholzer, C. T. Maravelias, I. Hermans, J. A. Dumesic and G. W. Huber, *Faraday Discuss.*, 2017.

2 S. H. Krishna, D. J. McClelland, Q. A. Rashke, J. A. Dumesic and G. W. Huber, *Green Chem.*, 2017, **19**, 1278â€“1285.

# 497020 High-Efficiency Photo-Electrochemical Chloro-Alkali Production

**Abstract:**

The increased penetration of photovoltaic (PV) technologies has led to significant challenges on the management of the electrical grid. This has made evident the need for energy storage technologies that can bridge the temporal mismatch between electricity production and demand imposed by intermittent solar power sources. Alternatively, the implementation of PVs for large-scale off-grid chemical processes could further accelerate the deployment of clean energy technologies. Electrochemical processes, such as the ones implemented for chlor-alkali production, account for more than 1% of the global energy consumption and could be adapted to directly operate with solar energy sources. In this study, we demonstrate a solar chloro-alkali reactor that operates at unprecedented solar-to-chemical efficiencies as high as 25.1%. The reactor implements high efficiency multi-junction InGaP/GaAs/InGaAsNSb solar cells within a novel planar solar concentrator which maintains a ± 40° acceptance angle with minimal tracking requirements. The concentrated PV system provides the required potential ($V_{OC}$ = 3.16 V) to drive a customized electrochemical cell fabricated via additive manufacturing. The electrochemical cell is composed of a Nickel cathode and a dimensionally stable anode based on a Titanium substrate coated with an Iridium and Ruthenium oxide blend. The two electrodes are separated by a Nafion® 117 membrane which allows for the collection of nearly pure $H_2$ and $Cl_2$ gas streams as well as a NaOH enriched aqueous stream in the cathode side. Under operation conditions the faradaic efficiency towards chlorine production of the anode was measured to be between 96-99%. The demonstration presented here provides insights and foundational design principles for the implementation of solar chemical processes. Related developments for the production of solar-Hydrogen and solar-textiles will be discussed.

# 500094 Gold Nanoparticle Catalysis: Colloidal Versus Supported Heterogeneous Catalysis and Methods for Colloidal Nanoparticle Recovery

**Abstract:**

Gold nanoparticles (AuNPs) have attracted enormous attention due to their unique catalytic activities. Potential advantages of colloidal AuNPs over supported AuNPs is the greater catalytic activity and selectivity. Colloidal catalysts also catalyze reactions under mild conditions and are very effective for chiral catalysis. Recovery of colloidal catalysts involve tedious methods like pH or temperature induced aggregation, solvent extraction, etc. The advantage of supported catalyst lies in its reusability and recovery by simple methods like centrifugation, filtration, sedimentation, etc. Supported catalysts are preferred over colloidal catalyst in industries because they can be used in a continuous reactor system. For colloidal systems, surface functionalization with ligands is a prerequisite to prevent aggregation. Surface passivation causes significant reduction in catalytic activity and selectivity. For supported catalysts on metal oxide supports, a certain portion of the active surface area of the catalyst is lost due to the immobilization process and simultaneously new edges at metal-metal oxide interfaces are created. In this work, we have attempted to characterize the catalytic activity and active sites of both colloidal and supported catalysts systems by employing a ligand adsorption based method in aqueous media. An in-depth study of the effect of nanoparticle stabilizing ligands such as thiolated polyethylene glycol (PEG) and 11-mercaptoundecanoic acid (MUA) on the nanoparticle catalytic activity has been carried out and we have quantified the loss of active surface sites due to the ligand stabilization. Decreasing PEG chain length and increasing surface coverage of PEG on AuNP reduces the catalytic activity. Moreover, the functionalization of AuNP with MUA completely passivates the surface and the AuNPs show no catalytic activity. Although, high MUA coverage on the AuNP are detrimental to the catalytic activity, it renders the AuNPs easily recoverable from the solution by pH induced aggregation and redispersion. To maintain both catalytic properties and reversible pH induced recovery, we have considered partial surface coverages of the AuNP with MUA. This work also reports the synthesis of thiolated poly (acrylic acid) (PAA) functionalized AuNPs and explored its application as a recoverable catalyst where the PAA provides pH responsive dispersibility in aqueous media. Thus, PAA-AuNPs are easily and completely recovered from the reaction mixture and reused in subsequent reactions. It was found that the AuNP-PAA catalyst was highly active and reusable with modest reductions in the reaction rate with up to four catalyst recycles. We have also supported â€˜bareâ€™ colloidal AuNPs (trisodium citrate stabilized) on metal oxides and quantified the loss of active surface area. We report the comparison of loss of surface area due to ligand stabilization and metal oxide supports and compare the effectiveness and recoverability of both system of catalysts. A direct consequence of being able to characterize active surface area of both colloidal and supported system simultaneously is the ability to measure the effect of the support on the catalytic activity of the AuNPs. We have quantified the metal oxide support effect on the catalytic activity of AuNPs for various metal oxide-AuNP catalysts.

# 500274 Isosorbide Production from Aqueous Sorbitol Solution over Silica-Aluminas

**Abstract:**

Isosorbide is an important chemical for the production of fuels, plastics, and pharmaceuticals. Current isosorbide production is from plant biomass. Most current studies have been conducted in non-aqueous systems or pure sorbitol. However, processing plant biomass usually involves water. Thus, isosorbide production in aqueous systems is more practical. In the present study, we investigated the conversion of aqueous sorbitol to isosorbide using amorphous silica-alumina (ASA) with various SiO2/Al2O3 ratios. The isosorbide yield of ~ 56% was achieved at 220ËšC and 22 h. Surface area/pore volume, densities of acid sites, and total acidity of ASA from Brunauer-Emmett-Teller (BET), Fourier transform infrared spectroscopy (FT-IR)-pyridine adsorption and NH3-Temperature Programmed Desorption (TPD), respectively, were found to be responsible for this high isosorbide selectivity in the aqueous system.

# 498148 Simple One-Step Reaction to Achieve Cross-Linked Polymer and Polymer Composites: Models for Rubber Tires That Can Recycled Multiple Times with Robust Recovery of Properties Related to Cross-Link Density

**Abstract:**

Conventional polymer networks cannot be recycled into high-value products because of permanent, covalent crosslinks. Rubber tires illustrate well the issues ranging from major economic loss to environmental problems that arise with spent, crosslinked polymer. Here, we will discuss a simple, one-step reaction strategy using nitroxide-mediated polymerization (NMP) that we developed to synthesize recyclable crosslinked polymer with dynamic covalent bonds from any monomers or polymers that contain carbon-carbon double bonds that are amenable to free radical polymerization. (See Advanced Materials 2016, 28, 6746.) Using this reaction strategy, we have designed network polymers using polybutadiene and styrene monomer as models for recyclable tires, yielding full property recovery after multiple melt-reprocessing steps. No previous study has achieved direct formation of recyclable networks from commercial monomer/polymer without presynthesis or prefunctionalization and demonstrated full recovery of crosslinks after multiple reprocessing steps. We have likewise demonstrated the effectiveness of this strategy with commercial natural (uncrosslinked) rubber, leading to crosslinked natural rubber that can be effectively recycled and reprocessed multiple times with full recovery of properties associated with crosslink density. Thus, the reaction strategy can be employed with exactly the same polymers currently used in the tire industry and with reaction temperatures comparable to those employed in tire molding operations. As a result, this strategy provides for green reaction engineering not with the use on biobased feedstock but instead by effective, high-value recycling of synthetic materials that have been highly engineered for demanding uses and that are unlikely to be replaced by biobased feedstock. Importantly, we have also shown that this strategy can work well with polymer composites, e.g. networks incorporating carbon black filler exactly of the type used in rubber tires. Additionally, we have also used one-step NMP to produce polymer networks from monomers with very good reprocessabiltiy and relative uniformity of chains between crosslinks. Examples of other novel material development using simple, one-step NMP will also be discussed, time permitting.

# 486988 Impacts of $CO_2$ Supply Systems for Algal-Based Biorefineries on Biofuel Life Cycle Assessments

**Abstract:**

Biofuels have great potential as low-carbon alternatives to fossil-based transportation fuels. This talk will focus on utilization of $CO_2$ in an advanced biofuel system and the integration of that system with $CO_2$ generation sources. The biofuel system is the Algenol Direct to Ethanol® technology, which provides an efficient, cyanobacteria-based system for producing ethanol and bio-crude. The Algenol process consumes $CO_2$ in a process that yields a biofuel with much lower greenhouse gas emissions than gasoline.[1-3] In this talk, different options for $CO_2$ capture and utilization (CCU) are considered and their impact on the overall system operation evaluated in terms of carbon footprint reduction. The boundary of the life cycle analysis includes the $CO_2$ source. Overall 13 scenarios are described including an artificial reference case with $CO_2$ arriving at the boundary limits of a 2000 acre, model biorefinery with zero carbon footprint burden and zero $CO_2$ cost at a 10% IRR. Most scenarios involve combined heat and power (CHP) systems with no power supplied from the grid. Both natural gas and cellulosic fuel sources are considered for the CHP system. Five of the scenarios are "stand-alone" in that they do not require co-location with an anthropogenic $CO_2$ source. This is achieved by generating enough $CO_2$ on site to supply the needs of the biorefinery and selling excess power back to the grid. Four of the five stand-alone systems have carbon footprints and economics that are fully competitive with the commonly considered coal flue gas case. Overall carbon footprint reductions range from 23% to 116%. Equivalent $CO_2$ costs (based on comparisons to the reference case) range from \$35/-\$96/tonne with coal flue gas being \$45/tonne.

[1]D. Luo et al, "Lifecycle Energy and Greenhouse Gas Emissions from Ethanol Produced by Algae", *Environmental Science and Technology* **44**, 8670 (2010).

[2]Notice of EPA pathway approval: http://epa.gov/otaq/fuels/renewablefuels/new-pathways/approved-pathways.htm

[3]R. P. Lively et al, "Anthropogenic $CO_2$ as a feedstock for the production of algal-based biofuels", *Biofuels, Bioproducts, and Biorefining* **9**, 72-81 (2015)

# 498698 Carbon Fixation By Rubisco-Nanostructure Complex to Produce 3-Phosphoglyceric Acid: A Life Cycle Assessment


**Abstract:**

As an effort to eliminate greenhouse gases (GHGs), biochemists, microbiologists, and engineers are working toward systems that mimic the plant cellular environment's ability to convert GHGs into higher value products [1]. In the Calvin cycle of photosynthesis, ribulose-1,5-bisphosphate carboxylase oxygenase (RubisCO), which is an enzyme present in chloroplasts, catalyzes the conversion of ribulose 1,5-bisphosphate (RuBP) to 3-phosphoglyceric acid (3-PGA) while incorporating atmospheric $CO_2$ into the organic molecule [1,2]. It has been said that RubisCO is the most abundant protein in the world and accounts for most of the biological $CO_2$ fixed on earth [3]. Satagopan et al. have developed multienzyme-nanostructure complexes that catalyze cascade reactions in the Calvin cycle to enhance the catalytic properties of enzymes [1]. The complexes are made of three enzymes that include Rubisco and a nanostructure to support the enzymes.

In this work, a carbon footprint and other environmental impacts of 3-PGA produced through carbon fixation by the multienzyme-nanostructure complexes are evaluated using the Life Cycle Assessment (LCA) approach [4]. Two types of nanostructures for the complex are investigated as follows: Camptothecin (CPT)-dipeptide nanotube [1,5] and fluorenylmethyloxycarbonyl (Fmoc) tetrapeptide nanofiber. To examine the effectiveness of carbon fixation by those emerging technologies, another route that doesn't fix $CO_2$ to prepare 3-PGA is investigated [6] and the carbon footprint of each route is compared. Also, the potential use of 3-PGA and its impacts are examined to account for a cradle-to-grave life cycle boundary. From the results of life cycle impact assessment, processes that show the highest contribution to the footprint are identified.

Many challenges exist in applying the LCA to emerging technologies due to the nascent nature of these technologies [7]. If any life cycle inventory (LCI) data are not available from existing database [8,9], the data are estimated from laboratory experiments or obtained from literatures and patents. To account for the uncertainty of those data, a sensitivity analysis is performed. Three cases are considered based on the amount of fugitive emissions and the recyclability of unreacted reactants and solvents. The reusability of the complexes is also taken into consideration through the sensitivity analysis because they function as catalysts for the carbon fixation. Also, most of energy flows are excluded from the inventory analysis because most of biological processes take place around room temperature. If product yields are unknown, they are determined by a stoichiometry calculation.

CPT is extracted from the stem of a specific tree species, *Camptotheca acuminate* [10]. Since its extraction yield from the tree and the subsequent product yield of the CPT-dipeptide nanotube are low, the total carbon footprint is affected significantly depending on whether biogenic $CO_2$ emissions are included in the analysis or not. In this work, therefore, the effects of the biogenic $CO_2$ emissions on the footprint are examined and discussed.

Even though this work accounts for some data uncertainty via sensitivity analysis, other sources of uncertainty are still left for a more robust analysis. Some data, such as labor, equipment, and other unavailable data, are excluded from the analysis boundary due to the lack of information. Also, inventories from different LCI database [8,9] often base on different location. Moreover, since many unavailable data from the LCI database are obtained from literatures and patents, there may be errors in the life cycle network as well as the inventory data. Those potential uncertainty and errors are mostly attributed to the nature of emerging technologies. If a more complete analysis that accounts for those issues is developed, it would present significant benefits for evaluating the impacts of emerging technologies.


References


[1] Sriram Satagopan, Yuan Sun, Jon R. Parquette, and F. Robert Tabita. Self-assembled nanostructures for capturing atmospheric $CO_2$. Unpublished manuscript.

[2] Spencer M. Whitney, Robert L. Houtz, and Hernan Alonso. Advancing our understanding and capacity to engineer nature's $CO_2$-sequestering enzyme, Rubisco. Plant Physiology. 155(1): 27-35, 2011.

[3] John A. Raven. Rubisco: still the most abundant protein of Earth? New Phytologist. 198(1): 1-3, 2013.

[4] ISO. Environmental management—Life cycle assessment—Principles and framework (ISO 14040:2006). ISO. Geneva, Switzerland, 2016.

[5] Se Hye Kim et al. The self-assembly of anticancer camptothecin–dipeptide nanotubes: A minimalistic and high drug loading approach to increased efficacy. Chemistry - A European Journal. 21(1): 101-105, 2015.

[6] Ines Mandl and Carl Neuberg. Preparation of D(-)-3-phosphoglyceric acid. Methods in Enzymology. 3:208-214, 1957.

[7] Marcelle C. McManus et al. Challenge clusters facing LCA in environmental decision-making–what we can learn from biofuels. The International Journal of Life Cycle Assessment. 20(10):1399-1414, 2015.

[8] U.S. Life Cycle Inventory Database. National Renewable Energy Laboratory, 2012.

[9] Rolf Frischknecht et al. The ecoinvent database: Overview and methodological framework. International Journal of Life Cycle Assessment, 10(1):3–9, 2005.

[10] J. George Buta and Michael J. Novak, Isolation of camptothecin and 10-methoxycamptothecin from camptotheca acuminate by gel permeation chromatography. Industrial and Engineering Chemistry Product Research and Development. 17(2): 160-161, 1978.

# 496672 Life Cycle Assessment of Biofuel and Bioenergy Production from Brown Algae through Biochemical Pathways


**Abstract:**

Brown algae *Laminaria Japonica* is the most cultivated type of seaweed with a large potential for sustainable biofuel and bioenergy production. This study aims to evaluate the environmental impacts of seaweed cultivation and utilization for biofuel production through biochemical conversion routes including the sugar platform (SP), methane platform (MP), and volatile fatty acids platform (VFAP). In the SP, bioethanol is produced through saccharification of carbohydrates and fermentation of produced sugars. In the MP and VFAP, methane and volatile fatty acids (VFA) are produced through anaerobic digestion of seaweed. In the MP, methane is utilized as fuel source in a boiler/turbogenerator for heat and power production. In VFAP, VFAs are recovered through extraction/distillation and then hydrogenated to mixed alcohols. All three processes are simulated in Aspen Plus v.8.8 to obtain heat and material balances. Simulation results along with data from literature and Ecoinvent V3 databases for seaweed cultivation, chemicals, nutrients, and waste treatment operation are used to obtain life cycle inventory. Then, life cycle models are developed in SimaPro V8.2 software for each process. The environmental impact categories considered include global warming potential, acidification, eutrophication, fresh water consumption, and non-renewable energy use. Results are used to identify bottlenecks, limitations, and main contributors to the environmental profile of each route and assess the potentials to reduce the emissions and enhance the sustainability of the production process.

# 497439 Integrated Design and Analysis of Chemical Production from Biomass Feedstocks

**Abstract:**

Integrated Design and Analysis of Chemical Production from Biomass Feedstocks


Abhay Athaley[1], Praneeth Annam[1], Basudeb Saha[2], Marianthi Ierapetritou[1]

[1] Department of Chemical and Biochemical Engineering, Rutgers - The State University of New Jersey

[2.] Catalysis Center for Energy Innovation & Department of Chemical & Biomolecular Engineering, University of Delaware



This work principally concentrates on the integrated design, techno-economic and the life-cycle analysis of the production of chemicals from the lignocellulose biomass. The priority for the development of variate fine chemicals from biomass feedstock stepped-up over the changing time. The idea of bio-refinery has been proposed that uses different conversion technologies to produce multiple products. The success of bio-refinery is highly dependent on its mature technology and its applicability to produce not only high volume low cost fuels but also to produce low volume high cost chemicals.[1]

Bio-based products acceptance in the market depends on the competitiveness of economics and sustainability when compared to oil-based chemicals and products. In our previous work, [2]we investigated a novel hydrolysis process[3] for the productions of sugars and the subsequent production of para-Xylene and compared it with two other hydrolysis process (Concentrated acid hydrolysis and Dilute acid hydrolysis) using techno-economic and life cycle assessment. In our present work, we integrate the novel hydrolysis process to production of other high value chemicals such as surfactants,[4] butadiene, 2-methyl styrene etc. and high value jet fuels and combine it with the production process of para-Xylene and phthalic anhydride.[5] The aforementioned chemicals are lucrative and also it furthers our goal of designing an integrated bio-refinery.

This work uses techno-economic analysis and life cycle assessment to design and evaluate the production alternatives. The detailed process flowsheet is developed and simulated using Aspen Plus. The steps for the production of different chemicals and fuels are integrated with the novel hydrolysis process into a single flowsheet. Next, heat integration and economic analysis is carried out to calculate the minimum selling price of each of the product. Furthermore, sensitivity analysis is performed to analyze the bottlenecks of entire process and to compare the cost of the chemicals with fluctuating cost of raw materials and capacity of the plant. It is also seen that the overall cost of raw materials increases considerably for the production of surfactants as compared to other chemicals such as butadiene or phthalic anhydride. To check the competitiveness of the above process in terms of sustainability Life Cycle Analysis is carried out on SimaPro®. The results with respect to carbon production and water consumption are shown and compared to the current oil based processes.


References:


1. Lin, Z., V. Nikolakis, and M. Ierapetritou, *Alternative Approaches for p-Xylene Production from Starch: Techno-Economic Analysis.* Industrial & Engineering Chemistry Research, 2014. **53**(26): p. 10688-10699.

2. Athaley A., S.B., Ierapetritou M.,, *TECHNO-ECONOMIC ANALYSIS OF DIFFERENT TYPES OF HYDROLYSIS PROCESS FOR THE PRODUCTION OF P-XYLENE.* 2017.

3. Sadula, S., Athaley, A., Zheng, W., Ierapetritou, M., Saha, B., *Process Intensification for Cellulosic Biorefineries.* ChemSusChem, 2017. **10**.

4. Park D., J.K., Koehle M., Krumm C., Ren L., Damen J., Shete M., Lee H., Zuo X., Lee B., Fan W., Vlachos., Lobo R., Tsapatsis M., Dauenhauer P.,, *Tunable Oleo-Furan Surfactants by Acylation of Renewable Furans.* ACS Catalysis, 2016. **2**: p. 820-824.

5. Lin, Z., Ierapetritou, M., Nikolakis, V.,, *Phthalic anhydride production from hemicellulose solutions: Technoeconomic analysis and life cycle assessment.* AIChE Journal, 2015. **61**: p. 3708-3718.

# 499590 Life-Cycle Analysis (LCA) of Bio-Derived Terephtalic Acid (TPA) and Bio-Derived-Hexamethylenediamine (HMDA)

**Abstract:**

Increasing efforts to reduce environmental impacts while boosting U.S chemical production have been an important driver in the growing bioeconomy. These efforts not only stimulate economic development but also promote competitiveness in the bioeconomy. Under the analysis led by the Clean Energy Manufacturing Analysis Center (CEMAC), screening of the impact of initial 170 chemicals derived from biomass were conducted, resulting in three relevant bioproducts: 3- hydroxypropionic acid (HP), terephtalic acid (TPA), and hexamethylenediamine (HMDA). These bio-derived chemicals can be used in the production of superabsorbent polymers, coatings, adhesives, plastics, fibers, paints, etc. In this work, we use the Greenhouse gases, Regulated Emissions and Energy use in Transportation (GREET®) model to conduct a life-cycle analysis (LCA) on the last two bioproducts: TPA and HMDA. We evaluate the key drivers of the GHG emissions and water consumption. We present thermochemical and biochemical pathways as alternative production pathways for these chemicals. These biomass-derived chemicals can displace fossil-derived products potentially and reduce GHG emissions, water and fossil fuel consumption as compared to their fossil-derived counterparts.

# 495175 Sustainability Analysis of Condensed-Phase n-Butanol Production from Ethanol

**Abstract:**

Bioethanol is one of the most valuable replacements for or additives to fossil fuels. However, ethanol is not without its limitations that include low energy density, corrosivity, and contamination issues when mixed with water. In contrast, n-butanol has a notable advantage in these aspects, as it is more miscible in gasoline, has higher energy density, and has an air to fuel ratio comparable to that of gasoline. As a result, there has been growing interest in the production of biobutanol and higher alcohols as biofuels as well as for use in the chemical industry.

The condensation of ethanol to butanol and higher alcohols is known as Guerbet chemistry [1]. The reactions are well known and have been substantially investigated. The path to Guerbet chemistry commercialization, however, has been hindered by substandard yields, catalyst stability concerns, and high capital and energy costs.

Our research group has studied the condensed-phase Guerbet reaction of ethanol to higher alcohols [2], [3]. Recently, we have optimized a condensed phase continuous flow reactor using multi-metallic catalysts on $\gamma$-$Al_2O_3$ support modified with $La_2O_3$. We have achieved higher alcohols selectivity of 80% at ethanol conversions of 40%. These encouraging results have prompted us to explore the economic feasibility and environmental impact of industrial-scale higher alcohol production using Guerbet chemistry.

An economic and environmental assessment of a first-generation facility producing 25 million gallons of n-butanol per year has been performed for a range of butanol and higher alcohol selectivities. Simulation of the catalytic reactor and the recovery and purification processes have been conducted using AspenPlus. Economic parameters such as capital costs, operating costs, annual profit, and return on investment have been determined for several scenarios. The results obtained show n-butanol required selling prices are reasonably competitive for conversion and selectivity combinations achieved in laboratory reactions. Interestingly, analysis of the separation train indicate significant economic benefit by carrying out distillations at elevated pressures. Environmental impacts such as climate change, resource depletion, water use, and eutrophication are also assessed.

References

[1] D. Gabriëls, W. Y. Hernández, B. Sels, P. Van Der Voort, and A. Verberckmoes, "Review of catalytic systems and thermodynamics for the Guerbet condensation reaction and challenges for biomass valorization," *Catal. Sci. Technol.*, vol. 5, pp. 3876–3902, 2015.

[2] T. L. Jordison, L. Peereboom, and D. J. Miller, "Impact of Water on Condensed Phase Ethanol Guerbet Reactions," *Ind. Eng. Chem. Res.*, vol. 55, no. 23, pp. 6579–6585, 2016.

[3] T. L. Jordison, C. T. Lira, and D. J. Miller, "Condensed-Phase Ethanol Conversion to Higher Alcohols," *Ind. Eng. Chem. Res.*, vol. 54, no. 44, pp. 10991–11000, 2015.

# 503130 Ionic Liquids for Absorption and Separation of Gases: An Extensive Database and a Systematic Screening Method

**Abstract:**

**Ionic liquids for absorption and separation of gases: an extensive database and a systematic screening method**

[a,b]Yongsheng Zhao, [c]Rafiqul Gani, [a,b]Raja Muhammad Afzal, [a,b]Suojiang Zhang[a,b,*]

[a]*Beijing Key Laboratory of Ionic Liquids Clean Process, State Key Laboratory of Multiphase Complex System,*
[b]*Institute of Process Engineering, Chinese Academy of Sciences, Beijing 100190, China*

[c]*Department of Chemical & Biochemical Engineering, Technical University of Denmark, DK 2800 Kgs. Lyngby, Denmark*

**Corresponding author: sjzhang@ipe.ac.cn*

**Keywords:** Ionic Liquids, Gas absorption and separation, Database, Systemic screening method

In modern society, with increased focus on energy and environment, absorption and separation of gases have become a very important research topic in both academic and industrial communities. In generally, solvent absorption method is one of the most common methods in the industry. There are, however, some problems with the use of commercial absorption solvents, such as alkanolamine solutions, which require high energy consumption, associated to high economic cost, cause high VOC (volatile organic chemical) issues, and caustic byproducts. Therefore, to overcome these disadvantages, alternative and greener solvents with better separation performance need to be considered. Recently, ionic liquids (ILs) have attracted much attention because of their unique properties, especially the negligible vapour pressure and tunable properties[1]. The negligible vapour pressure make ILs have low volatility and thus would not cause contamination of product and environment. The tunable property can makes us design special ILs for various purposes through proper combinations of cations and anions from a current chemical database. However, because of the tunable properties feature, there are also many possible ILs that could be synthesized. From a practical point of view, it is not feasible to measure all of the target (tunable) properties of synthesized ILs by experimental techniques, making their screening time consuming and economically costly. Therefore, it is necessary to design and select ILs using the computer aided techniques.

The aim of this study is to build a data-rich and extensive database of the Henryâ€™s law constants of twelve gases ($CO_2$, $NH_3$, $SO_2$, $H_2S$, $C_2H_2$, COS, $C_2H_4$, CO, $H_2$, $N_2$, $O_2$, and $CH_4$) in more than ten thousand of ILs (13585 ILs), which are calculated by the COSMO-RS method. That is, more than one hundred thousand data values of the Henryâ €™s law constants will be provided for academic and industry communities in this work. Furthermore, the objective is to develop a new, systematic, efficient and reliable screening method for selection of the appropriate IL from the 13585 ILs that are available in the database for specific absorption and separation of gases considering a set of identified tunable properties such as toxicity, bioaccumulation potential, viscosity, melting point, molality and molarity absorption capacity and separation performance. Finally, two important gases separation problems involving $CO_2/CH_4$ and $C_2H_2/C_2H_4$ will be employed as case studies to validate the developed screening method.

References:

[1] X. Zhang, X. Zhang, H. Dong, Z. Zhao, S. Zhang, Y. Huang, Carbon capture with ionic liquids: overview and progress, Energy & Environmental Science, 2012, 5: 6668-6681.

[2] R.D. Rogers, K.R. Seddon. Ionic liquids-solvents of the future? Science. 2003, 302: 792-793.

# Adsorption of polar molecules onto electrodes driven by the combined effects of dielectric inhomogeneity and electrostatic correlation

*Hongbo Chen[1,2] and Issei Nakamura[2]*

[1] *Polymer Physics and Chemistry, Changchun Institute of Applied Chemistry, University of Chinese Academy of Sciences.*

[2] *Department of Physics, Michigan Technological University*

## Introduction

- The adsorption of water in ionic liquids onto the electrodes is demonstrated using the molecular dynamics simulations

G. Feng and et al., ACS Nano, 8, 11685 (2014)

$\sigma = +0.16\ C/m^2$

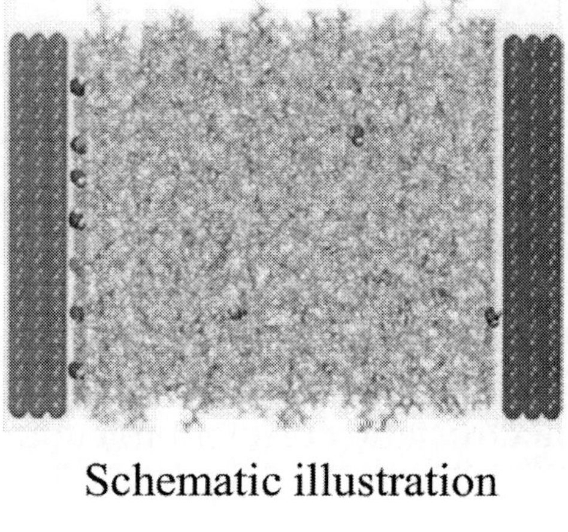

Schematic illustration

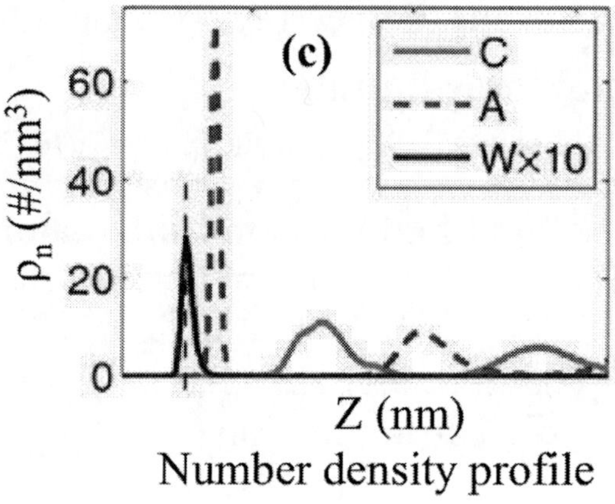


Number density profile

## Motivation

• **What are the key factors that account for the adsorption of water molecules onto the electrodes in ionic liquids?**

**Molecular volume, dielectric inhomogeneity, electronic polarization, and strong electrostatic correlation?**

## Key ingredients in our theory

• **Strong electrostatic correlation in ionic liquids**

M. Z. Bazant and et al., Phys. Rev. Lett. 106, 046102 (2011).

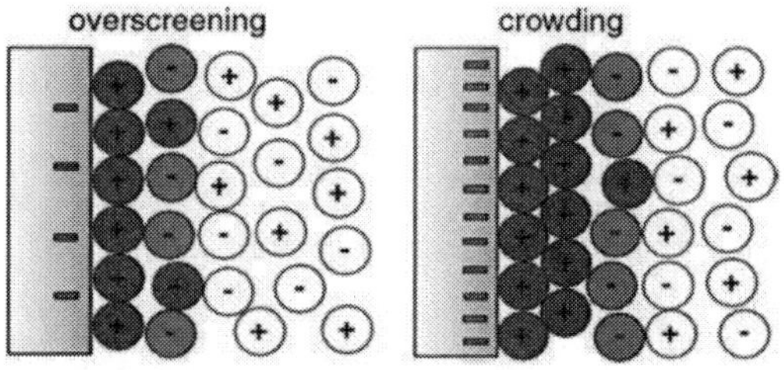


• **Different dielectric response between water molecules and ionic liquids**

I. Nakamura, J. Phys. Chem. C, 119, 7086 (2015)

---

- The modified Poisson Boltzmann equation:

$$\nabla\cdot[\varepsilon_r(r)\nabla\psi(r)] - l_c^2\nabla^4\psi(r) = -4\pi l_0\rho_f(r) - 4\pi l_0\sum_{s=\pm} q_s n_s(r)$$

local dielectric function

$l_c$ is the electrostatic correlation length

$$\varepsilon_r(r) = 1 + 4\pi l_0 \sum_{s=\pm} p_s^2 n_s(r)\left[\frac{L(p_s|\nabla\psi(r)|)}{p_s|\nabla\psi(r)|}\right] + 4\pi l_0 p_w^2 n_w(r)\left[\frac{L(p_w|\nabla\psi(r)|)}{p_w|\nabla\psi(r)|}\right]$$

## Density profile between charged plates

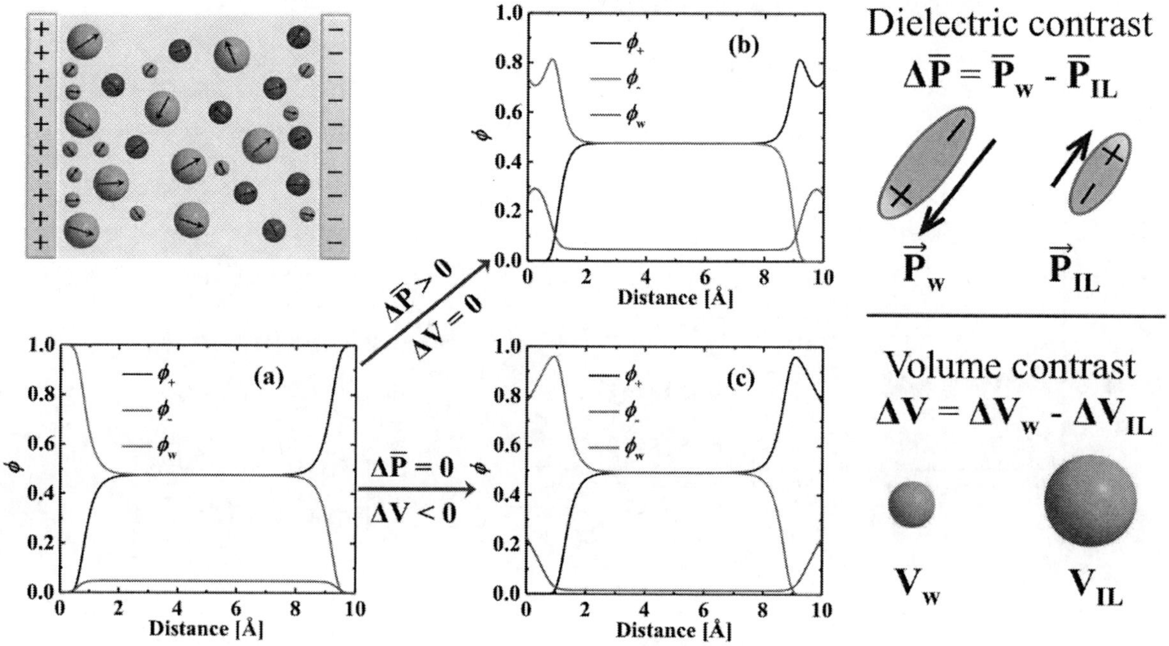


**Asymmetry in both the dielectric response and molecular volume results in the adsorption of solvent molecules onto electrodes.**

## Effects of the asymmetry between the cation and anion

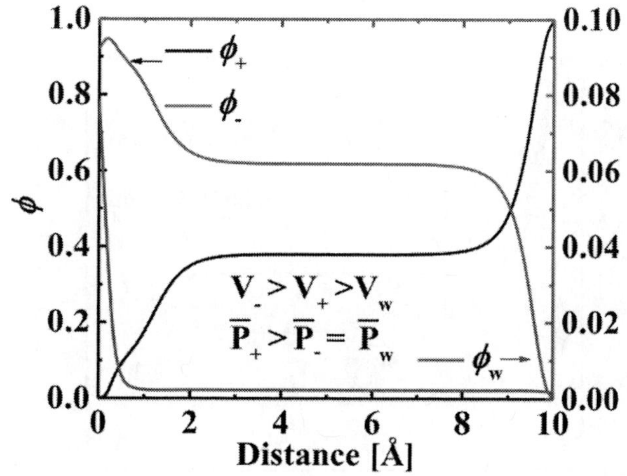


**Effects of the asymmetry between the cation and anion may account for the deviation of the water adsorption in the MD simulation.**

# Capacitance

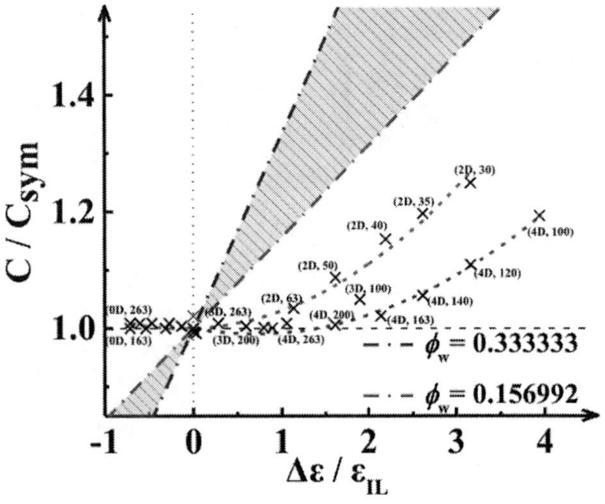


is deduced from the linear dielectric theory:

$$C/C_{sym} = \varepsilon_r / \varepsilon_{IL} = 1 + \frac{(\varepsilon_w - \varepsilon_{IL})}{\varepsilon_{IL}} \phi_w$$

$$\varepsilon_r = \varepsilon_{IL}\phi_{IL} + \varepsilon_w \phi_w$$

-·-·- the maximum $\phi_w$

-·-·- the minimum $\phi_w$

-·-·- the fitting curves

-·-·- depend on the $P_w$

**The linear dielectric theory may not account for the capacitance of the electrolytes with ionic liquid and water.**

# Effects of electronic polarization

- Polarizability contrast: $\mathbf{\Delta\alpha = \alpha_w - \alpha_{IL}}$

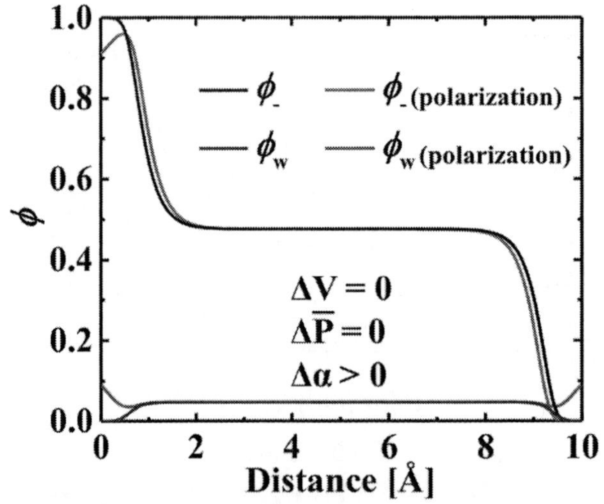


**The effects of electronic polarization affect wetting of the charged plates.**

# Effects of strong electrostatic correlation

- Without strong correlation
- With strong correlation

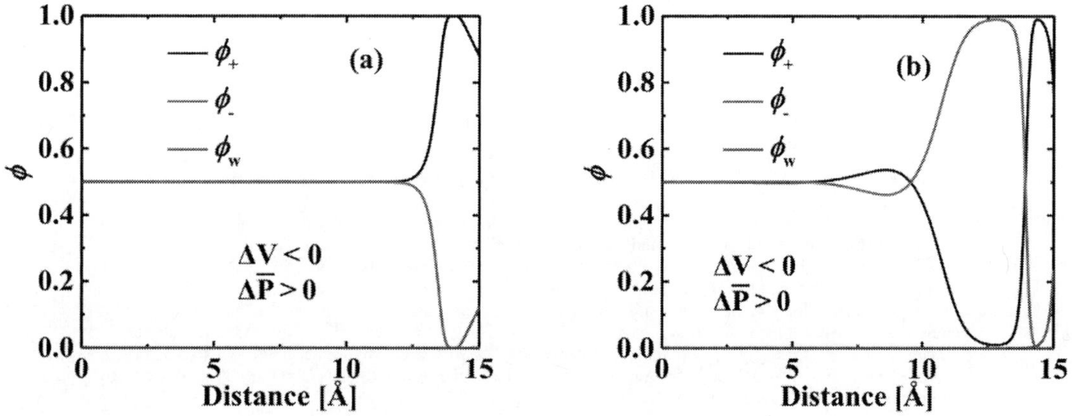


**The electrostatic correlation can be critical to adsorb water molecules onto the charged plates.**

# Summary

- Our coarse-grained mean-field theory accounts for the results of the MD simulations.
- Key factors on the water adsorption onto the electrodes:
  - ✓ Dielectric contrast between the species
  - ✓ Asymmetry in the molecular volume
  - ✓ Electronic polarization
  - ✓ Strong electrostatic correlations

***Effects of the molecular volume, the dielectric inhomogeneity, the electronic polarization, and the electrostatic correlation can be equally important to account for the adsorption of water in ionic liquid onto the electrodes.***

# 503048 Extraction of Carboxylic Acid from Aqueous Solution By Strong Hydrogen Bond Basicity Ionic Liquids

**Abstract:**

**Extraction of carboxylic acid from aqueous solution by strong hydrogen bond basicity ionic liquids**


Yinge Bai[1,2], Ruiyi Yan[1], Jianguo Qian[1,2], Xiangping Zhang[1]*, Suojiang Zhang[1]*

[1]*Beijing Key Laboratory of Ionic Liquids Clean Process, State Key Laboratory of Multiphase Complex Systems, Key Laboratory of Green Process and Engineering, Institute of Process Engineering, Chinese Academy of Sciences, 100190, Beijing, P. R. China*

[2]*School of Chemistry and Chemical Engineering, University of Chinese Academy of Sciences, 100190, Beijing, P. R. China*

**corresponding author:* xpzhang@ipe.ac.cn *(X. Zhang),* sjzhang@ipe.ac.cn *(S. Zhang)*



Carboxylic acids are important raw material for food, chemical and pharmaceutical industries [1]. The recovery of carboxylic acids from dilute aqueous solutions (< 10 wt%) is commonly encountered in petrochemical process and fermentation processes [2, 3]. For the liquid-liquid extraction, which is the commonly method used to separate carboxylic acid from aqueous solutions, the core task is to attain a green and highly efficient extractant. In this work, methacrylic acid (MAA) was employed as a model carboxylic acid. MAA in dilute aqueous solution (1 wt%) was recovered by quaternary ammonium salt ionic liquids (ILs) with strong hydrogen bond basicity (HBB). The partition coefficients of MAA were used to evaluate the extraction ability of ILs (Fig. 1). The extraction parameters such as extraction time, extraction temperature, concentrations of MAA in feed solution, concentration of IL were studied (Fig. 2). An internal mechanism of the extraction of MAA by ILs was revealed by combining solvatochromic study and quantum chemical calculations (Fig. 3). The results show that the ILs with strong hydrogen bond basicity has the high partition coefficient of MAA, which provides guidance for ILs molecular design to attain the high efficient extractive separation of carboxylic acids from dilute aqueous solution.


**Figure 1.** The partition coefficients of MAA between IL phase and water phase (Columns); the hydrogen bond basicity of [A336]TFA, [A336]Oxa, [A336]Suc and [A336]TCA (Line).

**Figure 2.** the partition coefficients of MAA for [A336]TCA at different concentration of MAA feed solution.

Figure 3. optimized structure of Suc-MAA complex at B3LYP/6-31++G(d,p) level.


[1] M.D. Waghmare, K.L. Wasewar, S.S. Sonawane, et al, Separation and Purification Technology, 120 (2013) 296-303.

[2] H. Uslu, D. Datta, H.S. Bamufleh, Industrial & Engineering Chemistry Research, 55 (2016) 3659-3667.

[3] A. Bekatorou, A. Dima, P. Tsafrakidou, et al, Bioresource technology, 220 (2016) 34-37.

# 503058 Unlocking Biomass Recalcitrance Using Lignin Based Ionic Liquids

**Abstract:**

Certain types of ionic liquids (ILs) have shown great solvation power for biopolymers, resulting in remarkable biomass pretreatment efficiency for production of renewable fuels and chemicals.Â To date, the best performing IL cations such as imidazolium and cholinium, are derived from petroleum and/or natural gas.Â ILs derived from inexpensive and renewable sources are thus highly desirable. Lignin is Earthâ€™s most abundant, naturally-occurring aromatic polymer, and replacement of petroleum-based platforms with those derived from lignin represents a virtuous challenge for chemists.

Protic ILs prepared from benzaldehydes and furfural have demonstrated efficacy as pretreatment solvents, yielding over 85% fermentable sugars from switchgrass. We herein present the first synthesis and evaluation of quaternary benzylammonium ILs from vanillin and syringaldehyde, two major products of oxidative lignin depolymerization. Reductive amination with dialkylamines ($C_1$-$C_6$) followed by methylation with dimethyl carbonate was used to prepare a series of ILs, which were characterized and evaluated for their ability to pretreat switchgrass. Under mild and dilute conditions, certain ILs gave excellent pretreatment yields of fermentable sugars.

# 503083 Ionic Liquids Based Consolidated Bioprocessing for Biofuel Production from Lignocellulosic Biomass

**Abstract:**

**Ionic Liquids based Consolidated Bioprocessing for Biofuel Production from Lignocellulosic Biomass**

Jian Sun, NVSN M. Konda, Tanmoy Dutta, Blake Simmons, Seema Singh

Deconstruction Division, Joint BioEnergy Institute, Emeryville, CA, USA

Biological and Engineering Sciences Center, Sandia National Laboratories, Livermore, CA, USA

## Abstract

Lignocelluloses are rich sugar precursors, which can be converted to biofuels with the help of efficient combination of enzymes and microbes. Pretreatment technologies aiming to increase fermentable sugar yields by overcoming recalcitrance of lignocellulosic biomass is highly essential. The use of ionic liquids (ILs) in biomass pretreatment has received considerable attention recently because of their effectiveness at decreasing biomass recalcitrance to subsequent enzymatic hydrolysis. There is a clear and unmet need for a consolidated bioprocessing technology that can process a wide-range of sustainable feedstocks and produce high yields of fermentable sugars and biofuels with minimal intervention between unit operations. However, the integration of IL based pretreatment with enzymatic saccharification and microbial fermentation is highly challenging due to the toxicity of the ILs as well as high pH issue currently used for pretreatment, requiring extensive water washes, pH adjustment or the development of engineered IL tolerant enzymes and microbes.

Herein, an integrated one-pot biomass conversion technology has been developed, for the first time, using cholinium based ionic liquids (ILs) for advanced fuel production. The process integration eliminates the need for separation and purification after biomass pretreatment, which simplifies the conversion process and reduces the operating costs. The used renewable ILs have been tested in terms of pretreatment efficiency and the compatibility with enzymes and microorganisms. Significant progress has been made to enhance the process efficiency, lower the water usage, and increase the fuel titer by using high-gravity fed-batch operations.[1] An innovative strategy using $CO_2$ as a reversible method of controlling pH was developed to avoid the addition of extra chemicals thus making the recycle of IL much easier and viable.[2] A preliminary technoeconomic analysis indicates that the above integrated approach could reduce production costs by 40–65% compared to previous IL biomass conversion methods studied. The established approaches indicate that a consolidated process for the biofuel production from biomass that eliminates the need for intermediate washing and/or separation steps can be enabled by ionic liquids.

## References

1. F. Xu, J. Sun, N. V. S. N. M. Konda, J. Shi, T. Dutta, C. D. Scown, B. A. Simmons, S. Singh, *Energy Environ. Sci.*, 2016, 9, 1042-1049

1. J. Sun, N. V. S. N. M. Konda, J. Shi, R. Parthasarathi, T. Dutta, F. Xu, C. D. Scown, B. A. Simmons, S. Singh, *Energy Environ. Sci.*, 2016, 9, 2822-2834.

# 488487 A Novel Coarse-Grained Model with Explicit Hydrogen Bond for Imidazolium-Based Ionic Liquids

**Abstract:**

Our recent research discovered that it was mainly the Z-bond interaction in ionic liquids(ILs) that determines the properties of ILs which are inherently different from the conventional molten salt and aqueous solution. A new, non-polarizable force field model with explicit hydrogen bond forms for imidazolium-based ionic liquids has been developed. The hydrogen bond interactions were refined to reproduce the potential surface by the density functional theory with a new explicit hydrogen bonds. The advantages of the proposed methods with explicit hydrogen bonds are clarity, simplicity, and flexibility, and it can reproduce experimental heats of vaporization, diffusion coefficients, viscosities of ILs very well. It can be easily improved the micro structures in the area with hydrogen bonds. We developed a novel coarse-grained(CG) model for 1-alkyl-3-methyl-imidazolium chloride ([Cnmim][Cl](n =2, 4, 6, 8, 10, 12)) with explicit hydrogen in the imidazolium ring for adding hydrogen bond partition. Many properties such as density, radial distribution functions, self-diffusion coefficient, mean square displacement of ILs were compared to verify the correctness of this CG model. Simulation results revealed that CG simulation results of [Cnmim][Cl] show good agreement with all-atom simulation results and experimental data. The novel CG model for the imidazolium-based ionic liquids would promote the development of large-scale simulation of IL system.

ACS Sustainable Chemistry & Engineering

Letter

pubs.acs.org/journal/ascecg

# Efficient Synthesis of Cyclic Carbonates from Atmospheric $CO_2$ Using a Positive Charge Delocalized Ionic Liquid Catalyst

Zhiguo Zhang,* Fangjun Fan, Huabin Xing, Qiwei Yang, Zongbi Bao, and Qilong Ren

Key Laboratory of Biomass Chemical Engineering of Ministry of Education, College of Chemical and Biological Engineering, Zhejiang University, Zheda Road 38, Hangzhou 310027, China

Supporting Information

ABSTRACT: The chemical fixation of $CO_2$ into value-added products has recently received much attention. One of the most well-known methods is the coupling of $CO_2$ with epoxides to form cyclic carbonates. Although this field has progressed considerably, it is still a challenge to perform the reaction with atmospheric $CO_2$ or flue gas. Herein we described the development of 4-(dimethylamino)pyridine hydrobromide ([DMAPH]Br) as a highly efficient and recyclable catalyst for the formation of cyclic carbonates from atmospheric $CO_2$ and epoxides. In the presence of 1 mol % of [DMAPH]Br, excellent conversions and selectivities for a broad range of terminal epoxides were achieved under solvent-free conditions. Furthermore, the catalyst could be recycled over five times without appreciable loss of catalytic activity. The excellent catalytic performance of [DMAPH]Br is attributed to the enhanced synergistic interplay of acidic proton and bromide to epoxides and $CO_2$ through positive charge delocalization on its cation.

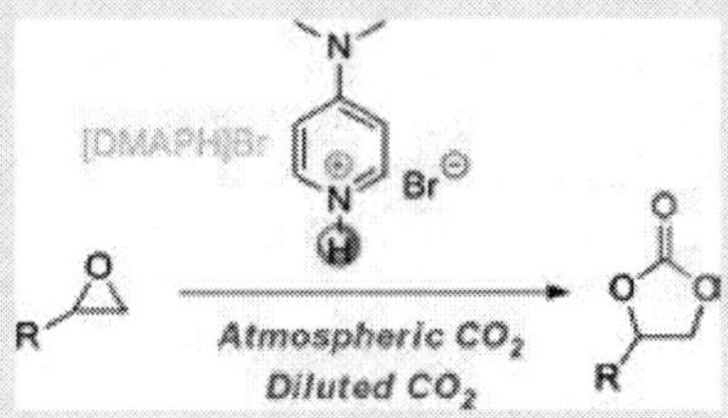


KEYWORDS: *Carbon dioxide, Cyclic carbonates, DMAP, Protic ionic liquids, Hydrogen bonding*

## INTRODUCTION

Carbon dioxide ($CO_2$) is arguably one of the most well-known greenhouse gases and its continuous emission into the atmosphere has caused increasing environmental concerns. For this reason, the development of efficient chemical processes for capture and utilization of $CO_2$ (CUC) is currently of intensive study.[1−3] In this context, the chemical fixation of $CO_2$ into value-added chemicals has seen an explosion of interest, largely due to the potential of $CO_2$ as a renewable and cheap C1 feedstock for the chemical industry.[4−8] As a consequence, a number of methodologies have been developed for the chemical fixation of $CO_2$. Of these, the coupling of epoxides with $CO_2$ to form five-membered cyclic carbonates (Scheme 1) has long been considered to be one of the most promising routes for the utilization of $CO_2$,[9−14] as cyclic carbonates have a range of applications such as aprotic polar solvents, electrolytes for lithium ion batteries, monomers for polymer synthesis, and intermediates for chemicals and pharmaceuticals.[3,9,15−17] Even though the synthesis of cyclic carbonates from epoxides and $CO_2$ is a highly exothermic reaction, however, owing to the inherent thermodynamic stability of $CO_2$, a catalytic system is generally required to activate the epoxides and/or $CO_2$.

**Scheme 1. Synthesis of Cyclic Carbonates from $CO_2$ and Epoxides**

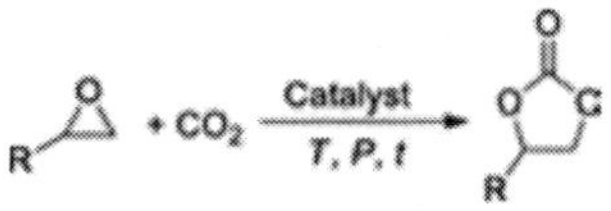


As a result, a number of effective catalysts have been developed for this transformation including transition metal complexes,[18−20] alkali metal salts,[21−24] metal oxides,[25−27] ionic liquids (ILs)[13,28−33] as well as organocatalysts.[14,34−38] Nevertheless, most of the protocols still operate at high temperatures and pressures with highly purified $CO_2$,[11,39] which lead the synthesis of cyclic carbonates a net carbon dioxide emitter rather than consumer. Thus, from a practical point of view, considerable effort has been devoted to the development of efficient catalytic systems for carrying out this transformation at ambient conditions. In this respect, the combination of a well-designed Lewis acid catalyst with a quaternary ammonium halide cocatalyst has proved to be a viable strategy on promoting this transformation at ambient conditions even with diluted carbon dioxide.[40−46] Although $CO_2$ can be efficiently transformed using these catalytic systems, the potential metal contamination sometimes has to be paid attention to. In particular, it is also difficult to recover and reuse these catalysts. Meanwhile, organic molecules based on pyridinium,[47,48] imidazolium,[3249,] ammonium,[50,51] and phosphonium salts[52,53] as well as *N*-heterocyclic carbenes[54] have been successfully equipped into the arsenal for catalyzing the synthesis of cyclic carbonates. In this context, bifunctional strategies have been often taken for the design of efficient organocatalysts, upon which the incorporation of hydrogen donors into ionic liquids can greatly improve the catalytic activities based on a putative mechanism that epoxides ring

Received: February 17, 2017
Revised: March 12, 2017
Published: March 13, 2017

opening is facilitated via a synergistic effect of hydrogen bonding and nucleophilic anions.[55,56] Although significant advances have been achieved in this organocatalytic field, it is still a challenge to perform the reaction at ambient $CO_2$ pressure or with flue gas.

4-(Dimethylamino)pyridine (DMAP) is well-known as an acyl-transfer catalyst for esterification and probably the most frequently used nucleophilic catalyst for various synthetically useful transformations.[57−59] In addition, the use of DMAP as a nucleophilic cocatalyst in combination with H-bond donors[60,61] or Lewis acids[62,63] has also been employed to accelerate the coupling of epoxides with $CO_2$. However, no DMAP-based protic ILs have been explored as a promoter for the title reaction. In view of several recent reports on ammonium ILs catalyzed fixation of $CO_2$ with epoxides[50] and the "privileged" structure of DMAP in positive charge delocalization (Scheme 2), we reasoned that DMAP derived protic ILs would serve as more effective catalysts for the fixation of $CO_2$ into cyclic carbonates than conventional ILs. The enhanced activity was largely ascribed to the weaker electrostatic cation/anion interactions and increased nucleophilicity of anions as well as the simultaneous presence of H-bond interaction with epoxides. Herein, we demonstrated for the first time that [DMAPH]Br was an efficient and recyclable organocatalyst for the synthesis of cyclic carbonates with atmospheric $CO_2$ as well as flue gas.

**Scheme 2. Three Mesomeric Forms of Protic DMAP**

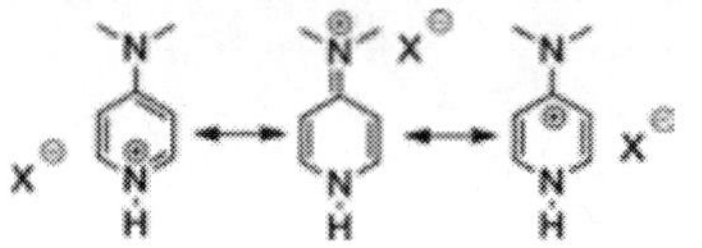

On the basis of these considerations, we commenced our investigations by first synthesis of a series of protonated DMAP derivatives (**1a−d**) for evaluation. Several known aliphatic protonated onium salts (**2a** [DBUH]Cl,[64] **2b** [DBUH]Br, **3** [HMIM]Br,[65] **4** [TMGH]Br, and **5** [DABCOH]Br) have also been prepared for comparison (Scheme 3).

## ■ RESULTS AND DISCUSSION

The initial experiment was performed using styrene oxide (SO) as a model substrate in the presence of 1 mol % of protic ILs depicted in Scheme 3 under 1 atm of $CO_2$, and the results are shown in Table 1. Control experiment demonstrated that no reaction occurred in the absence of any catalyst (Table 1, entry 1). The use of DMAP/NaBr as the catalyst gave a conversion of 48% under identical condition (Table 1, entry 2). Obviously, the cation structures of the investigated ILs have strong impact on the catalytic activities (entries 3−11) and [DMAPH]Br **1a** showed the best activity, giving the desired product in excellent conversion and selectivity (entry 3). The results obtained suggest that a synergetic interplay of protic acid and bromide is crucial for the enhanced catalytic activity. The variation of substituents at 4-position of **1a** to less electron-donating groups gave rise to less effective activity, and the yields decreased to 78%, 60%, and 47%, respectively (entries 4−6), presumably due to inefficient positive charge delocalization on these cations. For comparison, several known protic ILs derived from various bases were also tested for the title reaction, which generally showed inferior activity to **1a** (entries 7−11). This is in line with previous work using [DBUH]Cl[64] and [HMIM]Br[65] as

**Scheme 3. Compounds Prepared as Potential Catalysts for the Title Reaction**

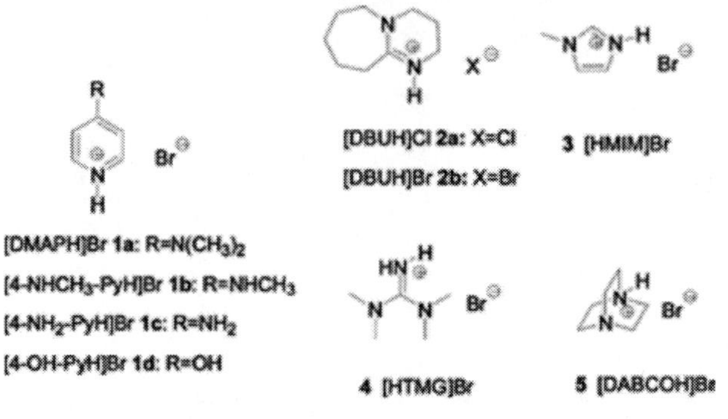


**Table 1. Synthesis of Styrene Carbonate (SC) Catalyzed with Different Protic ILs**

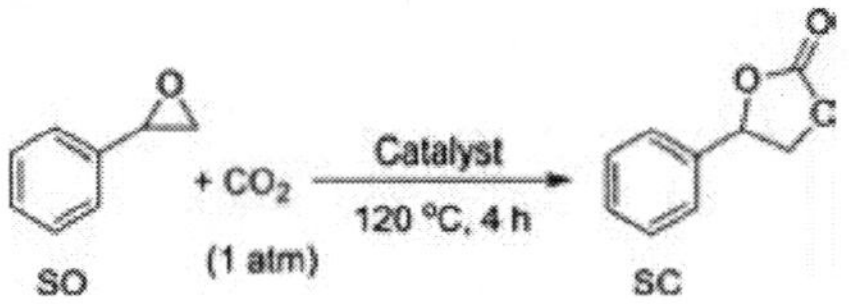


| entry[a] | catalyst | SO conv. [%][b] | SC selec. [%][b] |
|---|---|---|---|
| 1 | none | n.d. | n.d. |
| 2 | DMAP/NaBr | 48 | 99 |
| 3 | **1a** [DMAPH]Br | 96 | 99 |
| 4 | **1b** [4-NHCH$_3$-PyH]Br | 78 | 99 |
| 5 | **1c** [4-NH$_2$-PyH]Br | 60 | 99 |
| 6 | **1d** [4-OH-PyH]Br | 47 | 99 |
| 7 | **2a** [DBUH]Cl | 80 | 99 |
| 8 | **2b** [DBUH]Br | 85 | 99 |
| 9 | **3** [HMIM]Br | 84 | 99 |
| 10 | **4** [HTMG]Br | 71 | 99 |
| 11 | **5** [DABCOH]Br | 61 | 99 |

[a]Reaction conditions: SO (10 mmol), $CO_2$ (1 atm), catalyst (1 mol %), 120 °C, 4 h. [b]Determined by $^1$H NMR ($CDCl_3$, 400 MHz).

**Table 2. Optimization of Reaction Conditions**

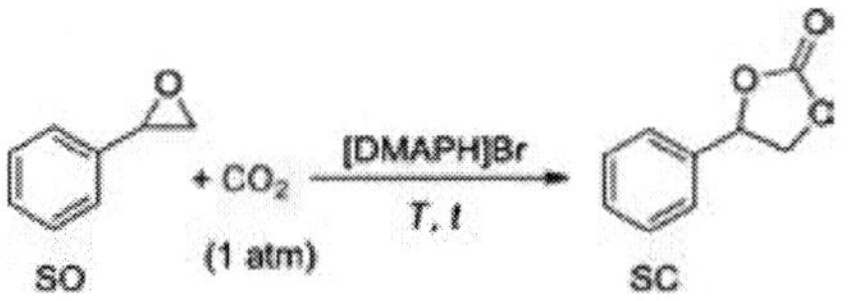


| entry[a] | catalyst loading [mol %] | $T$ [°C] | $t$ [h] | SO conv. [%][b] | SC selec. [%][b] |
|---|---|---|---|---|---|
| 1 | 0.1 | 120 | 4 | 38 (86)[c] | 99 (99) |
| 2 | 0.3 | 120 | 4 | 56 | 99 |
| 3 | 0.5 | 120 | 4 | 76 | 99 |
| 4 | 0.7 | 120 | 4 | 85 | 99 |
| 5 | 1.0 | 120 | 4 | 96 | 99 |
| 6 | 1.0 | 100 | 4 | 62 | 99 |
| 7 | 1.0 | 80 | 4 | 36 | 99 |
| 8 | 1.0 | 80 | 12 | 76 | 99 |
| 9[d] | 1.0 | 120 | 14 | 92 | 99 |

[a]Reaction conditions: SO (10 mmol), $CO_2$ (1 atm). [b]Determined by $^1$H NMR ($CDCl_3$, 400 MHz). [c]Reaction time 24 h, and the conversion is in parathesis. [d]Mixture of 15% $CO_2$ and 85% $N_2$ in volume.

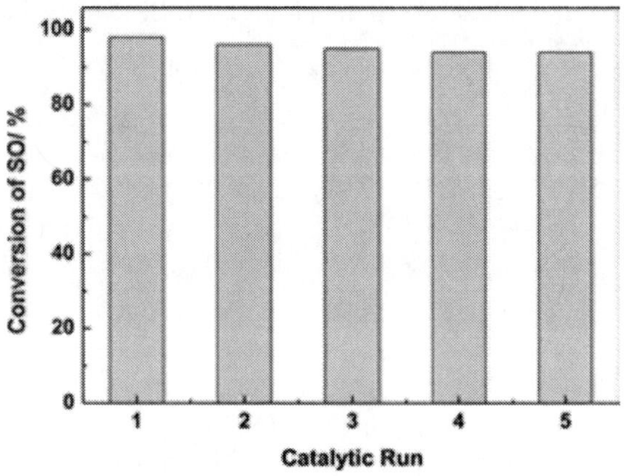


**Figure 1.** Catalyst recycling for cycloaddition reaction (reaction conditions similar to those of Table 1, entry 3).

**Scheme 4. Proposed Catalytic Mechanism**

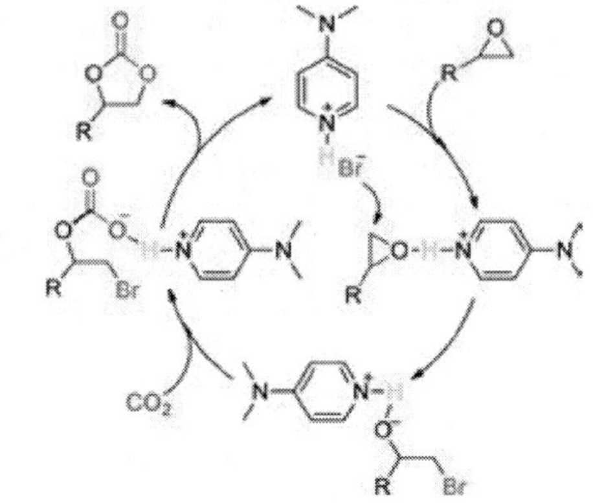


the catalysts, where the reaction had to undergo at high pressure of carbon dioxide.

**Table 3. Reactions of Various Epoxides with $CO_2$ using [DMAPH]Br as the Catalyst**

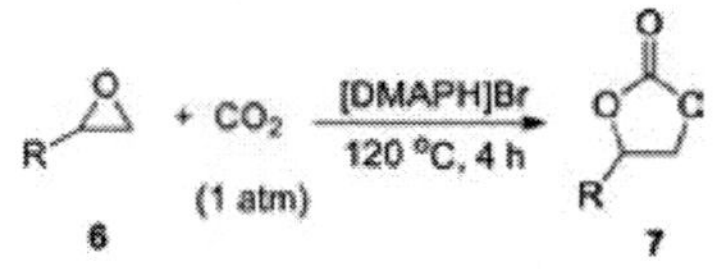


| Entry[a] | Substrate | Product | Conv. [%][b] | Selec. [%][b] |
|---|---|---|---|---|
| 1 |  6a | 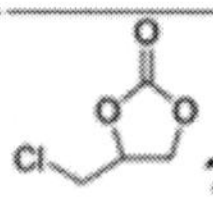 7a | 96 | 99 |
| 2 |  6b | 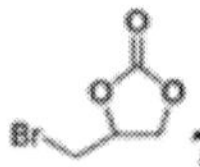 7b | 96 | 99 |
| 3 | 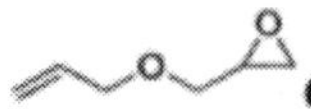 6c | 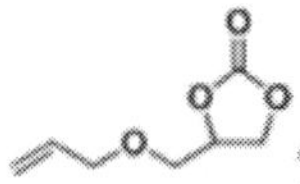 7c | 95 | 99 |
| 4 | 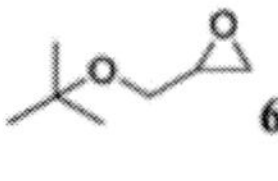 6d | 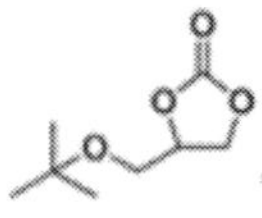 7d | 95 | 99 |
| 5 | 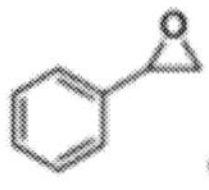 6e | 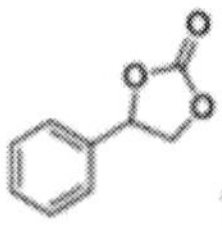 7e | 96 | 99 |
| 6 | 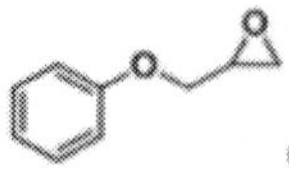 6f | 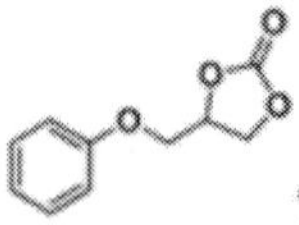 7f | 99 | 99 |
| 7[c] | 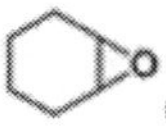 6g | 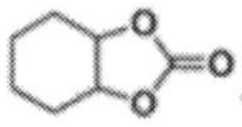 7g | 39 | 85 |

[a]Reaction conditions: 1 (10 mmol), $CO_2$ (1 atm), 120 °C, 1a (1 mol %), 4 h. [b]Determined by $^1$H NMR ($CDCl_3$, 400 MHz). [c]12 h.

After [DMAPH]Br **1a** was established as the optimal catalyst, we next investigated the effects of other reaction parameters on the reaction including catalyst loadings, reaction temperatures, and concentrations of $CO_2$ (Table 2). In general, the conversions of SO were proportional to the catalyst loadings and 1 mol % of catalyst loading was optimal for this reaction (entries 1–5). Of note, decreasing the catalyst loading to 0.1 mol %, a decent conversion of SO can also be obtained albeit with a prolonged reaction time (24 h) (Table 2, entry 1). Furthermore, the SO conversions were strongly affected by the reaction temperature, which grew steadily from 38% to 96% when the reaction temperature was increased from 80 to 120 °C (entries 5–8). The results in Table 2 display that 120 °C was a suitable temperature. It is particularly worth mentioning that under optimal conditions and using a mixture of 15% $CO_2$ and 85% $N_2$ to mimic flue gas in industry, catalyst **1a** was able to convert SO into SC with excellent conversion (entry 9), which, to our knowledge, represents the first example of protic ILs catalyzed synthesis of cyclic carbonate with diluted carbon dioxide.

To test catalyst reusability, the reaction was carried out in the presence of a catalytic amount of [DMAPH]Br under the optimal condition. After the reaction was completed, 5.0 mL of water was added into the mixture, which was then subjected to centrifugal separation. The aqueous phase was collected and the catalyst was recovered after removal of water by rotary evaporation. The recovered catalyst was used for next run without further purification. The results shown in Figure 1 demonstrate that the catalyst could be recycled over five times without appreciable loss of catalytic activity.

Under the optimal reaction conditions, the substrate scope of this protocol was also examined and the results are summarized in Table 3. Generally, terminal epoxides bearing with both electron-withdrawing and electron-donating groups could be transformed to the corresponding cyclic carbonates with excellent conversions and selectivities (entries 1–6), but for internal cyclohexene oxide, a relatively low conversion and selectivity was obtained even when the reaction time was prolonged to 12 h (entry 7), presumably due to the steric hindrance of cyclohexene oxide.

On the basis of previous reports[31,33,34,56,65] and our experimental results, a plausible mechanism was proposed for [DMAPH]Br-catalyzed cyclization of epoxides with atmospheric $CO_2$ (Scheme 4). First, epoxide was activated via the hydrogen bonding between the proton of [DMAPH]Br and the oxygen of the epoxide, which resulted in the polarization of the C–O bond and therefore facilitated its ring opening by nucleophilic attack of the bromide. The formed bromo-alkoxide can be stabilized by coordination to the cation part through hydrogen bonding, and then reacts with $CO_2$ and cyclizes to give the cyclic carbonate with the regeneration of [DMAPH]Br.

## CONCLUSION

In summary, [DMAPH]Br **1a** has been developed as a highly efficient and recyclable catalyst for the synthesis of cyclic carbonates from epoxides and atmospheric carbon dioxide. Various terminal epoxides bearing with both alkyl and aryl substituents are tolerated with this protocol, giving the corresponding product in excellent conversions and selectivities. It is particularly worth mentioning that using a mixture of 15% $CO_2$ and 85% $N_2$ to mimic flue gas in industry, catalyst **1a** was also able to convert SO into the product with excellent conversion. Moreover, the catalyst can be recyled over five times without considerable loss in activties. Further exploitation of [DMAPH]Br to other transformations is currently under investigation in our laboratory.

## ASSOCIATED CONTENT

### Supporting Information

The Supporting Information is available free of charge on the ACS Publications website at DOI: 10.1021/acssuschemeng.7b00513.

Preparations of ILs, NMR and FT-IR of ILs, NMR of products, procedure of the cycloaddition reaction (PDF)

## AUTHOR INFORMATION

**Corresponding Author**

*Z. Zhang. E-mail: zhiguo.zhang@zju.edu.cn.

**ORCID**

Zhiguo Zhang: 0000-0003-1681-4853

Qiwei Yang: 0000-0002-6469-5126

**Notes**

The authors declare no competing financial interest.

## ACKNOWLEDGMENTS

We are grateful for financial support from the MOST (2016YFA0202900), the National Natural Science Foundation of China (21376212, 21436010), the Fundamental Research Funds for the Central Universities (2016FZA4019) and the Natural Science Foundation of Zhejiang Province, China (LY13B060001).

## REFERENCES

(1) Zhang, X.; Zhang, X.; Dong, H.; Zhao, Z.; Zhang, S.; Huang, Y. Carbon capture with ionic liquids: overview and progress. *Energy Environ. Sci.* **2012**, *5*, 6668.

(2) Yu, K. M.; Curcic, I.; Gabriel, J.; Tsang, S. C. Recent advances in $CO_2$ capture and utilization. *ChemSusChem* **2008**, *1*, 893.

(3) Markewitz, P. Worldwide Innovations in the development of carbon capture technologies and the utilization of $CO_2$. *Energy Environ. Sci.* **2012**, *5*, 7281.

(4) Gibson, D. H. The organometallic chemistry of carbon dioxide. *Chem. Rev.* **1996**, *96*, 2063.

(5) Sakakura, T.; Choi, J. C.; Yasuda, H. Transformation of carbon dioxide. *Chem. Rev.* **2007**, *107*, 2365.

(6) Aresta, M.; Dibenedetto, A. Utilisation of $CO_2$ as a chemical feedstock: opportunities and challenges. *Dalton T.* **2007**, *28*, 2975.

(7) Yu, K. M.; Curcic, I.; Gabriel, J.; Tsang, S. C. Recent advances in $CO_2$ capture and utilization. *ChemSusChem* **2008**, *1*, 893.

(8) Omae, I. Recent developments in carbon dioxide utilization for the production of organic chemicals. *Coord. Chem. Rev.* **2012**, *256*, 1384.

(9) Shaikh, A. A. G.; Sivaram, S. Organic carbonates†. *Chem. Rev.* **1996**, *96*, 951.

(10) Fukuoka, S.; Kawamura, M.; Komiya, K.; Tojo, M.; Hachiya, H.; Hasegawa, K.; Aminaka, M.; Okamoto, H.; Fukawa, I.; Konno, S. A novel non-phosgene polycarbonate production process using by-product $CO_2$ as starting material. *Green Chem.* **2003**, *5*, 497.

(11) North, M.; Pasquale, R.; Young, C. Synthesis of cyclic carbonates from epoxides and $CO_2$. *Green Chem.* **2010**, *12*, 1514.

(12) Schäffner, B.; Schäffner, F.; Verevkin, S. P.; Börner, A. Organic carbonates as solvents in synthesis and catalysis. *Chem. Rev.* **2010**, *110*, 4554.

(13) He, Q.; O'Brien, J. W.; Kitselman, K. A.; Tompkins, L. E.; et al. Synthesis of cyclic carbonates from $CO_2$ and epoxides using ionic liquids and related catalysts including choline chloride-metal halide mixtures. *Catal. Sci. Technol.* **2014**, *4*, 1513.

(14) Fiorani, G.; Guo, W.; Kleij, A. W. Sustainable conversion of carbon dioxide: the advent of organocatalysis. *Green Chem.* **2015**, *17*, 1375.
(15) Clements, J. H. Reactive applications of cyclic alkylene carbonates. *Ind. Eng. Chem. Res.* **2003**, *42*, 663.
(16) Yoshida, M.; Ihara, M. Novel methodologies for the synthesis of cyclic carbonates. *Chem. - Eur. J.* **2004**, *10*, 2886.
(17) Vidal, M.; Domínguez, J. Making plastics from carbon dioxide: salen metal complexes as catalysts for the production of polycarbonates from epoxides and $CO_2$. *Chem. Rev.* **2007**, *107*, 2388.
(18) Cokoja, M.; Bruckmeier, C.; Rieger, B.; Herrmann, W. A.; Kuhn, F. E. Transformation of carbon dioxide with homogeneous transition-metal catalysts: A molecular solution to a Global Challenge? *Angew. Chem., Int. Ed.* **2011**, *50*, 8510.
(19) Comerford, J. W.; Ingram, I. D. V.; North, M.; Wu, X. Sustainable metal-based catalysts for the synthesis of cyclic carbonates containing five-membered rings. *Green Chem.* **2015**, *17*, 1966.
(20) Martín, C.; Fiorani, G.; Kleij, A. W. Recent advances in the catalytic preparation of cyclic organic carbonates. *ACS Catal.* **2015**, *5*, 1353.
(21) Kihara, N.; Hara, N.; Endo, T. Catalytic activity of various salts in the reaction of 2,3-epoxypropyl phenyl ether and carbon dioxide under atmospheric pressure. *J. Org. Chem.* **1993**, *58*, 6198.
(22) Barkakaty, B.; Morino, K.; Sudo, A.; Endo, T. Amidine-mediated delivery of $CO_2$ from gas phase to reaction system for highly efficient synthesis of cyclic carbonates from epoxides. *Green Chem.* **2010**, *12*, 42.
(23) Huang, J. W.; Shi, M. Chemical fixation of carbon dioxide by $NaI/PPh_3/PhOH$. *J. Org. Chem.* **2003**, *68*, 6705.
(24) Song, J.; Zhang, B.; Zhang, P.; Ma, J.; Liu, J.; Fan, H.; Jiang, T.; Han, B. Highly efficient synthesis of cyclic carbonates from $CO_2$ and epoxides catalyzed by KI/lecithin. *Catal. Today* **2012**, *183*, 130.
(25) Yamaguchi, K.; Ebitani, K.; Yoshida, T.; Yoshida, H.; Kaneda, K. Mg–Al mixed oxides as highly active acid–base catalysts for cycloaddition of carbon dioxide to epoxides. *J. Am. Chem. Soc.* **1999**, *121*, 4526.
(26) Yano, T.; Matsui, H.; Koike, T.; Ishiguro, H.; Fujihara, H.; Yoshihara, M.; Maeshima, T. Magnesium oxide-catalysed reaction of carbon dioxide with an epoxide with retention of stereochemistry. *Chem. Commun.* **1997**, 1129.
(27) Tomishige, K.; Yasuda, H.; Yoshida, Y.; Nurunnabi, M.; Li, B.; Kunimori, K. Catalytic performance and properties of ceria based catalysts for cyclic carbonate synthesis from glycol and carbon dioxide. *Green Chem.* **2004**, *6*, 206.
(28) Peng, J.; Deng, Y. Cycloaddition of carbon dioxide to propylene oxide catalyzed by ionic liquids. *New J. Chem.* **2001**, *25*, 639.
(29) Kawanami, H.; Sasaki, A.; Matsui, K.; Ikushima, Y. A rapid and effective synthesis of propylene carbonate using a supercritical $CO_2$-ionic liquid system. *Chem. Commun.* **2003**, 896.
(30) Xie, Y.; Zhang, Z.; Jiang, T.; He, J.; Han, B.; Wu, T.; Ding, K. $CO_2$ cycloaddition reactions catalyzed by an ionic liquid grafted onto a highly cross-linked polymer matrix. *Angew. Chem., Int. Ed.* **2008**, *39*, 7255.
(31) Sun, J.; Zhang, S.; Cheng, W.; Ren, J. Hydroxyl-functionalized ionic liquid: a novel efficient catalyst for chemical fixation of $CO_2$ to cyclic carbonate. *Tetrahedron Lett.* **2008**, *49*, 3588.
(32) Ghazali-Esfahani, S.; Song, H.; Paunescu, E.; Bobbink, F. D.; Liu, H.; Fei, Z.; Laurenczy, G.; Bagherzadeh, M.; Yan, N.; Dyson, P. J. Cycloaddition of $CO_2$ to epoxides catalyzed by imidazolium-based polymeric ionic liquids. *Green Chem.* **2013**, *15*, 1584.
(33) Xu, B.; Wang, J.; Sun, J.; Huang, Y.; Zhang, J.-P.; Zhang, X.-P.; Zhang, S.-J. Fixation of $CO_2$ into cyclic carbonates catalyzed by ionic liquids: a multi-scale approach. *Green Chem.* **2015**, *17*, 108.
(34) Cokoja, M.; Wilhelm, M. E.; Anthofer, M. H.; Herrmann, W. A.; Kuhn, F. E. Synthesis of cyclic carbonates from epoxides and carbon dioxide by using organocatalysts. *ChemSusChem* **2015**, *8*, 2436.
(35) Sopeña, S.; Fiorani, G.; Martín, C.; Kleij, A. W. Highly Efficient organocatalyzed conversion of oxiranes and $CO_2$ into organic carbonates. *ChemSusChem* **2015**, *8*, 3248.
(36) Kohrt, C.; Werner, T. Recyclable bifunctional polystyrene and silica gel-supported organocatalyst for the coupling of $CO_2$ with epoxides. *ChemSusChem* **2015**, *8*, 2031.
(37) Whiteoak, C. J.; Nova, A.; Maseras, F.; Kleij, A. W. Merging sustainability with organocatalysis in the formation of organic carbonates by using $CO_2$ as a feedstock. *ChemSusChem* **2012**, *5*, 2032.
(38) Lan, D. H.; Fan, N.; Wang, Y.; Gao, X.; Zhang, P.; Chen, L.; Au, C.-T.; Yin, S.-F. Recent advances in metal-free catalysts for the synthesis of cyclic carbonates from $CO_2$ and epoxides. *Chin. J. Catal.* **2016**, *37*, 826.
(39) Yoshida, M.; Ihara, M. Novel methodologies for the synthesis of cyclic carbonates. *Chem. - Eur. J.* **2004**, *10*, 2886.
(40) Castro-Osma, J. A.; Lamb, K. J.; North, M. Cr(salophen) complex catalyzed cyclic carbonate synthesis at ambient temperature and pressure. *ACS Catal.* **2016**, *6*, 5012.
(41) Ma, R.; He, L.; Zhou, Y. An efficient and recyclable tetraoxo-coordinated zinc catalyst for the cycloaddition of epoxides with carbon dioxide at atmospheric pressure. *Green Chem.* **2016**, *18*, 226.
(42) Lang, X. D.; Yu, Y. C.; He, L. N. Zn-salen complexes with multiple hydrogen bonding donor and protic ammonium bromide: Bifunctional catalysts for $CO_2$ fixation with epoxides at atmospheric pressure. *J. Mol. Catal. A: Chem.* **2016**, *420*, 208.
(43) Yu, S.; Liu, X.; Ma, J.; Niu, Z.; Cheng, P. A new catalyst for the solvent-free conversion of $CO_2$ and epoxides into cyclic carbonate under mild conditions. *J. CO2 Util.* **2016**, *14*, 122.
(44) Liu, L.; Zhang, J.; Fang, H.; Chen, L.; Su, C.-Y. Metal-organic gel material based on UiO-66-$NH_2$ nanoparticles for improved adsorption and conversion of carbon dioxide. *Chem. - Asian J.* **2016**, *11*, 2278.
(45) North, M.; Wang, B.; Young, C. Influence of flue gas on the catalytic activity of an immobilized aluminium(salen) complex for cyclic carbonate synthesis. *Energy Environ. Sci.* **2011**, *4*, 4163.
(46) Barthel, A.; Saih, Y.; Gimenez, M.; Pelletier, J. D. A.; Kuhn, F. E.; D'Elia, V.; Basset, J.-M. Highly integrated $CO_2$ capture and conversion: direct synthesis of cyclic carbonates from industrial flue gas. *Green Chem.* **2016**, *18*, 3116.
(47) Motokura, K.; Itagaki, S.; Iwasawa, Y.; Miyaji, A.; Baba, T. Silica-supported aminopyridinium halides for catalytic transformations of epoxides to cyclic carbonates under atmospheric pressure of carbon dioxide. *Green Chem.* **2009**, *11*, 1876.
(48) Ochiai, B.; Endo, T. Polymer-supported pyridinium catalysts for synthesis of cyclic carbonate by reaction of carbon dioxide and oxirane. *J. Polym. Sci., Part A: Polym. Chem.* **2007**, *45*, 5673.
(49) Sun, J.; Wang, J.; Cheng, W.; Zhang, J.; Li, X.; Zhang, S.; She, Y. Chitosan functionalized ionic liquid as a recyclable biopolymer-supported catalyst for cycloaddition of $CO_2$. *Green Chem.* **2012**, *14*, 654.
(50) Foltran, S.; Mereau, R.; Tassaing, T. Theoretical study on the chemical fixation of carbon dioxide with propylene oxide catalyzed by ammonium and guanidinium salts. *Catal. Sci. Technol.* **2014**, *4*, 1585.
(51) Zhao, Y.; Yao, C.; Chen, G.; Yuan, Q. Highly efficient synthesis of cyclic carbonate with $CO_2$ catalyzed by ionic liquid in a microreactor. *Green Chem.* **2013**, *15*, 446.
(52) Büttner, H.; Steinbauer, J.; Werner, T. Synthesis of cyclic Carbonates from epoxides and carbon dioxide by using bifunctional one-component phosphorus-based organocatalysts. *ChemSusChem* **2015**, *8*, 2655.
(53) Toda, Y.; Komiyama, Y.; Kikuchi, A.; Suga, H. Tetraarylphosphonium salt-catalyzed carbon dioxide fixation at atmospheric pressure for the synthesis of cyclic carbonates. *ACS Catal.* **2016**, *6*, 6906.
(54) Yang, L.; Wang, H. Recent advances in carbon dioxide capture, fixation, and activation by using N-heterocyclic carbenes. *ChemSusChem* **2014**, *7*, 962.
(55) Chen, W.; Zhong, L.; Peng, X.; Sun, R.; Lu, F.-c. Chemical fixation of carbon dioxide using a green and efficient catalytic system based on sugarcane bagasse—an agricultural waste. *ACS Sustainable Chem. Eng.* **2015**, *3*, 147.

(56) Cheng, W.; Su, Q.; Wang, J.; Sun, J.; et al. Ionic Liquids: The synergistic catalytic effect in the synthesis of cyclic carbonates. *Catalysts* **2013**, *3*, 878–901.
(57) Steglich, D. W.; Höfle, G. N,N-Dimethyl-4-pyridinamine, a very effective acylation catalyst. *Angew. Chem., Int. Ed. Engl.* **1969**, *8*, 981.
(58) Höfle, G.; Steglich, W.; Vorbrüggen, H. 4-Dialkylaminopyridines as highly active acylation catalysts. *Angew. Chem., Int. Ed. Engl.* **1978**, *17*, 569.
(59) Scriven, E. F. V. 4-Dialkylaminopyridines: Super acylation and alkylation catalysts. *Chem. Soc. Rev.* **1983**, *12*, 129.
(60) Shen, Y. M.; Duan, W. L.; Shi, M. Phenol and organic bases co-catalyzed chemical fixation of carbon dioxide with terminal epoxides to form cyclic carbonates. *Adv. Synth. Catal.* **2003**, *345*, 337.
(61) Roshan, R. K.; Achuthan, R. P.; Kathalikkattil, A. C.; Babu, R.; Mathai, G.; Lee, H.-S.; Park, D.-W. A computational study into the mechanistic insights of base catalysed synthesis of cyclic carbonates from $CO_2$: Bicarbonate anion as active species. *Catal. Sci. Technol.* **2016**, *6*, 1030.
(62) Paddock, R. L.; Hiyama, Y.; McKay, J. M.; Nguyen, S. T. Co(III) porphyrin/DMAP: an efficient catalyst system for the synthesis of cyclic carbonates from $CO_2$ and epoxides. *Tetrahedron Lett.* **2004**, *45*, 2023.
(63) Ramidi, P.; Munshi, P.; Gartia, Y.; Pulla, S.; Biris, A. S.; Paul, A.; Ghosh, A. Synergistic effect of alkali halide and Lewis base on the catalytic synthesis of cyclic carbonate from $CO_2$ and epoxide. *Chem. Phys. Lett.* **2011**, *512*, 273.
(64) Yang, Z.; He, L.; Miao, C.; Chanfreau, S. Lewis Basic Ionic Liquids-Catalyzed Conversion of Carbon Dioxide to Cyclic Carbonates. *Adv. Synth. Catal.* **2010**, *352*, 2233.
(65) Xiao, L.; Su, D.; Yue, C.; Wu, W. Protic ionic liquids: A highly efficient catalyst for synthesis of cyclic carbonate from carbon dioxide and epoxides. *J. CO2 Util.* **2014**, *6*, 1.

# 493435 New Catalytical System in Ionic Liquid for Small Molecule Conversion

**Abstract:**

Ionic liquids (ILs) have gained practical and fundamental interests as replacements of conventional electrolytes in modern electrochemical devices (e.g., advanced batteries, fuel cells, super-capacitors, sensors, and electrochromic displays). All electrochemical applications require fundamental understanding of the interface reactions, interface structure, and properties which can be obtained through in situ characterization of the dynamic interface processes. In this work, we discuss new catalytical system detrived from ionic liquid and ILâ€“electrode interface for methane, CO, methanol and et.al conversion, which were conducted in our lab.Â The related in situ characterization techniques including electrochemical impedance spectroscopy (EIS), electrochemical quartz crystal microbalance (EQCM), and in situ infrared spectroscopy (IR) for studying ILâ€“electrode interface will be presented. This task-specific ILs and their surface immobilization on solid supports as a replacement of conventional recognition elements and electrolytic media could be widely applied in green energy and enviromental monitoring.

# 496951 Ionic Liquids for Separation of Aromatics and Aliphatics: Extraction and Regeneration of Solvent Using $CO_2$

**Abstract:**

Ionic liquids have been shown to be effective extraction solvents for the extraction of aromatic compounds from aliphatic compounds, as many ionic liquids are nearly insoluble in the raffinate phase and have good selectivity towards aromatics. One of the most important steps in an extraction process is regeneration of the solvent, which is usually accomplished by distillation or stripping. These methods often require high temperatures and high energy consumption, leading to a desire for a greener alternative. It has been shown that near supercritical $CO_2$ can be used as an antisolvent gas to induce a liquid-liquid phase split in some mixtures of ionic liquids and organic compounds. Previous work has shown the organic-rich liquid phase contains negligible ionic liquid, making the gas antisolvent process a very effective separation technique, as $CO_2$can be removed easily by lowering pressure.

Solubility has been measured in mixtures of ionic liquids and aromatic compounds, where a liquid+vapor to liquid+liquid+vapor phase transition is observed, demonstrating the ability of $CO_2$ to regenerate the ionic liquid. To be an energy efficient process, a low phase transition pressure with a low composition of $CO_2$ in the liquid phases is desired. Additionally, the ionic liquid must have high selectivity and capacity for the aromatic in the liquid-liquid extraction process. Various ionic liquids have been tested both for extraction and regeneration with $CO_2$ by measuring ternary liquid-liquid equilibrium of ionic liquid, aromatic, and heptane and high pressure phase behavior of ionic liquid, aromatic, $CO_2$ systems. Properties such as excess enthalpy and excess volume have also been studied to gain insight into the underlying thermodynamics to help guide process optimization.

# 500469 Coculture of Methanotrophs and Microalgae â€“ a Flexible Platform for Biological $CH_4/CO_2$ Co-Utilization

**Abstract:**

Recent findings published in Nature[1] and Nature Geoscience[2] suggest that the coupling of methane oxidation and oxygenic photosynthesis are prevalent in nature and reduce $CH_4$ emissions and reuse $CO_2$. In addition, a 2015 ISME Journal paper reported[3] that methane removal in anoxic waters (freshwater lakes depleted of dissolved $O_2$) is due to true aerobic oxidation of methane, rather than anaerobic oxidation, fueled by *in situ* $O_2$ produced from photosynthetic algae. These findings suggest that the coculture of methanotrophs and microalgae represents not only a feasible, but also a highly promising strategy for simultaneous conversion of biogas (both $CH_4$ and $CO_2$) into useful products, feed, energy or other products.

Such coupling has been partially validated in laboratory settings. (1) It was reported that coculture of *Methylocystis parvus* (methanotroph) and *Scenedesmus sp.* (microalgae) can achieve total microbial conversion of both $CH_4$ (60%) and $CO_2$ (40%) in a synthetic biogas without external $O_2$ supply[4]; (2) coculture of *Methylomicrobium alcaliphilum* (methanotroph) and *Synechococcus* PCC 7002 (cyanobacteria) exhibit robust growth on diverse gas mixtures including raw biogas and synthetic natural gas[5]. However, both published research were conducted under continuous illumination, and there has not been any systematic study on the factors that could affect the stability and performance of the cocultures.

In this work, we hypothesize that the coculture of methanotrophs and microalgae offers a flexible platform for simultaneous biological $CH_4/CO_2$ co-utilization, and cocultures of different methanotroph and microalgae strains can be stabilized if their growth conditions are compatible. To validate our hypothesis, we have established three pairs of methanotroph-microalgae coculture, with one pair preferring high pH and high salt condition, while the other two pairs prefer neutral pH and low salt condition. Multiple transfers of the coculture to new media confirm the stability of the three coculture pairs, all under light-dark cycles (12:12 or 16:8) of the illumination. In addition, one pair of the model coculture was used to gain fundamental understandings on the interactions between the methanotroph and microalgae strains. Specifically, inoculation ratio, gas phase composition and illumination regimes were examined using batch experiments. Individual and total cell concentrations, $CH_4$ consumption rate, overall $CO_2$ production/consumption rate, and overall $O_2$ production/consumption rate, were measured over time to assess the effect of different factors, and evaluate the conversion performance of the coculture under different conditions.

(1) Raghoebarsing, A. A.; Smolders, A. J.; Schmid, M. C.; Rijpstra, W. I. C.; Wolters-Arts, M.; Derksen, J.; Jetten, M. S.; Schouten, S.; Damsté, J. S. S.; Lamers, L. P. Methanotrophic Symbionts Provide Carbon for Photosynthesis in Peat Bogs. *Nature* **2005**, *436*, 1153.

(2) Kip, N.; van Winden, J. F.; Pan, Y.; Bodrossy, L.; Reichart, G.-J.; Smolders, A. J.; Jetten, M. S.; Damsté, J. S. S.; den Camp, H. J. O. Global Prevalence of Methane Oxidation by Symbiotic Bacteria in Peat-Moss Ecosystems. *Nat. Geosci.* **2010**, *3*, 617.

(3) Milucka, J.; Kirf, M.; Lu, L.; Krupke, A.; Lam, P.; Littmann, S.; Kuypers, M. M.; Schubert, C. J. Methane Oxidation Coupled to Oxygenic Photosynthesis in Anoxic Waters. *ISME J.* **2015**.

(4) Van der Ha, D.; Nachtergaele, L.; Kerckhof, F.-M.; Rameiyanti, D.; Bossier, P.; Verstraete, W.; Boon, N. Conversion of Biogas to Bioproducts by Algae and Methane Oxidizing Bacteria. *Environ. Sci. & Technol.* **2012**, *46*, 13425.

(5) Hill, E. A.; Chrisler, W. B.; Beliaev, A. S.; Bernstein, H. C. A Flexible Microbial Co-Culture Platform for Simultaneous Utilization of Methane and Carbon Dioxide from Gas Feedstocks. *Bioresour. Technol.* **2017**.

# 497573 Identification of Antibiotic Resistome in Urban Watersheds Via High-Throughput Screening

**Abstract:**

Urban watersheds from point sources are potential reservoirs of antibiotic resistance genes (ARGs). However, few studies have investigated urban watersheds of non-point source, which can provide information on the background level of antibiotic resistance in a bacterial community. Antibiotic resistance levels of two watersheds (R1, R3) were examined by employing heterotrophic plate counts (HPC) as a culturing method to obtain counts of bacteria resistant to seven antibiotics belonging to different classes (erythromycin, kanamycin, lincomycin, norfloxacin, sulfanilamide, tetracycline and trimethoprim). From the HPC study, 239 antibiotic resistant bacteria were characterized for resistance to more antibiotics. Furthermore, ARGs and antimicrobial biosynthesis genes were identified using GeoChip version 5.0 to elucidate the resistomes of surface waters in R1 and R3.Â GeoChip revealed similar ARGs in both watersheds, but with significantly higher intensities for *tetX* and Î²-lactamase B genes in R1 compared to R3. The genes with the highest average normalized intensities in R1 and R3 were tetracycline *(tet)* and fosfomycin (*fosA)* resistance genes, respectively. This study demonstrates that the use of a DNA microarray is useful for identifying naturally occurring resistomes and for comparing bacterial communities in urban watersheds of different land-use profiles.

# 490216 Influence of Subsurface Storage on the Microbiological and Physicochemical Quality of Surface Water


**Abstract:**

### Background

In land-scarce countries such as Singapore, storage of surface water in underground caverns is being considered as a potential solution to ensure water security for the future. However, long-term subsurface storage of the surface water is expected to undergo quality changes, as the conditions in underground caverns would differ markedly from those in surface water reservoirs. The objective of this study was to investigate the effect of subsurface storage conditions on the quality of surface water in model underground caverns.

### Methods

Bench-top simulated underground caverns were constructed by drilling cylindrical wells of diameter 150 mm and depth 50 mm in a granite slab and half of the wells were covered with opaque black acrylic sheets. The remaining wells were used as control and covered with transparent acrylic sheets for the light from a white LED source (Wavelength - 6000 K, Power - 10 W, Intensity ~ 1 $W/m^2$) to pass through. Raw water from three local reservoirs was mixed in equal proportion and 600 ml of the mixed water was added to each well. Microbiological water quality parameters were monitored periodically for six months by CFU counts to estimate bacteria and fungi, and by chlorophyll-A measurement to estimate total algae. Physicochemical water quality parameters including pH, conductivity, redox potential, chloride, total organic carbon (TOC), total nitrogen (TN), nitrate, ammonia and phosphate were measured either using water quality test kits or by water quality sensor probes. Leaching of heavy metal ions from granite was quantified using inductively coupled plasma coupled with mass spectrometry (ICP-MS). When needed, sodium tungstate at a final concentration of 10 mM was used to inhibit denitrification. Total RNA extracted from the water stored in cavern condition and control was sequenced by 2×300 bp paired end sequencing on MiSeq platform. Data were processed and analyzed using open source tools.

### Results

Heterotrophic plate count (HPC) in the planktonic phase reduced significantly ($p<0.05$) in the control ($2.5\pm0.9 \times 10^4$ CFU/ml) as well as in the cavern condition ($1.1\pm0.8 \times 10^5$ CFU/ml) as compared to the initial HPC ($3.0\pm0.3 \times 10^5$ CFU/ml) of the raw water. On the contrary, total algae increased in the light and decreased in the dark. There was no appreciable change in the pH, conductivity, redox potential, chloride concentration and TOC of the water. In addition, leaching of metal ions from granite was found to be highly dependent on the source of the granite. Under the simulated bench cavern condition used in this study, arsenic and cadmium leached out from the granite and reached a concentration of 8±1 mg/L and 41±2 mg/L, respectively, within a month. Intriguingly, the concentration of TN in the water under the cavern condition increased steadily from 0.8±0.1 mg/L to 2.4±0.2 mg/L over 6 months, which was not observed in the control. Further analyses showed that most of the N accumulated in the cavern condition was in the form of nitrate. In the presence of sodium tungstate (a denitrification inhibitor), nitrate accumulation was observed in the control with light, suggesting that the nitrate accumulation in the cavern condition was likely due to the inhibition of denitrification. Taxonomic and functional analyses of the RNA sequencing data support the theory that denitrification is inhibited in the cavern condition.

### Conclusions

Storage of surface water in the dark under cavern condition inhibits denitrification and leads to the accumulation of nitrate in the water. Leaching of metal ions from the rocks interacting with the water can also be a concern depending on the geochemical property of the storage system. The findings draw our attention to the potential areas of concern that should be considered in the design and operation of underground caverns for water storage.

## 501868 Use *Parageobacillus Thermoglucosidasius*: Save the Planet

**Abstract:**

Water, food, and energy are the key requirements for maintaining the growing population and sustainable growth. To fulfil the growing energy demands corn and sugarcane, which are also food sources, are being used for producing alternative fuel resources example bioethanol. The increase in use of corn or sugarcane for bioethanol production, and hence the cultivable land for their production, has brought an imbalance to the food, and water resources, surfacing food vs fuel, and drought like situations. Hence, the current study was done with an objective to utilize municipal solid waste (food waste, and paper and cardboard waste) as a substitute of corn and sugarcane for ethanol production, and reduce the overall water consumed in the bioprocess. *Parageobacillus thermoglucosidasius* was used as the fermentative microbe in the work. *P. thermoglucosidasius* is a thermophilic facultative anaerobe, which can produce ethanol from a wide variety of carbon sources. Different food (bread, tortillas, rice, potato etc.), and paper wastes (newspaper, packaging cardboard, and paper towel etc.) were used as carbon sources for ethanol production. In the initial studies, the maximum ethanol concentration yield was 4.0g/L, using paper towels as the waste feedstock. A mixture of organic acids (lactate, acetate, and formate) was also obtained from the paper towel waste. Similar product mixtures were obtained from the other waste materials during fermentation. To direct the metabolic flux towards ethanol production, future studies will involve increasing the ethanol production by removing the genes for lactate and formate production. Other objectives will include using higher waste concentrations in a continuous mode. The developed bioprocess can reduce the pressure on the cultivable land to grow corn or sugarcane, decrease the overall water requirement for biofuel production, and provide a sustainable solution for managing solid waste processing and energy resources.

# Efficient Preparation of β-O-4’ Dilignols and Their Oxidative Coupling Studies

Chang Peng
pengchang@dicp.ac.cn

Division of Biotechnology, Dalian Institute of Chemical Physics, Chinese Academy of Sciences

Minneapolis, October 30, 2017

## Outline

- 01 | Introduction
- 02 | Synthesis of β-O-4’ dilignols
- 03 | Oxidative coupling of β-O-4’ dilignols
- 04 | Conclusions

2

# Lignin: Renewable aromatic polymer

- Lignin has a heterogeneous, cross-linked structure derived biosynthetically from three phenylpropane (C9) units.

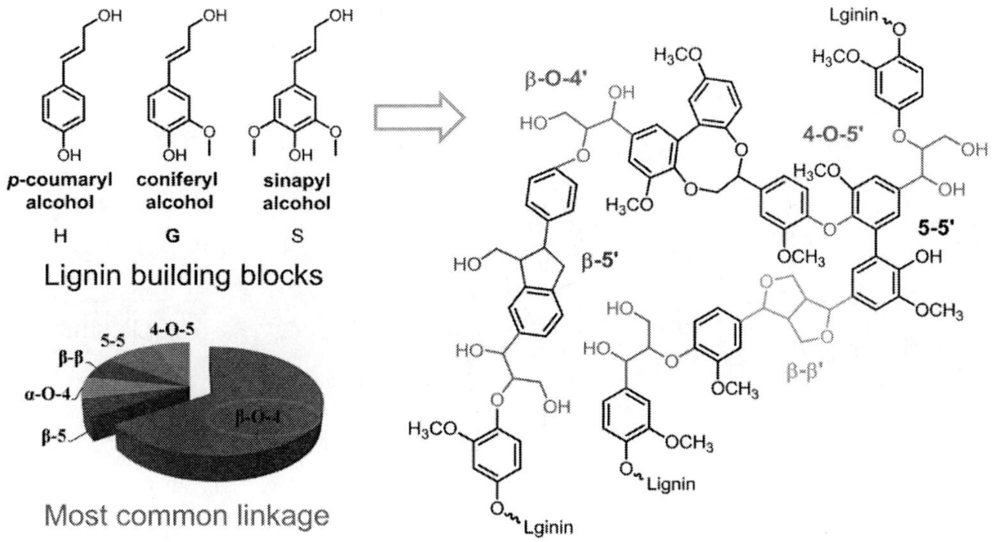


Chem. Rev. , 2010, 110, 3552; Chem. Rev. 2015, 115, 11559.

3

# Lignin to chemical and fuel

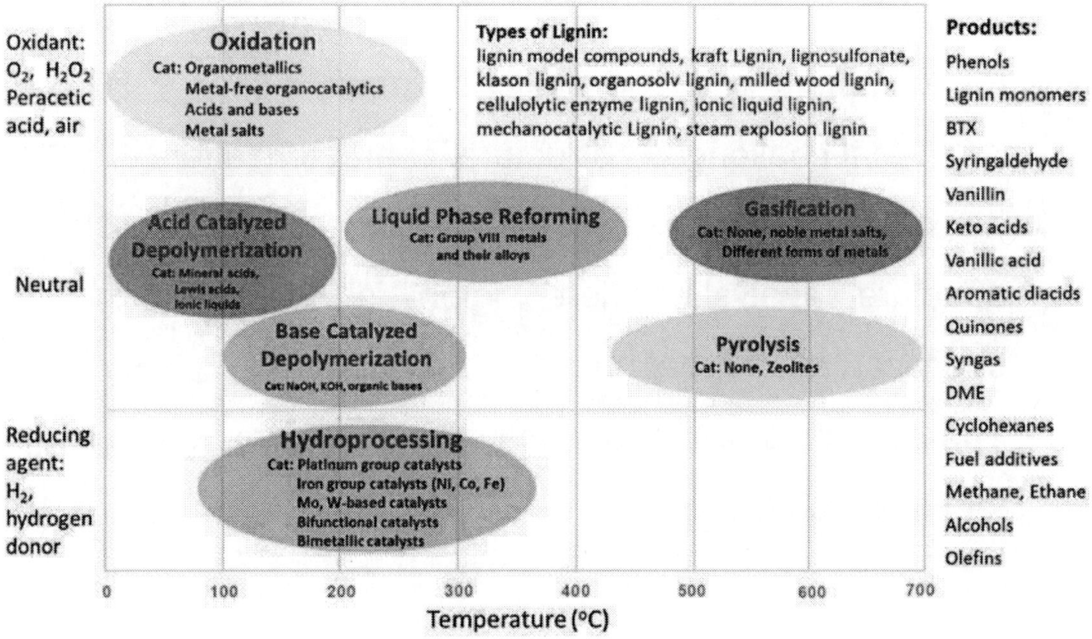


Chem. Rev. 2015, 115, 11559.

4

# Lignin to material

Lignin modification

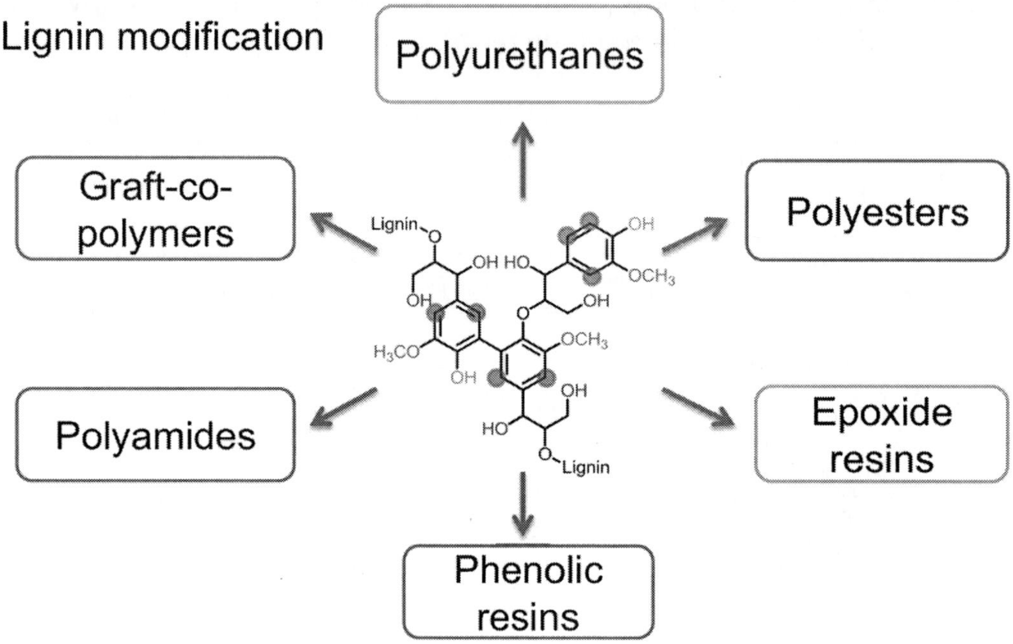


5

# Challenges of lignin utilization

- Heterogeneity and Complexity of lignin.
- Analytical challenges: structural characterization
- It is challenging to explain the chemistry and establish structure analysis where lignin is used for the chemical transformation.
- The synthesis of model compounds is highly desired for lignin conversion.

6

## Commonly used β-O-4' dimers

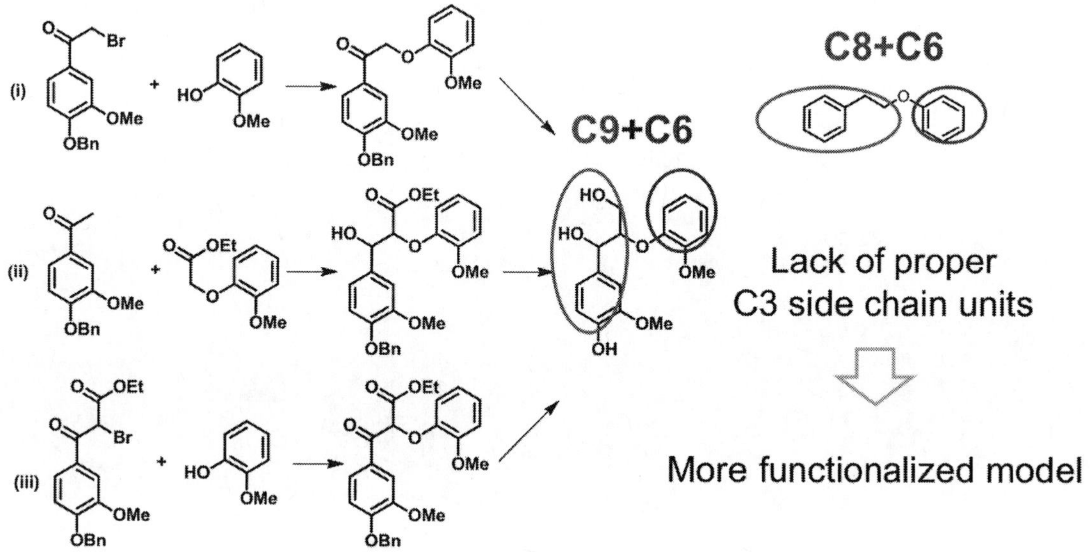


Chem. Rev. 2015, 115, 11559; Acta. Chem. Scand., 1955, 9, 341; Chem Eur. J. 2011, 17, 13877; Holzforschung, 1975, 29, 165.

**7**

## Our study

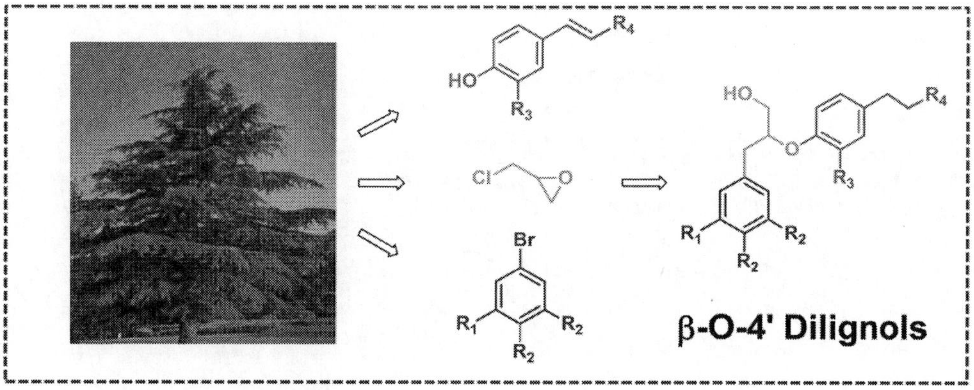


**8**

# Synthesis of β-O-4′ dilignols

### Synthesis of protected intermediates

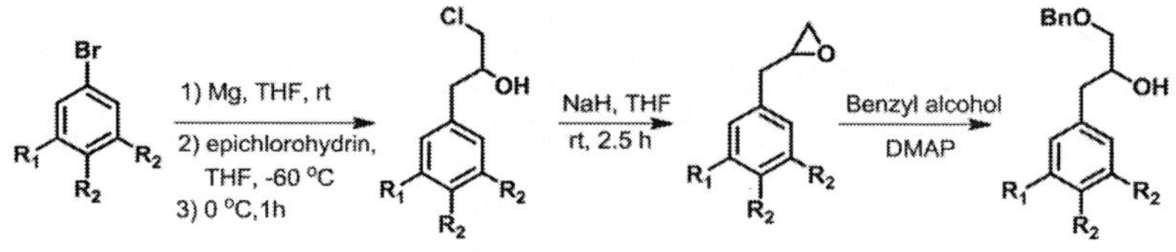


**1a**: $R_1$ = H, $R_2$ = $OCH_3$
**1b**: $R_1$ = $R_2$ = $OCH_3$

**2a-b**
**2a**-95%
**2b**-68%

**3a-b**
**3a**-92%
**3b**-89%

**4a-b**
**4a**-72% Coniferyl alcohol
**4b**-74% Sinapyl alcohol

### Construction of β-O-4′ linkage

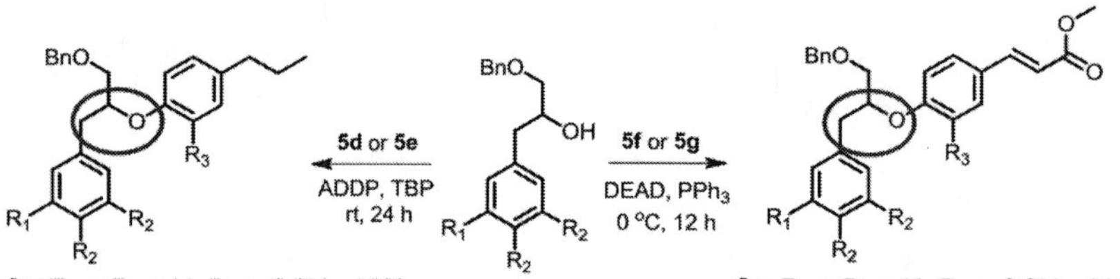


**6a**: $R_1$ = $R_3$ = H, $R_2$ = $OCH_3$, 52%
**6b**: $R_1$ = $R_2$ = $OCH_3$, $R_3$ = H, 44%
**7a**: $R_1$ = H, $R_2$ = $R_3$ = $OCH_3$, 62%
**7b**: $R_1$ = $R_2$ = $R_3$ = $OCH_3$, 47%

**4a-b**
**4a**-72%
**4b**-74%

**8a**: $R_1$ = $R_3$ = H, $R_2$ = $OCH_3$, 88%
**8b**: $R_1$ = $R_2$ = $OCH_3$, $R_3$ = H, 77%
**9a**: $R_1$ = H, $R_2$ = $R_3$ = $OCH_3$, 84%
**9b**: $R_1$ = $R_2$ = $R_3$ = $OCH_3$, 77%

**9**

# Synthesis of β-O-4′ dilignols

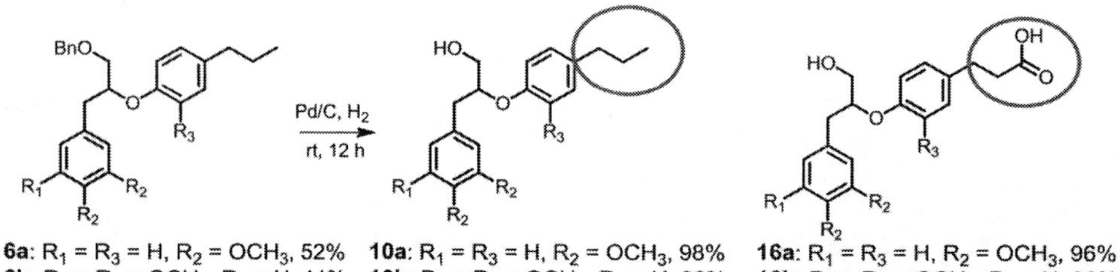


**6a**: $R_1$ = $R_3$ = H, $R_2$ = $OCH_3$, 52%
**6b**: $R_1$ = $R_2$ = $OCH_3$, $R_3$ = H, 44%
**7a**: $R_1$ = H, $R_2$ = $R_3$ = $OCH_3$, 62%
**7b**: $R_1$ = $R_2$ = $R_3$ = $OCH_3$, 47%

**10a**: $R_1$ = $R_3$ = H, $R_2$ = $OCH_3$, 98%
**10b**: $R_1$ = $R_2$ = $OCH_3$, $R_3$ = H, 96%
**11a**: $R_1$ = H, $R_2$ = $R_3$ = $OCH_3$, 97%
**11b**: $R_1$ = $R_2$ = $R_3$ = $OCH_3$, 93%

**16a**: $R_1$ = $R_3$ = H, $R_2$ = $OCH_3$, 96%
**16b**: $R_1$ = $R_2$ = $OCH_3$, $R_3$ = H, 96%
**17a**: $R_1$ = H, $R_2$ = $R_3$ = $OCH_3$, 93%
**17b**: $R_1$ = $R_2$ = $R_3$ = $OCH_3$, 93%

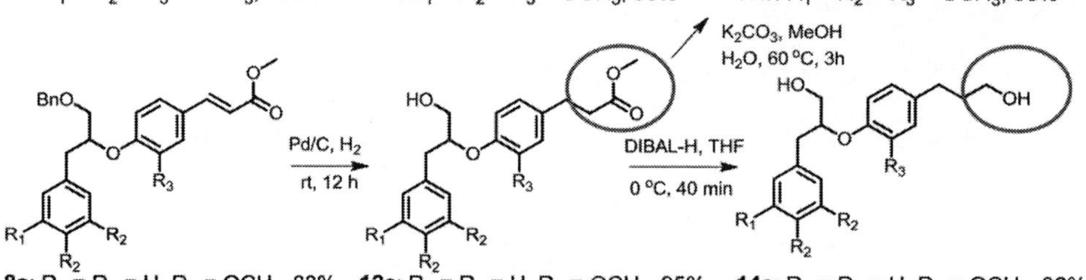


**8a**: $R_1$ = $R_3$ = H, $R_2$ = $OCH_3$, 88%
**8b**: $R_1$ = $R_2$ = $OCH_3$, $R_3$ = H, 77%
**9a**: $R_1$ = H, $R_2$ = $R_3$ = $OCH_3$, 84%
**9b**: $R_1$ = $R_2$ = $R_3$ = $OCH_3$, 77%

**12a**: $R_1$ = $R_3$ = H, $R_2$ = $OCH_3$, 95%
**12b**: $R_1$ = $R_2$ = $OCH_3$, $R_3$ = H, 94%
**13a**: $R_1$ = H, $R_2$ = $R_3$ = $OCH_3$, 94%
**13b**: $R_1$ = $R_2$ = $R_3$ = $OCH_3$, 98%

**14a**: $R_1$ = $R_3$ = H, $R_2$ = $OCH_3$, 96%
**14b**: $R_1$ = $R_2$ = $OCH_3$, $R_3$ = H, 99%
**15a**: $R_1$ = H, $R_2$ = $R_3$ = $OCH_3$, 97%
**15b**: $R_1$ = $R_2$ = $R_3$ = $OCH_3$, 99%

β-O-4′ dilignols with alkyl, carboxyl group, ester and hydroxyl group were synthezied

**10**

# Application of β-O-4′ dilignols

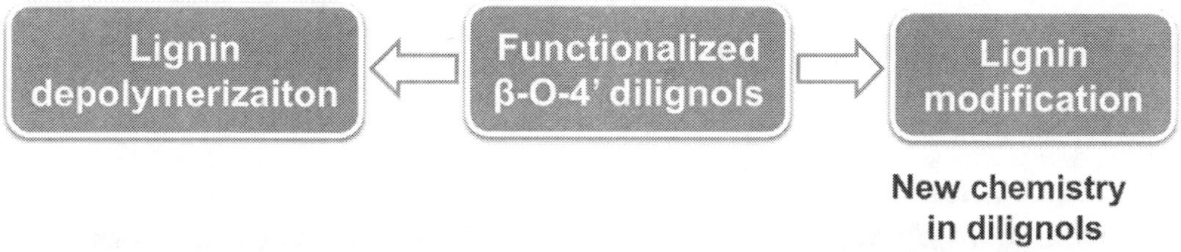


- Hypervalent iodine reagent (III) have been widely used for C-H modification and coupling reaction of aromatic compounds.
- PIFA and PIDA have received much attention due to their low toxicity, ready availability, easy handing and similar reactivity to those of heavy metal reagents.

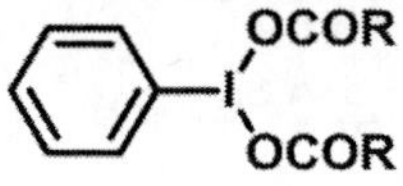


R= $CH_3$ (PIDA)
R= $CF_3$ (PIFA)

Chem. Rec. 2015, 15, 886

11

# Oxidative coupling of β-O-4′ dilignols

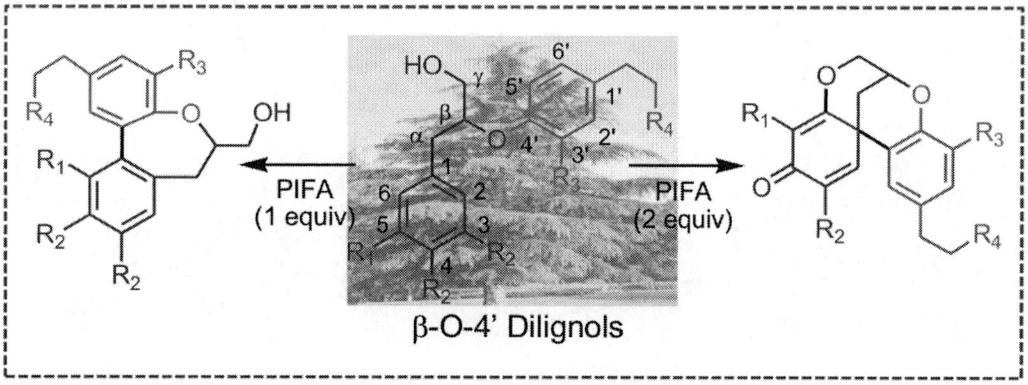


12

# Oxidative coupling of β-O-4′ dilignols

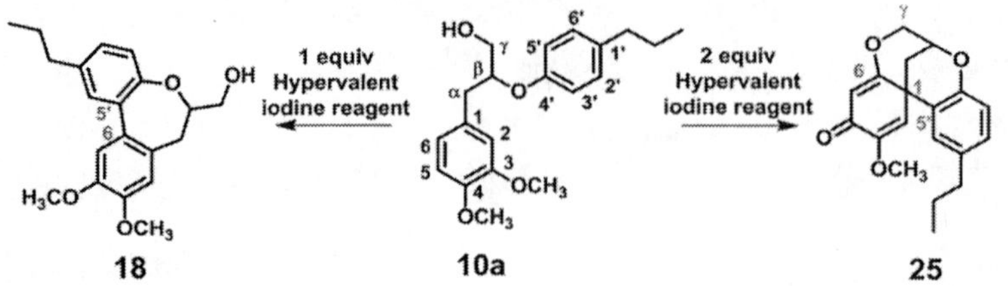


| Entry | Reaction Condition | Product | Yield/% |
|---|---|---|---|
| 1[a] | PIFA (1.05 equiv), $BF_3$•$Et_2O$, DCM, -40 °C, 1 h | **18** | 67.1 |
| 2 | PIFA (1.05 equiv), HFIP, 0 °C, 2 h | **18** | 74.2 |
| 3[b] | PIFA (1.05 equiv), MK10, HFIP, 0 °C, 2 h | **18** | 88.7 |
| 4[b] | PIDA(1.05 equiv), MK10, HFIP, 0 °C, 2 h | **18** | 59.5 |
| 5[c] | PIFA (2 equiv), MK10, HFIP, 0 °C, 2 h | **25** | 81.1 |

[a] $BF_3$•$Et_2O$ (2.1 equiv); [b] MK10 (montmorillonite K10, 0.5 g/mol), HFIP (hexafluoroisopropanol); [c] MK10 (1 g/mol).

Tetrahedron, 2007, 63, 4052

**13**

# Crystal structure

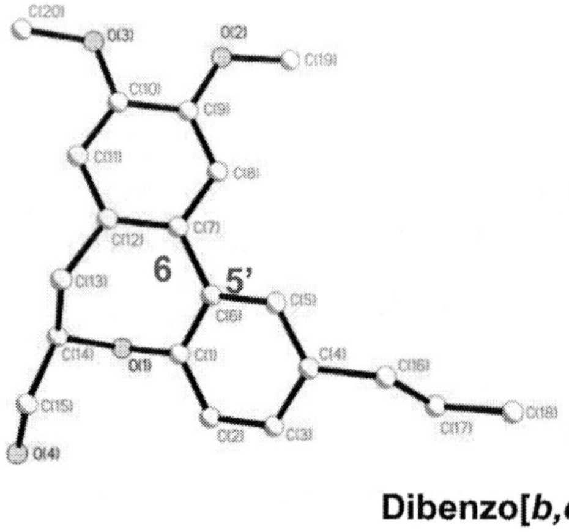

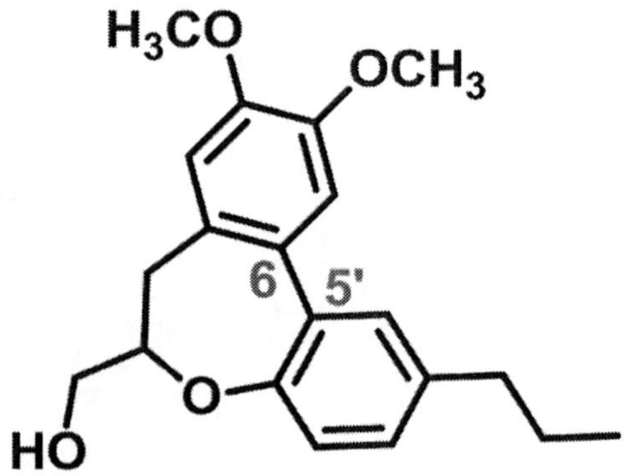


**Dibenzo[*b,d*]oxepine**

Crystal data: $C_{20}H_{24}O_4$, M = 328.39, *a* = 14.4834 (14), *b* = 5.5205 (5), *c* = 21.6449 (19) Å, *U* = 1719.3 (3) Å$^3$, Mo-Ka radiation, Z = 4, space group P21/c, *F*(000) = 704, *Dx* = 1.268 g cm$^{-3}$, *μ* = 0.087 mm$^{-1}$.

**Figure 2.** Crystal structure of **18**

**14**

# Oxidative coupling of β-O-4′ dilignols

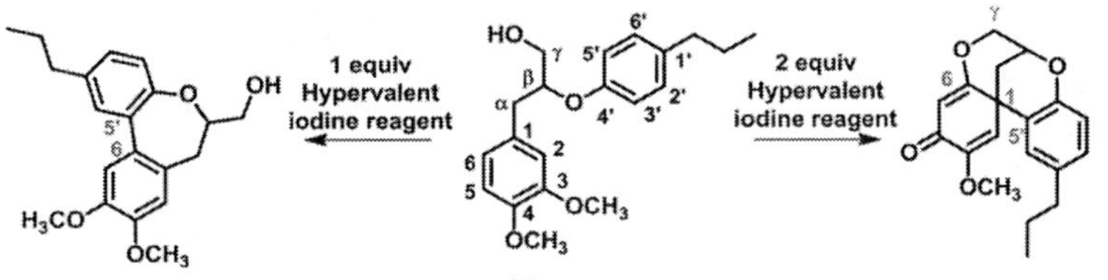


**18** **10a** **25**

**Dibenzo[*b,d*]oxepine** **2,7-dioxabicyclo[3,3,1]nonane**

| Entry | Reaction Condition | Product | Yield/% |
|---|---|---|---|
| 1[a] | PIFA (1.05 equiv), $BF_3 \cdot Et_2O$, DCM, -40 °C, 1 h | **18** | 67.1 |
| 2 | PIFA (1.05 equiv), HFIP, 0 °C, 2 h | **18** | 74.2 |
| 3[b] | PIFA (1.05 equiv), MK10, HFIP, 0 °C, 2 h | **18** | 88.7 |
| 4[b] | PIDA (1.05 equiv), MK10, HFIP, 0 °C, 2 h | **18** | 59.5 |
| 5[c] | PIFA (2 equiv), MK10, HFIP, 0 °C, 2 h | **25** | 81.1 |

[a] $BF_3 \cdot Et_2O$ (2.1 equiv); [b] MK10 (montmorillonite K10, 0.5 g/mol), HFIP (hexafluoroisopropanol); [c] MK10 (1 g/mol).

Tetrahedron, 2007, 63, 4052

**15**

# Crystal structure

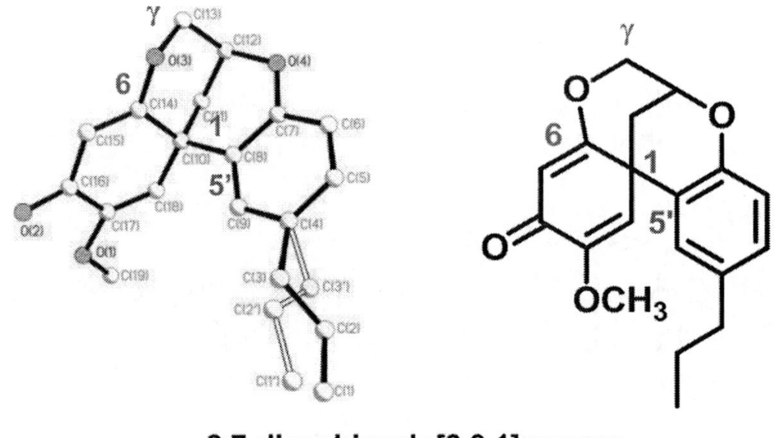

**2,7-dioxabicyclo[3,3,1]nonane**

Crystal data: $C_{19}H_{20}O_4$, M = 312.35, *a* = 5.8007 (4), *b* = 14.3745 (10), *c* = 19.066 (16) Å, *U* = 1612.2 (2) Å$^3$, Mo-Ka radiation, Z = 4, space group P21/c, *F*(000) = 664, *Dx* = 1.287 g cm$^{-3}$, $\mu$ = 0.090 mm$^{-1}$.

**Figure 3.** Crystal structure of **25**.

**16**

# Oxidative coupling of β-O-4′ dilignols

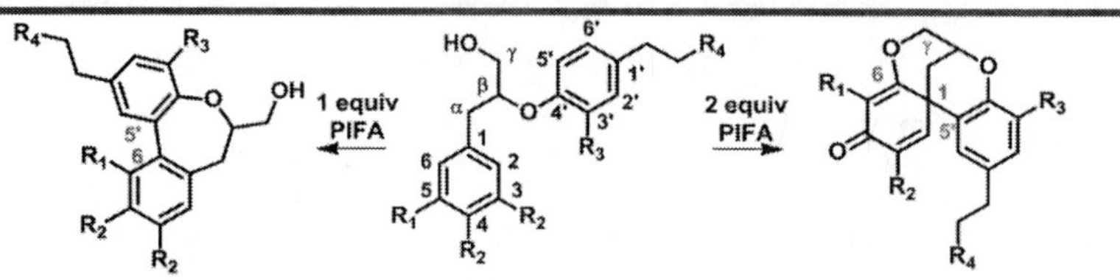


18-24 | 10, 11, 12,14, 16 | 25-30

10a: $R_1 = R_3 = H$, $R_2 = OCH_3$, $R_4 = CH_3$
10b: $R_1 = R_2 = OCH_3$, $R_3 = H$, $R_4 = CH_3$
11a: $R_1 = H$, $R_2 = R_3 = OCH_3$, $R_4 = CH_3$
11b: $R_1 = R_2 = R_3 = OCH_3$, $R_4 = CH_3$
12a: $R_1 = R_3 = H$, $R_2 = OCH_3$, $R_4 = COOCH_3$
14a: $R_1 = R_3 = H$, $R_2 = OCH_3$, $R_4 = CH_2OH$
16a: $R_1 = R_3 = H$, $R_2 = OCH_3$, $R_4 = COOH$

| Entry | Substrate | 1 equiv PIFA | | 2 equiv PIFA | |
|---|---|---|---|---|---|
| | | Product | Yield/% | Product | Yield/% |
| 1 | **10a** | **18** [a] | 88.7 | **25** [b] | 81.1 |
| 2 | **10b** | **19** [a] | 88.7 | **26** [b] | 41.2 |
| 3 | **11a** | **20** [a] | 92.3 | **27** [b] | 36.1 |
| 4 | **11b** | **21** [a] | 79.5 | **ND** [b] | - |
| 5 | **12a** | **22** [a] | 78.8 | **28** [b] | 73.1 |
| 6 | **14a** | **23** [a] | 47.0 | **29** [b] | 23.5 |
| 7 | **16a** | **24** [a] | 81.1 | **30** [b] | 89.1 |

[a] PIFA (1.05 equiv), HFIP, MK10 (0.5 g/mol), 0 °C, 2 h.
[b] PIFA (2 equiv), HFIP, MK10 (1.0 g/mol), 0 °C, 2 h. ND: not detected.

17

# Comparison with other oxidative coupling

Lignin biosynthesis or Biomimetic oxidative coupling

β-O-4′, β-5′, β-β′, β-1′, 5-5′, et al.

Hypervalent iodine reagent (III) Catalyzed oxidative coupling

Rare linkages: 6-5′, 1-5′, γ-O-6

- This process may be applied to expand our ability in designing new chemical compounds from lignin as well as modified lignin contained rare cross linkages.

18

# Possible reaction mechanism

**SET** | **Dienone/phenol-type rearrangement**

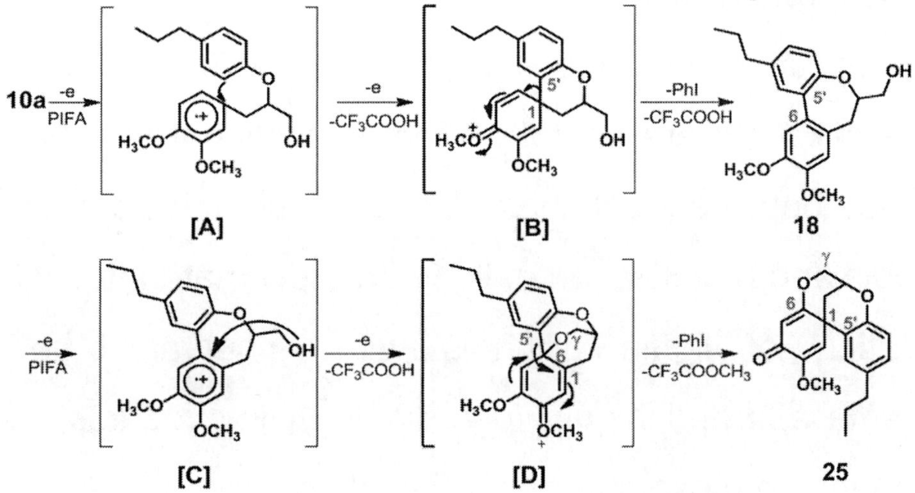


**aryl group migration**

Treatment of **18** with one equiv of PIFA afforded **25** in 81.5% yield.

Chem. Eur. J. 2002, 8, 5377.

**19**

# Cytotoxic activity of dibenzo[b,d]oxepine

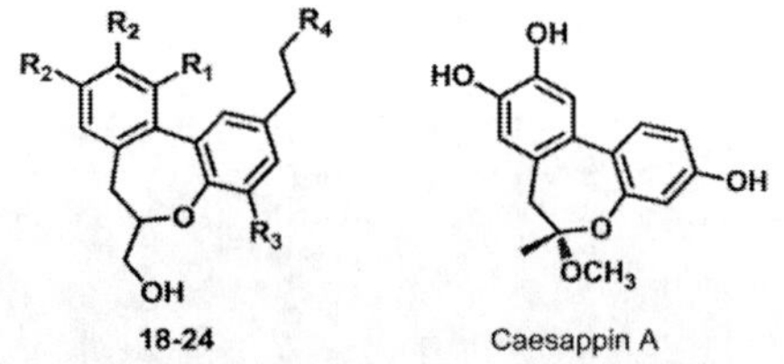


A novel protosappanin

| | Compounds | $IC_{50}$ (μM) | |
|---|---|---|---|
| | | MCF7 | A549 |
| **S type** | **18** | > 100 | > 100 |
| | **19** | 42.7±6.4 | > 100 |
| | **20** | > 100 | > 100 |
| | **21** | 62.9±7.5 | 97.6±4.4 |
| | **22** | > 100 | > 100 |
| | **23** | > 100 | > 100 |
| | **24** | > 100 | > 100 |
| | Caesappin A | > 100 | 54.2 |

Fitoterapia, 2014, 92, 280

**20**

## Conclusion

- Practical routes were devised ensuring an easy access to more functionalized models β-O-4' dilignols in good yields.
- These dilignols are useful as more functionalized models to study both destruction of lignin into chemicals such as phenols and modification of lignin for new materials.
- Oxidation of lignin dilignols with hypervalent iodine reagents afforded highly fused polycyclic products, such as dibenzo[b,d]oxepine and 2,7-dioxabicyclo[3,3,1]nonane derivatives.

21

## Acknowledgment

**Professor**
**Zongbao (Kent) Zhao**

$$$ Dalian Institute of Chemical Physics, CAS
$$$ National Natural Science Foundation of China

22

# Thank you!

# 491883 Development of Unique Trimetallic Copper Iron Manganese Oxygen Carrier for Chemical Looping Combustion

**Abstract:**

**Development of Unique Trimetallic Copper Iron Manganese Oxygen Carrier for Chemical Looping Combustion**

William Benincosa[1,2,3] Ranjani Siriwardane[1]* Hanjing Tian[1,2] Jarrett Riley[1,2]

[1] *U.S. Department of Energy, National Energy Technology Laboratory, 3610 Collins Ferry Road, P.O. Box 880, Morgantown, WV 26507-0880, USA*

[2] *Department of Chemical Engineering, West Virginia University, Morgantown, WV 26505, USA*

[3] *Oak Ridge Institute for Science and Education, P.O. Box 117, Oak Ridge, TN 37831, USA*

*Corresponding Author: Ranjani.Siriwardane@netl.doe.gov

Chemical looping combustion (CLC) is a novel combustion method that uses oxygen from an oxygen carrier (OC), or metal oxide, to react with a fuel source and produce power/heat and a sequestration ready stream of carbon dioxide. It requires the use of a reactive, robust, inexpensive, and environmentally benign OC. As a result, synthetic OC development has gained interest to improve performance by increasing reactivity while maintaining chemical and mechanical stability.

In our previous studies, a synergetic effect among copper, iron, and manganese mixed metal oxides has demonstrated improved reactivity compared with pure single metal oxides for chemical looping combustion. This observed synergy led to the discovery and development of trimetallic copper iron manganese oxide ($CuFeMnO_4$). This unique oxygen carrier has demonstrated excellent reactivity and stability during multi-cycle reduction and oxidation experiments using thermogravimetric analysis with online mass spectrometry (TGA-MS). It also possesses excellent resistance to attrition. In addition, this oxygen carrier was synthesized from metal oxide precursors, which is the most economical route to commercialization, and the formation of the dominant reactive phase has been distinguished from its single and bimetallic counterparts. Thus, $CuFeMnO_4$is an excellent OC candidate for CLC. To gain a deeper understanding of the enhanced reactivity due to the formation of this unique phase, a comprehensive study was conducted to determine the oxygen carrierâ€™s reduction behavior. Experimental TGA-MS data was evaluated with conventional solid-state modeling approaches to identify and compare the reduction mechanism that is attributing to the enhanced performance of the trimetallic OC with its single and bi-metallic counterparts. The kinetic parameters and reduction model obtained from this study were also evaluated in process simulations to obtain information for process scale-up. This work aims to highlight the important milestones that were achieved in the design and development of the novel trimetallic oxygen carrier.

# 500556 Dopant Modified Iron Based Materials for Fuel Chemical Looping Combustion and Reforming Applications

**Abstract:**

The chemical looping technologies have multiple applications in fuel combustion, as well as methane, coal and biomass reforming. With advantages of chemical looping process design, it can result in a dramatic reduction of capital cost required for the process and improvement of carbon utilization efficiency. Oxygen carriersâ€™ properties are the most significant part of the success of chemical looping fuel combustion and reforming processes. The Ohio State University utilizes iron-based materials for oxygen carriers of chemical looping process. Low cost, multiple phases and environmental friendly nature of iron oxide leads to the industrial possibility of chemical looping technology utilization. However, the moderate reactivity and selectivity of iron oxide with fuel reaction still impedes the success of the commercialization of chemical looping process. In details, the low ionic diffusion of oxygen, less active sites on the surface and low adsorption energy of iron oxide hindered its reactivity and selectivity. Dopants on the surface can modify iron oxide redox reactivity by providing active sites for fuel gas adsorption and aliovalent dopants inside the bulk can alter the iron oxide oxygen ionic diffusion by creating schottky defects. In our study, we examine at the nanoscale to characterize undoped, single-dopant and binary-dopant modified iron oxides that undergo redox reactions with fuel gas, such as hydrogen, carbon monoxide and methane. Doped iron oxides are fabricated by sol gel or solid-state synthesis. Afterwards, X-ray Diffractometry (XRD), Scanning electronic microscopy (SEM)/ Energy-dispersive X-ray spectroscopy (EDS) and X-ray photoelectron spectroscopy (XPS) were utilized to determine the structural and morphological modification of iron oxide. The lattice parameters and surface functional groups of iron oxide variation indicate the existence and functions of these dopants. The redox reaction and temperature programming reaction (TPR) are conducted with fuel gases by thermogravimetric analysis (TGA) with Mass spectrometry (MS). The difference among the undoped iron oxide and single / binary dopant modified iron oxides will be discussed in detail. Meanwhile, density functional theory (DFT) calculation studies are used as a guideline for searching for appropriate dopants. Moreover, the understanding of fuel to syngas or fuel combustion mechanism using dopant modification provides us a methodology to improve the reactivity and recyclability of oxygen carriers, which is a promising way to build the oxygen carriers with low cost and high reactivity and selectivity for commercialization success of chemical looping technology.

# 501673 Tuning Bulk and Surface Properties of Mixed Metal Oxides for Partial Oxidation Applications via Chemical Looping Schemes

**Abstract:**

Using metal oxide based oxygen carriers as the reaction intermediate, the chemical looping strategy intensifies carbonaceous fuel conversion processes via a cyclic redox scheme. Although conventional chemical looping combustion processes aim to fully oxidize carbonaceous fuels for carbon capture, a similar principle can be applied to generate value added products such as syngas and olefins. The product selectivity requirement calls for oxygen carriers, also known as redox catalysts, with tailored bulk and surface properties.

Monometallic oxides of Cu, Fe, or Ni are among the most frequently investigated oxygen carriers for chemical looping. However, they have limited applicability for selective product generations. The present study discusses the limitation of monometallic oxides for such applications as well as the potential advantages of mixed metal oxide based redox catalysts. Specifically, iron and manganese containing mixed oxides are exemplified for methane partial oxidation, oxidative dehydrogenation (ODH) of ethane, and water/$CO_2$-splitting. Redox catalyst selection criteria and the effect of its bulk and surface properties on redox activity and product selectivity are also presented. The underlying redox reaction pathways and mechanisms are also revealed by experiments and simulations. Compared to traditional heterogeneous catalysts, we demonstrate that mixed metal oxide based redox catalysts can potentially be more selective. Moreover, significant decrease in redox reaction temperatures can be achieved by tailoring the bulk and surface properties of the mixed oxides. In terms of methane partial oxidation, we report excellent redox activity and over 90% syngas selectivity at temperatures as low as 600 °C. In terms of ethane ODH, over 70% single-pass olefin yield was achieved by redox catalysts with tailored surface properties. Rationally designed, iron containing mixed metal oxides demonstrated close to 100% water and $CO_2$ splitting efficiencies.

# 489942 Continuous Large-Scale Production of Red Mud Oxygen Carriers with a Rotary Kiln for Chemical Looping Combustion

**Abstract:**

Chemical looping combustion (CLC) has the feature of inherently enriching $CO_2$ concentration by circulating solid metal oxides (called oxygen carrier or OC) between an air reactor and fuel reactor. The durability and cost of OC directly impacts the economy and competitiveness of the CLC technology. Unfortunately, current available OC production techniques require a series of fabrication steps in order to obtain necessary properties, particularly mechanical strength and reactivity. The complicated production process significantly increases the cost of OC preparation and makes it difficult to scale up, one of the biggest challenges for CLC. Looking for a cost-effective production technique for large-scale OC production is very meaningful for the development of CLC. In this work, red mud, byproduct from the alumina industry, was used as raw OC material. An electrical, lab-scale rotary kiln was adopted for continuously producing red mud-based OC. Different fabrication parameters, including firing temperature, feeding rate, rotation speed and kiln angle, were investigated for the purpose of optimizing the resulting clinker properties. Clinker product was evaluated based on its mechanical strength, density and reactivity. The product obtained at 1250°C from the rotary kiln showed a higher strength (2.7 N) than, but similar reactivity (3.86%/min from $Fe_2O_3$ to $Fe_3O_4$) to, the OC prepared by the lab-scale freeze granulation method (2.3 N, 3.87%/min from $Fe_2O_3$ to $Fe_3O_4$). Repeatability of OC fabrication with the rotary kiln was verified as well. The mineral phase formations at different firing temperatures were analyzed. $Fe_3Al_2(SiO_4)_3$ and $Fe_3(Al_{0.2}Fe_{1.8})(SiO_4)_3$, almandines, which contribute to the hardness of the clinker, were detected when firing at 1250°C and 1300°C, respectively. Fabrication costs associated with the rotary kiln was estimated to be around \$346/ton, or two orders less than conventional preparation methods.

# 497477 Correlation of the Physical and Solid State Chemistry Changes for a $CuO$-$Fe_2O_3$-$Al_2O_3$ Oxygen Carrier during Reduction with $H_2$ and CO for Chemical Looping Combustion Applications

**Abstract:**

Chemical-looping combustion (CLC) is a novel combustion technology for energy generation that uses a metal oxide oxygen carrier (OC) to transport oxygen to a fuel source, avoiding direct contact between fuel and air. Combusting the fuel with an oxygen carrier is advantageous because ideally only carbon dioxide ($CO_2$) and water vapor are generated. After condensing the water, sequestration-ready $CO_2$ is produced making the technology desirable for increased carbon capture efficiency with decreased energy penalties. Development of efficient oxygen carriers is essential to successful CLC systems and a promising mixed metal oxide oxygen carrier has been developed at NETL consisting of CuO-$Fe_2O_3$-$Al_2O_3$. The metal oxide combination forms a unique $Cu(Fe_{2\text{â``}x}Al_x)O_4$ spinel structure. The novel mixed metal OC possesses desirable properties including enhanced reactivity and physical stability. In addition, the OC also demonstrates exothermic behavior during reduction with fuels, which is essential to sustaining auto-thermal operation for energy generation applications. The oxygen carrier has been implemented successfully in pilot scale operation for CLC at NETL, showing that it is a proven candidate for CLC applications and merits further examination. A thorough understanding of the changes that occur during reduction and oxidation (redox) mechanisms are needed for implementation of modeling strategies and to deepen the understanding of materials used for this unique application.

The changes in physical and chemical properties during reduction and oxidation for the $Cu(Fe_{2\text{â``}x}Al_x)O_4$Â OC are unique and not thoroughly understood. The act of transferring oxygen to the fuel source induces the formation oxygen vacancies causing the structure to distort and ultimately change phase when enough oxygen has been depleted. The phase change to base metal and metal oxide components due to the stripping of lattice oxygen is often referred to as the reduction route, and comparably the oxidation route being the reverse process. With this noted, chemical properties can vary from solid-state solution changes including predominant phase, the reduction and oxidation routes, and component migration that can occur due to lattice distortion and reorientation. In addition, physical properties of the material can also be impacted including particle size changes during redox reactions, density changes associated with predominant phase, surface area, etc. This comprehensive study identifies and correlates these property changes with extent of solid oxygen carrier conversion for the development and implementation of accurate reduction and oxidation models for further use in process scale up and optimization strategies. A combination of various characterization techniques including in-situ X-ray diffraction, Scanning Electron Microscopy, Thermogravimetric analysis coupled with mass spectrometry and others are used to establish correlations between the property changes with the metal oxide during reduction with $CO/H_2$ and oxidation with $O_2$.

# 499519 Chemical Looping Partial Oxidation of Solid Fuels for High Purity Syngas Production

**Abstract:**

The Ohio State University is developing the chemical looping partial oxidation (CLPO) processes for the production of high-purity syngas from solid carbonaceous fuels including biomass and coal. Compared to conventional coal and biomass gasification processes, the CLPO processes eliminate the need for molecular oxygen from an air separation unit (ASU), and deliver increased cold gas efficiencies and decreased fuel consumptions. The CLPO processes utilize the unique co-current moving bed gas-solid reducer reactors for the partial oxidation of solid fuels to syngas with metal oxide oxygen carriers. The co-current moving bed reactor design provides a desirable gas-solid contacting while minimizing gas bypassing and solid back mixing, which result in high fuel conversion and close-to-equilibrium gas composition. In this study, active iron-titanium composite metal oxide (ITCMO) materials are used as the oxygen carrier in the CLPO processes. Theoretical analysis aided by a modified Ellingham Diagram illustrates that syngas production is thermodynamically favored in presence of ITCMO oxygen carrier. Fundamental reaction mechanism studies are performed using density functional theory (DFT) simulation to unveil the role of oxygen vacancy in partial oxidation reactions by metal oxide oxygen carrier. The experimental results from a bench-scale moving bed reactor have successfully shown the production of high purity syngas from coal and biomass with the combination of moving bed reducer and ITCMO oxygen carrier. Adjustable syngas compositions were achieved by the co-injection of steam or $CO_2$. A fully integrated sub-pilot scale test unit with non-mechanical gas sealing and solid circulation devices was designed, constructed, and tested for continuous syngas production. The bench and sub-pilot demonstrations confirmed the syngas produced from the CLPO processes are consistent with the prediction from thermodynamic equilibrium models.

# 504028 Low-Temperature Synthesis of $NH_3$ from Air and Water for Electrical Energy Storage

**Abstract:**

Direct synthesis of ammonia from air and water is attractive for chemical energy storage of renewable electricity and for production of ammonia fertilizers as well. However, the conversion unit must be compact and efficient enough for economical production, because renewable energies such as wind and solar power are distributed over a wide range of territory. Electrochemical conversion at low temperature and low pressure is a promising approach. The literature studies on the subject will be reviewed and analyzed. Critical challenges will be delineated. A new catalytic membrane reactor based on low-temperature alkaline membrane separator and nano-/micro-structured electrodes will be introduced to address these challenges.

# Metal Oxide/Metal Sulfate Water Splitting Cycle for Hydrogen Generation

**(Principal Author)**
**Rahul Bhosale**
Department of Chemical Engineering
College of Engineering
Qatar University, Doha, Qatar
rahul.bhosale@qu.edu.qa

**(Other Authors)**
**Gorakshnath Takalkar**
Department of Chemical Engineering
College of Engineering
Qatar University, Doha, Qatar

**Anand Kumar**
Department of Chemical Engineering
College of Engineering
Qatar University, Doha, Qatar

**Fares A. AlMomani**
Department of Chemical Engineering
College of Engineering
Qatar University, Doha, Qatar

Prepared for Presentation at
American Institute of Chemical Engineers
2017 Fall Meeting
Minneapolis, MN, USA
October 29 – November 3, 2017

# Metal Oxide/Metal Sulfate Water Splitting Cycle for Hydrogen Generation

**Rahul R. Bhosale, Anand Kumar, Gorakshnath Takalkar, Fares A. AlMomani**

Department of Chemical Engineering
College of Engineering
Qatar University, Doha, Qatar

**Keywords:** Metal Oxide, Metal Sulfate, Hydrogen, Water Splitting

## Abstract

This study reports the thermodynamic analysis of the cobalt oxide – cobalt sulfate (CoO – CoS) based water splitting cycle. CoO-CoS cycle consists of two steps: first step – exothermic oxidation of CoO via water splitting reaction producing $H_2$, and second step – the $CoSO_4$ is thermally decomposed into CoO, $SO_2$, and $O_2$. The CoO is recycled back to first step and can be used in multiple steps. This study is divided into two steps: 1. Thermodynamic equilibrium analysis, and 2. Second law energy and exergy analysis. At the end, the efficiency of CoO-CoS cycle is calculated and the results of the thermodynamic analysis are reported in detail.

## 1. Introduction

Among the various thermochemical cycles [1-27], the sulfur-iodine cycle and the hybrid sulfur cycle are more promising due to the requirement of lower operating temperatures. Following reaction chemistry is involved in both cycles:

Sulfur – Iodine Cycle

$H_2SO_4 \rightarrow SO_3 + H_2O$ (673K) (1)

$SO_3 \leftrightarrow SO_2 + {}^1/_2\, O_2$ (1123 – 1273K) (2)

$SO_2 + 2H_2O + I_2 \rightarrow H_2SO_4 + 2HI$ (393K) (3)

$2HI \rightarrow H_2 + I_2$ (573 – 723K) (4)

Hybrid - Sulfur Cycle

$H_2SO_4 \rightarrow SO_3 + H_2O$ (673K) (5)

$SO_3 \leftrightarrow SO_2 + {}^1/_2\, O_2$ (1123 – 1273K) (6)

$SO_2 + 2H_2O \rightarrow H_2SO_4$ (353 – 393K) (7)

Although these cycle seems promising, the need of expensive catalytic materials and sulfur poisoning of the catalyst are major limitations associated with these cycles. Therefore, recently, we are dealing with 'metal oxide – metal sulfate' based water-splitting cycle operated using concentrated solar energy. It is one of the alternatives to achieve $H_2$ production at moderate temperatures. The solar 'metal oxide – metal sulfate' cycle is represented by following equations.

$$MO + SO_2(g) + H_2O(g) \rightarrow MSO_4 + H_2(g) \tag{8}$$

$$MSO_4 \rightarrow MO + SO_2(g) + {}^1/_2\, O_2(g) \tag{9}$$

It is a two-step process in which the first step (non-solar) deals with exothermic oxidation of MO with the help of $SO_2$ and $H_2O$ producing metal sulfate ($MSO_4$) and $H_2$. In the second step (solar), the produced $MSO_4$ reduced into MO, $SO_2$, and $O_2$. Such that by recycling the MO, this cycle operated on a continuous basis.

In this paper, thermodynamic analysis of CoO-CoS thermochemical water splitting cycle was performed in detail. HSC Chemistry 7.0 software and its database was used for this purpose. This analysis includes the determination of equilibrium composition and solar to fuel conversion efficiency by following second law of thermodynamics. A typical CoO-CoS solar thermochemical water splitting cycle is shown below.

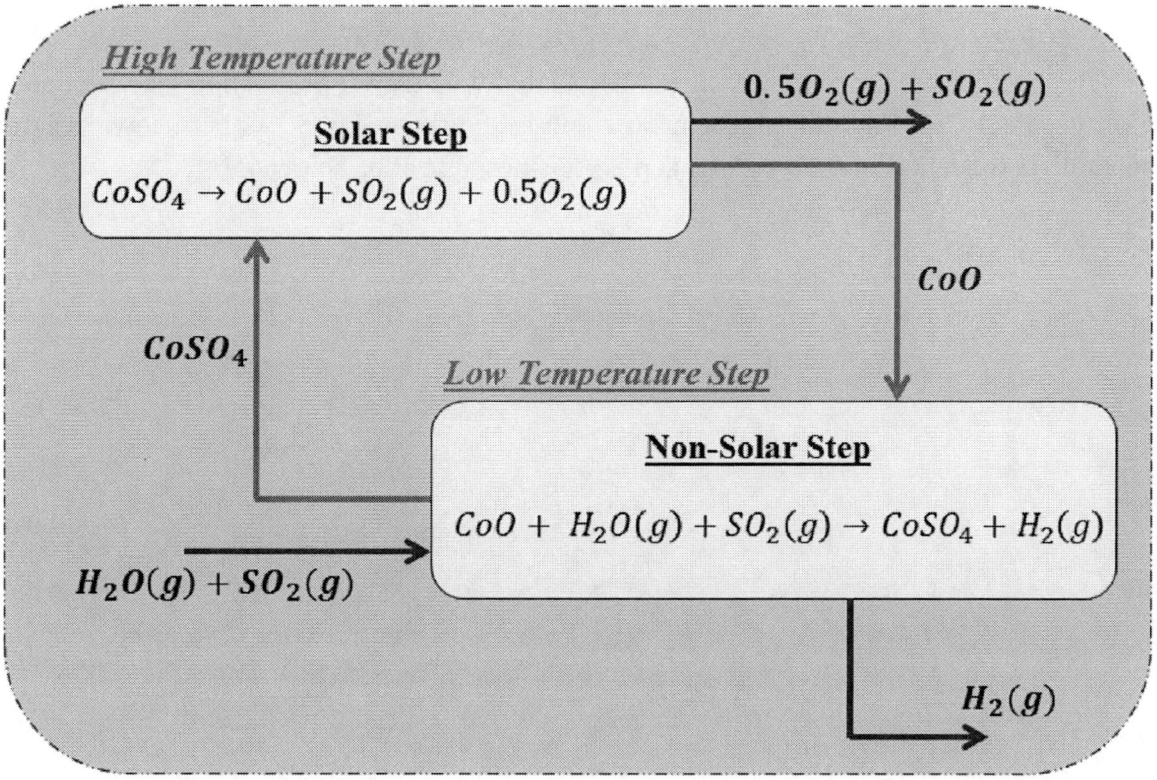


Fig. 1. A typical CoO-CoS solar thermochemical water splitting cycle.

## 2. Equilibrium Composition

As per Fig. 1, the thermodynamic analysis predicts that the complete thermal reduction of $CoSO_4$ into CoO, $SO_2$(g), and $O_2$(g) is feasible at or above 1400K (in absence of Ar).

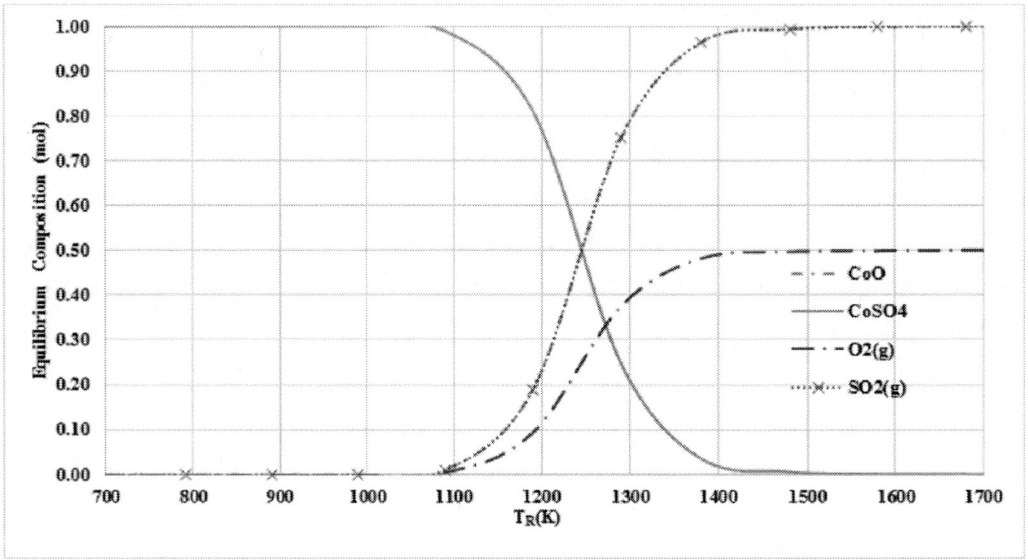


Fig. 2. Thermal decomposition of $CoSO_4$ in absence of Ar.

The equilibrium molar compositions associated with the $H_2$ production via water splitting reaction via CoO-CoS solar thermochemical cycle is shown in Fig. 3. As per shown results, $H_2$ production via oxidation of CoO by $SO_2$ and $H_2O$ is possible at or below 450 K.

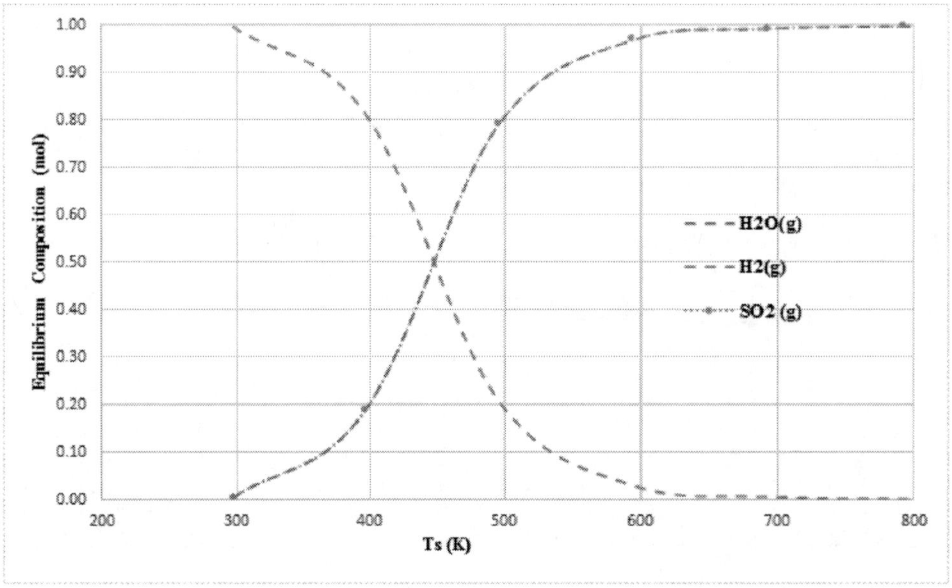


Fig. 3. Water splitting reaction using CoO-CoS cycle.

## 3. Efficiency Analysis

The energy and exergy analysis of the solar thermochemical CoO-CoS water splitting cycle was performed by following the second law analysis. The process flow diagram for the CoO-CoS cycle is presented in Fig. 4.

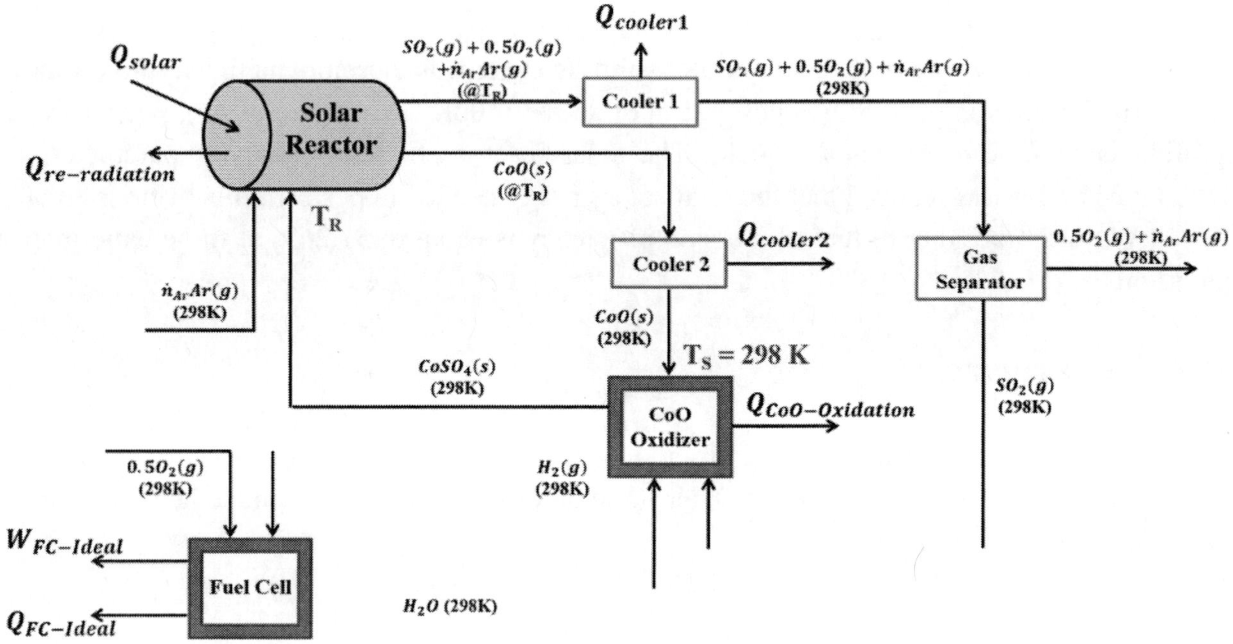


Fig. 4. The process flow diagram for the CoO-CoS cycle.

The cycle thermodynamic efficiency analysis is performed by applying the principles of second law of thermodynamics. The thermal reduction was carried out at 1525K whereas the $H_2O$ splitting is commenced at 298K. The values of the various parameters associated with the solar reactor efficiency analysis calculated by using the HSC Chemistry thermodynamic software are reported in Table 1.

Table 1. Solar reactor efficiency analysis for CoO-CoS cycle (I = 1000 W/m$^2$, C = 5000 suns, $T_H$ = 2528 K, $T_L$ = 500 K).

| Parameters | Values |
|---|---|
| $Q_{reactor-net}$ | 523.4 kW |
| $Q_{solar}$ | 557.6 kW |
| $Q_{cooler-1}$ | 93.9 kW |
| $Q_{cooler-2}$ | 76.1 kW |
| $Q_{CoO-Oxidizer}$ | 67.7 kW |
| $Irr_{reactor}$ | 0.263 kW/K |

| | |
|---|---|
| $Irr_{cooler-1}$ | 0.170 kW/K |
| $Irr_{cooler-2}$ | 0.109 kW/K |
| $Irr_{CoO-Oxidizer}$ | 0.104 kW/K |
| $\eta_{solar-to-fuel}$ | 51.26% |

## 4. Conclusions

The CoO-CoS thermochemical water splitting cycle was thermodynamically investigated. The thermal reduction of $CoSO_4$ is possible at or above 1400K (in absence of Ar), whereas water splitting is feasible at or below 450 K. The solar reactor efficiency analysis parameters are determined and it was realized that the total solar energy needed to perform this cycle is equal to 557.6 kW. Also, the solar to fuel conversion efficiency is equal to 51.26% at these experimental conditions.

## 5. Acknowledgments

This publication was made possible by the NPRP grant (NPRP8-370-2–154) from the Qatar National Research Fund (a member of Qatar Foundation). The statements made herein are solely the responsibility of author(s).

## References

[1] J.R. Scheffe, A. Steinfeld, Oxygen exchange materials for solar thermochemical splitting of $H_2O$ and $CO_2$: a review, Materials Today 17 (2014) 341-348.

[2] P.G. Loutzenhiser, A. Steinfeld, Solar syngas production from $CO_2$ and $H_2O$ in a two-step thermochemical cycle via Zn/ZnO redox reactions: thermodynamic cycle analysis, International Journal of Hydrogen Energy 36 (2011) 12141-12147.

[3] A. Stamatiou, P. Loutzenhiser, A. Steinfeld, Solar syngas production from $H_2O$ and $CO_2$ via two-step thermochemical cycles based on Zn/ZnO and FeO/$Fe_3O_4$ redox reactions: Kinetic analysis, Energy & Fuels 24 (2010) 2716-2722.

[4] R. R. Bhosale, A. Kumar, F. AlMomani, R. B. Gupta, Solar thermochemical ZnO/$ZnSO_4$ water splitting cycle for hydrogen production, International Journal of Hydrogen Energy (2017) in-press.

[5] P. Loutzenhiser, M. Gálvez, I. Hischier, A. Stamatiou, A. Frei, A. Steinfeld, $CO_2$ splitting via two-step solar thermochemical cycles with Zn/ZnO and FeO/$Fe_3O_4$ redox reactions II: Kinetic analysis, Energy & Fuels 23 (2009) 2832-2839.

[6] R. R. Bhosale, A. Kumar, van den Broeke, Leo JP, S. Gharbia, D. Dardor, M. Jilani, J. Folady, M.S. Al-Fakih, M.A. Tarsad, Solar hydrogen production via thermochemical iron oxide–iron sulfate water splitting cycle, International Journal of Hydrogen Energy 40 (2015) 1639-1650.

[7] P. Charvin, S. Abanades, F. Lemont, G. Flamant, Experimental study of $SnO_2$/SnO/Sn thermochemical systems for solar production of hydrogen, AIChE Journal 54 (2008) 2759-2767.

[8] R. R. Bhosale, A. Kumar, P. Sutar, Thermodynamic analysis of solar driven $SnO_2$/SnO based thermochemical water splitting cycle, Energy Conversion & Management 135 (2017) 226-235.

[9] R. R. Bhosale, A. Kumar, F. AlMomani, U. Ghosh, P. Sutar, G. Takalkar, A. Ashok, I. Alxneit, Effectiveness of Ni incorporation in iron oxide crystal structure towards thermochemical $CO_2$ splitting reaction, Ceramics International 43 (2017) 5150-5155.

[10] C.C. Agrafiotis, C. Pagkoura, A. Zygogianni, G. Karagiannakis, M. Kostoglou, A.G. Konstandopoulos, Hydrogen production via solar-aided water splitting thermochemical cycles: Combustion synthesis and preliminary evaluation of spinel redox-pair materials, International Journal Hydrogen Energy 37 (2012) 8964-8980.

[11] R. R. Bhosale. A. Kumar, F. AlMomani, I. Alxneit, Propylene oxide assisted sol–gel synthesis of zinc ferrite nanoparticles for solar fuel production, Ceramics International 42 (2016) 2431-2438.

[12] M. Roeb, N. Gathmann, M. Neises, C. Sattler, R. Pitz-Paal, Thermodynamic analysis of two-step solar water splitting with mixed iron oxides, International Journal of Energy Research 33 (2009) 893-902.

[13] R.R. Bhosale, R.V. Shende, J.A. Puszynski, Thermochemical water-splitting for $H_2$ generation using sol-gel derived Mn-ferrite in a packed bed reactor, International Journal of Hydrogen Energy 37 (2012) 2924-2934.

[14] J.R. Scheffe, J. Li, A.W. Weimer, A spinel ferrite/hercynite water-splitting redox cycle, International Journal of Hydrogen Energy 35 (2010) 3333-3340.

[15] R.R. Bhosale, A. Kumar, F. AlMomani, I. Alxneit, Sol–gel derived $CeO_2$–$Fe_2O_3$ nanoparticles: Synthesis, characterization and solar thermochemical application, Ceramics International 42 (2016) 6728-6737.

[16] R.R. Bhosale, A. Kumar, F. AlMomani, U. Ghosh, S. Al-Muhtaseb, R. Gupta, I. Alxneit, Assessment of $Ce_xZr_yHf_zO_2$ based oxides as potential solar thermochemical $CO_2$ splitting materials, Ceramics International 42 (2016) 9354-9362.

[17] J.R. Scheffe, A. Steinfeld, Thermodynamic analysis of cerium-based oxides for solar thermochemical fuel production, Energy & Fuels 26 (2012) 1928-1936.

[18] W.C. Chueh, C. Falter, M. Abbott, D. Scipio, P. Furler, S.M. Haile, A. Steinfeld, High-flux solar-driven thermochemical dissociation of $CO_2$ and $H_2O$ using nonstoichiometric ceria, Science 330 (2010) 1797-1801.

[19] P. Furler, J.R. Scheffe, A. Steinfeld, Syngas production by simultaneous splitting of $H_2O$ and $CO_2$ via ceria redox reactions in a high-temperature solar reactor, Energy & Environmental Science 5 (2012) 6098-6103.

[20] J.R. Scheffe, R. Jacot, G.R. Patzke, A. Steinfeld, Synthesis, Characterization, and Thermochemical Redox Performance of $Hf^{+4}$ , $Zr^{+4}$ , and $Sc^{+3}$ Doped Ceria for Splitting $CO_2$, The Journal of Physical Chemistry C 117 (2013) 24104-24114.

[21] J.R. Scheffe, D. Weibel, A. Steinfeld, Lanthanum–strontium–manganese perovskites as redox materials for solar thermochemical splitting of $H_2O$ and $CO_2$, Energy & Fuels 27 (2013) 4250-4257.

[22] A.H. McDaniel, E.C. Miller, D. Arifin, A. Ambrosini, E.N. Coker, R. O'Hayre, W.C. Chueh, J. Tong, Sr-and Mn-doped $LaAlO_{3-\delta}$ for solar thermochemical $H_2$ and CO production, Energy & Environmental Science 6 (2013) 2424-2428.

[23] M. Gálvez, R. Jacot, J. Scheffe, T. Cooper, G. Patzke, A. Steinfeld, Physico-chemical changes in Ca, Sr and Al-doped La–Mn–O perovskites upon thermochemical splitting of $CO_2$ via redox cycling, Physical Chemistry Chemical Physics 17 (2015) 6629-6634.

[24] R. R. Bhosale, A. Kumar, A. Ashok, P. Sutar, G. Takalkar, M. Khraisheh, F. AlMomani, U. Ghosh, La-Based Perovskites as Oxygen-Exchange Redox Materials for Solar Syngas Production, MRS Advances (2017) in-press.

[25] R. R. Bhosale, R. Khadka, J. Puszynski, R. V. Shende, $H_2$ generation from two-step thermochemical water-splitting reaction using sol-gel derived $Sn_xFe_yO_z$, Journal of Renewable & Sustainable Energy 3 (2011) 063104-1 – 063104-12.

[26] R. Bhosale, R. Shende, J. Puszynski, $H_2$ generation from thermochemical water splitting using sol-gel derived Ni-ferrite, Journal of Energy and Power Engineering 4 (2010) 27-38.

[27] R.R. Bhosale, R.V. Shende, J.A. Puszynski, $H_2$ Generation From Thermochemical Water-Splitting Using Sol-Gel Synthesized Zn/Sn/Mn-doped Ni-Ferrite, International Review of Chemical Engineering 2 (2010) 852-862.

# $sCO_2$ Power Cycles with Integrated Thermochemical Energy Storage Using an MgO-Based $sCO_2$ Sorbent in Direct Contact with Working Fluid for Grid Energy Storage Applications

**Andy Muto**

Kevin McCabe

Daniel Real

SOUTHERN RESEARCH 1941-2016 75 YEARS

Energy and Environment Division, Southern Research, Durham, North Carolina 27712

SR SOUTHERN RESEARCH

# Presentation Outline

- **Introduction to SunShot program and Southern Research**
- **Thermochemical energy storage fundamentals and system design**
- **Heat exchange reactor design**
- **Technoeconomic analysis**
- **$CO_2$ sorbent material engineering**
- **Summary and technical challenges**

amuto@southernresearch.org

SR SOUTHERN RESEARCH 2

# SunShot CSP Performance and Cost Targets

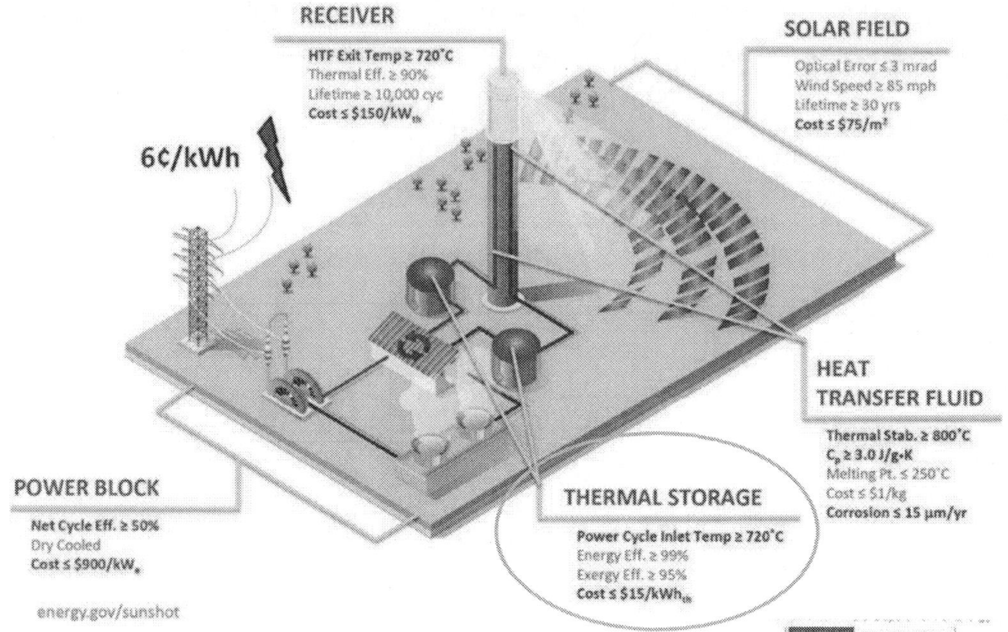


amuto@southernresearch.org SR SOUTHERN RESEARCH 3

SR SOUTHERN RESEARCH

- Non-profit, established in 1941
- Headquarters in Birmingham Alabama
- 450 Scientists, Engineers and Technicians
- Four operating divisions

# Energy Storage Projects at Southern Research

**SunShot thermochemical energy storage projects**

**CaO-based $CO_2$ sorbent**

**1) ELEMENTS (2 yrs):** "Regenerative Carbonate-Based Thermochemical Energy Storage System for Concentrating Solar Power", DE-EE0006535

**2) APOLLO (3 yrs):** "Demonstration of High-Temperature Calcium-Based Thermochemical Energy Storage System for use with Concentrating Solar Power Facilities", DE-EE0007116

**MgO-based $CO_2$ sorbent**

**3) EERE**, Combined water gas shift/$CO_2$ capture process for integrated gasification combined cycle systems, DE-FE0026388

**4) SunShot Technology to Market (2 yrs):** "$sCO_2$ power cycle with integrated thermochemical energy storage using an MgO-based $sCO_2$ sorbent in direct contact with working fluid", DE-EE0008126

amuto@southernresearch.org

SR SOUTHERN RESEARCH 5

# Types of Thermal Energy Storage

**Sensible Heat**

- Energy stored in vibrational modes of molecules (molten salts, sand, etc.)

**Latent Heat**

- Energy stored in media as it changes phase from solid to liquid (molten salts, Al-Si alloy)

**Thermochemical Energy**

- Reversible chemical reactions involving gases. Gases can be compressed and allow for a temperature swing, pressure swing or combination thereof to drive reaction.

Too Cold ↑

Ammonia sorbent
Hydrates

Carbonates
Metal hydrides
Ammonia decomposition
Hydrocarbon decomposition

Too Hot ↓

Metal oxides

SR SOUTHERN RESEARCH 6

# Balance of Plant CaO TCES example

*CaO-based TCES*

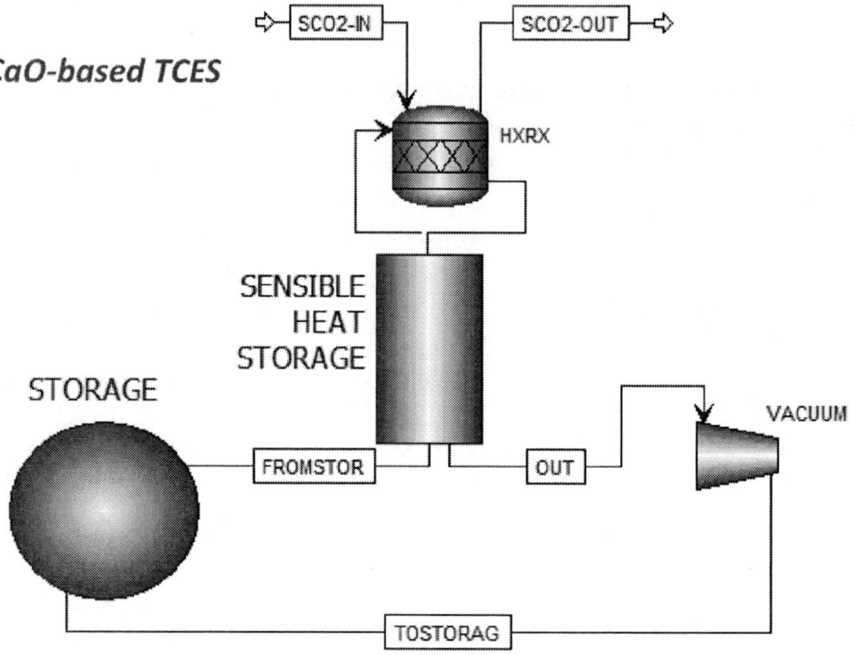


7

# MgO carbonation reaction offers potentially the fewest components of any TCES

*High pressure!*

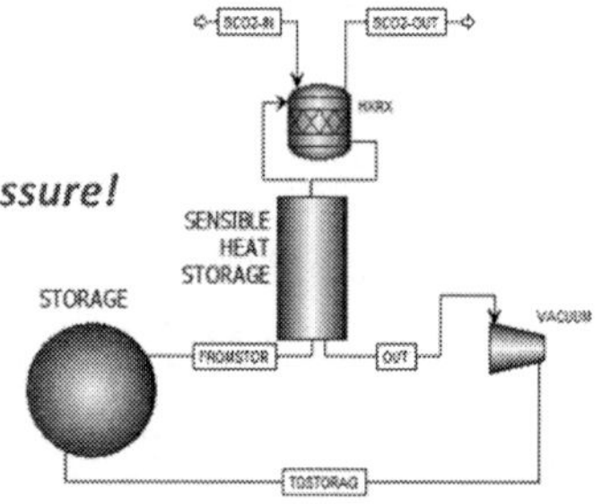


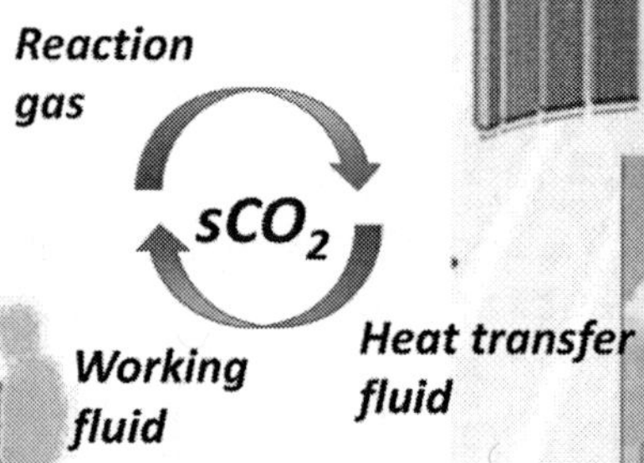

*Echogen's 10 MWe $sCO_2$ power turbine compared to a 10 MWe steam turbine.*

ECHOGEN power systems

BraytonEnergy

amuto@southernresearch.org

SR SOUTHERN RESEARCH

8

# Reaction Thermodynamics

$MgO + CO_2 <-> MgCO_3 + 107\ kJ$

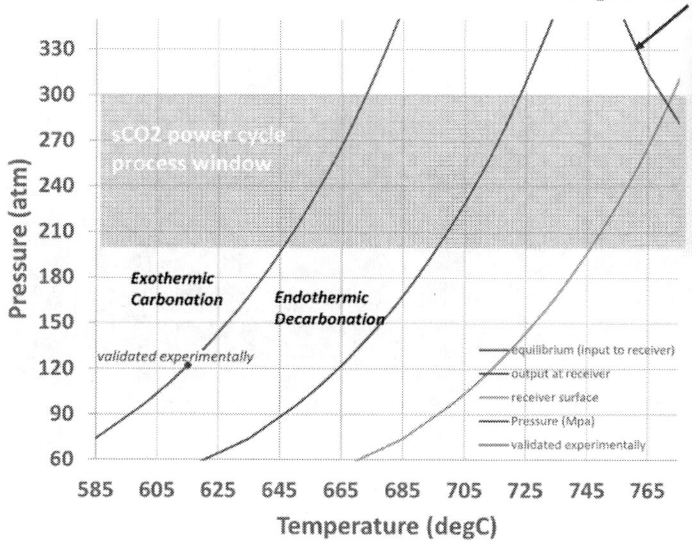


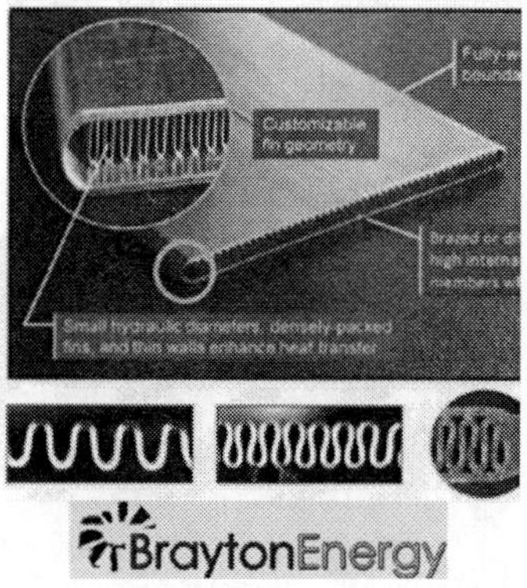


SR SOUTHERN RESEARCH 9

# Integrated $sCO_2$ Process Flow Diagram

An $sCO_2$ power cycle with "reheat" is utilized to accept thermal energy under a smaller temperature range: 50°C as opposed to 200°C

New major components:

1) heat exchange reactor
2) $sCO_2$ storage vessel
3) sensible heat storage system for the reaction gas.

Additional minor components include cyclone separators or filters (not pictured) to eliminate the transport of sorbent fines

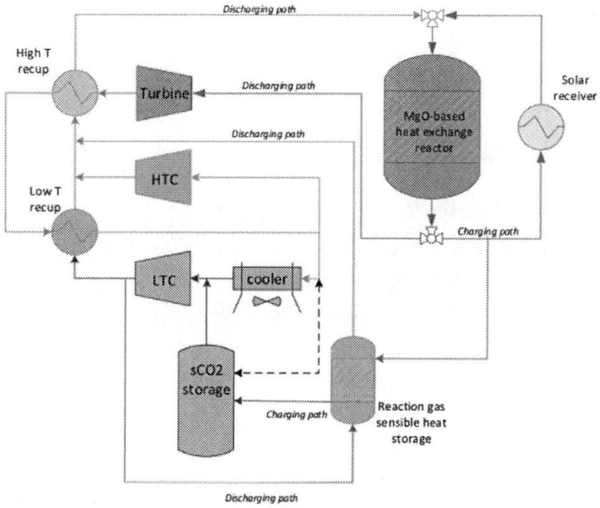


amuto@southernresearch.org

SR SOUTHERN RESEARCH 10

# Heat Exchange Reactor Conceptual Design

- T=575-685 degC, P= 70-300 atm
- All major TCES components are contained in a underground shaft. The surrounding earth is used as a cost-effective means of structural support.
- Possibly constructed using mine-shaft drilling technology: 16 m diameter, 114 m deep.
- Sorbent is loaded into parallel packed beds, with a fluid path length of 5 m and a pellet size of at least 200-650 μm.
- High pressure natural gas storage pilot plants have demonstrated similar volumes capable of >500 atm.

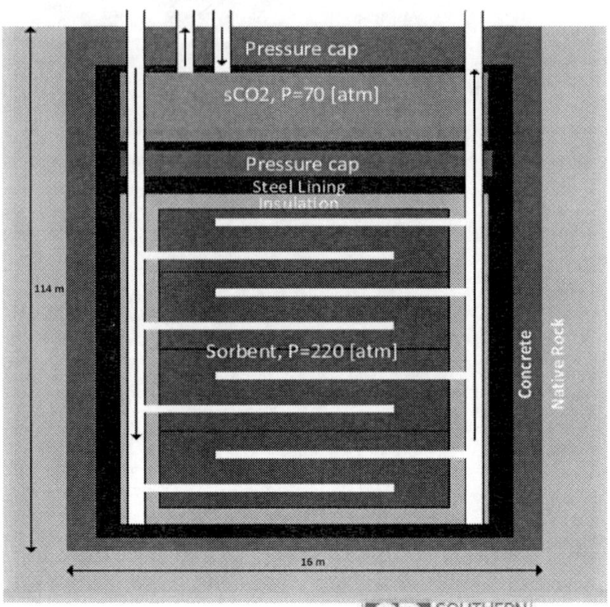


amuto@southernresearch.org

SOUTHERN RESEARCH 11

# Round-trip Exergy Efficiency

- Discharge (carbonation) temperature is held constant at 675°C, 300 atm.
- Charging (decarbonation) temperature is swept from 585°C, 73 atm to 675°C, 300 atm
- Chemical heat pump operation is possible but there is an efficiency penalty

**Reactor irreversibilities**

| | |
|---|---|
| dT_bed_approach | 2°C |
| dT_decarbonation | 50°C |
| dT_carbonation | 50°C |
| Efficiency_thermal_insulation | 0.99 |

**Balance of plant irreversibilities**

| | |
|---|---|
| 2nd_law_efficiency_power_cycle | 0.7 |
| dT_loss_sensible_heat_storage_reaction_gas | 50° C |
| 2nd_law_efficiency_compressor | 0.93 |

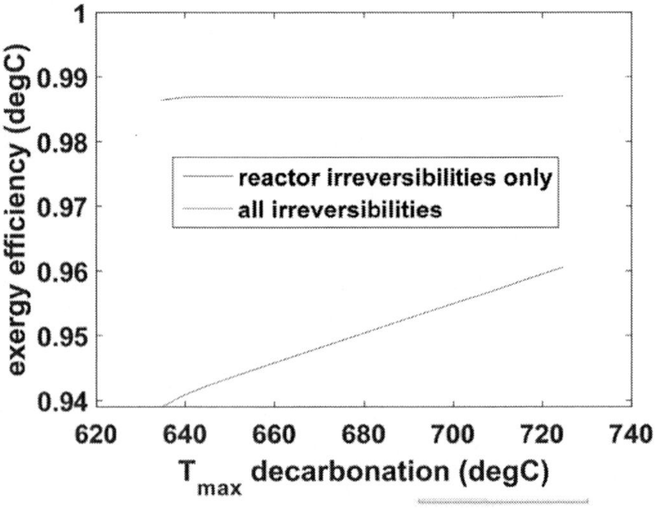

# Technoeconomic Analysis

**Major cost drivers:**

1) energy density of the MgO-based sorbent
2) Underground excavation + lining ($\$400/m^3$)

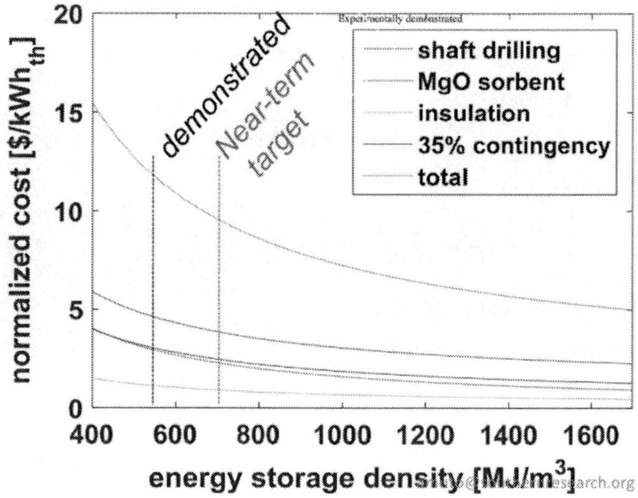


SOUTHERN RESEARCH

13

# Sorbent Durability Tests in TGA

**CaO-based stabilized $CO_2$ synthetic sorbent**
**38% wt gain over 100 cycles at 750°C**

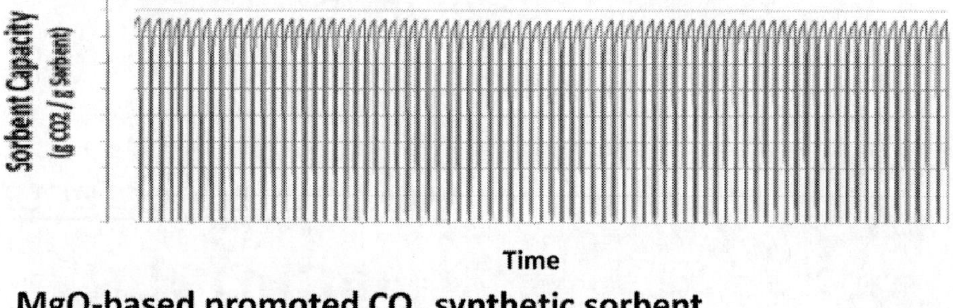


**MgO-based promoted $CO_2$ synthetic sorbent**
**25% wt gain over 150 cycles at 350°C**

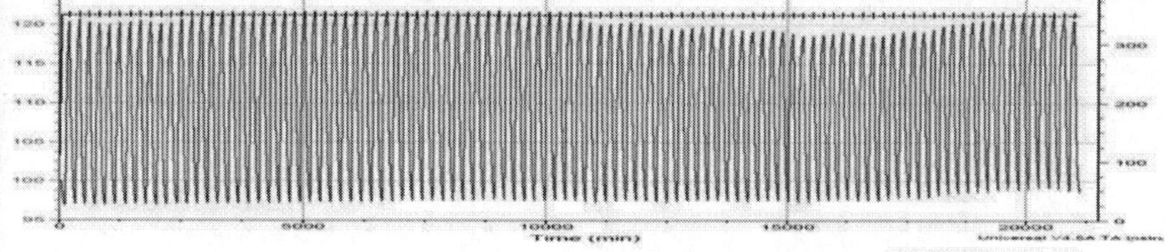

# Summary

- SR has engineered high performance CaO- and MgO-based $CO_2$ sorbents and thermal energy systems. Similar methods and approaches will be applied to development of the high pressure MgO-based TCES system.
- Custom instruments are being built to demonstrate durability of high pressure MgO-based $CO_2$ sorbents under an active SunShot project in collaboration with Echogen Power Systems.
- Identified a viable path to meet **SunShot performance targets**
  - Exergy efficiency ≥ 95 %
  - Capital cost ≤ \$15/$kWh_{th}$ for a commercial plant
  - Power cycle efficiency ≥ 50%

## *Technical Challenges*

- High pressure reactor detailed design
- Low cost, high pressure, underground excavation
- Sorbent total life-cycle costs
- Integration and controls of the receiver + thermal storage + power cycle

amuto@southernresearch.org

SR SOUTHERN RESEARCH 15

SR SOUTHERN RESEARCH

Solving The World's Hardest Problems

Acknowledgements

ECHOGEN power systems

BraytonEnergy

U.S. DEPARTMENT OF ENERGY

SunShot U.S. Department of Energy

## Concentrated solar power (CSP) problem statement and value proposition

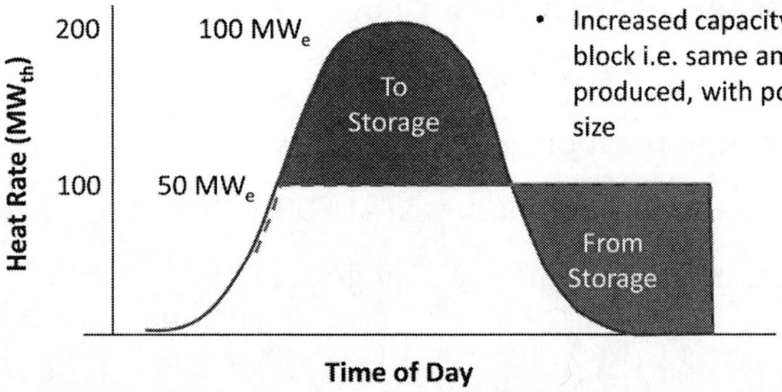


- Dispatchable energy
- Increased capacity factor of power block i.e. same amount of energy produced, with power block ½ the size

Storage system requirements (SunShot program, DOE)

- Exchange heat with a sCO2 cycle operating at **200-300 bar, 700-750°C**
- **10 hr** storage for 200 $MW_{th}$ plant
- Target: **Capital cost $15/kWh** for a commercial plant
- Target: **Exergy efficency>90 %**
- **Lifetime: 30 yr**

amuto@southernresearch.org

SR SOUTHERN RESEARCH 17

## Energy & Environment Project Portfolio

**LOW CARBON ENERGY SYSTEMS**

- carbon capture
- emissions control
- waste heat to power
- biomass to energy
- photovoltaics
- energy storage
- Gen IV nuclear

**WATER TREATMENT**

- management of industrial brines and waste waters
- engineered biology for contaminant removal from watersheds
- water quality monitoring

**SUSTAINABLE CHEMISTRY**

- catalyst and process development for cost competitive conversion of bio-derived raw materials to fuels and chemicals
- chemical process intensification

**RESOURCE RECOVERY**

- "mining" valuable metals (e.g., Li, Ge, Gd, Zn, other Rare Earth Elements) from industrial wastes
- nutrient recovery from agriculture waste streams

**INTENSIVE FOOD PRODUCTION**

- development of food protein production using engineered aquatic plant systems
- combined carbon capture, water treatment and food production

▪ New areas in last 24 months.

SR SOUTHERN RESEARCH

# Selecting a Thermochemical Reaction

- Reversible, near equilibrium, good kinetics

$$\Delta G = \Delta H - T\Delta S = 0$$

- High heat of reaction: $\Delta H$
- Reaction equilibrium occurs in at 600-800°C
- Equates to a large entropy change $\Delta S = \frac{\Delta H}{T}$

*Therefore decomposition reactions are favorable*

- Low cost, non-toxic storage media
- Minimal required balance of plant

amuto@southernresearch.org

SOUTHERN RESEARCH 19

# Engineering $CO_2$ Sorbent

***Primary challenge is to resist sintering and maintain pore structure,*** inert stabilizing phases are required therefore trade-offs in capacity vs. durability

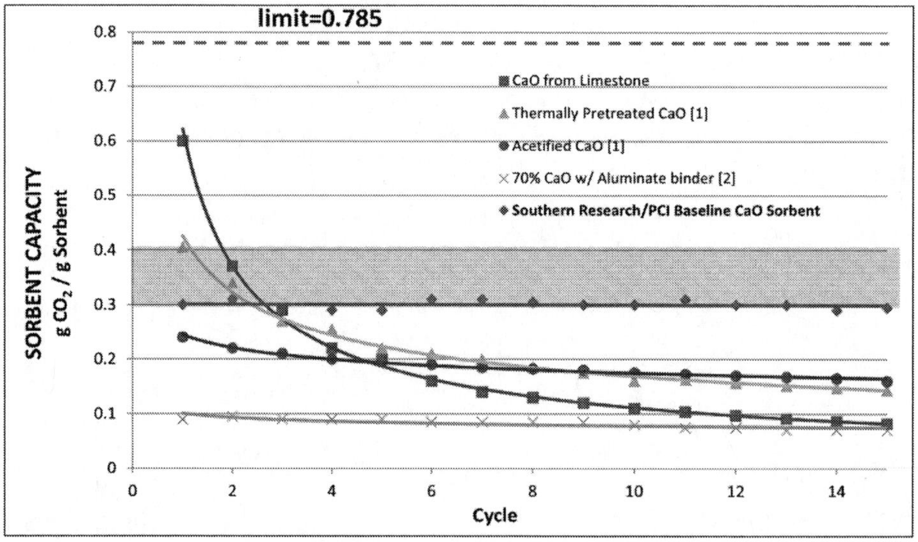


20

# High Pressure Reactors for MgO experiments

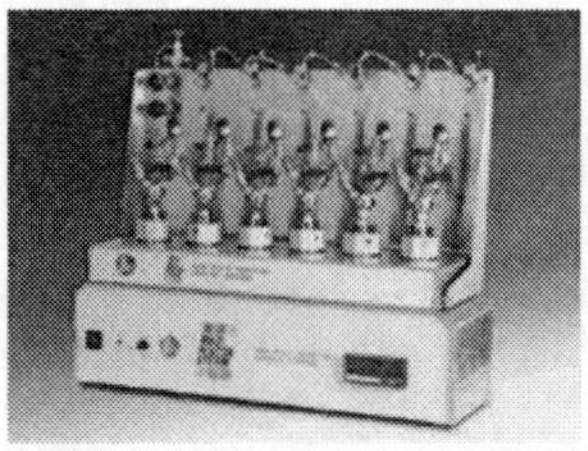

21

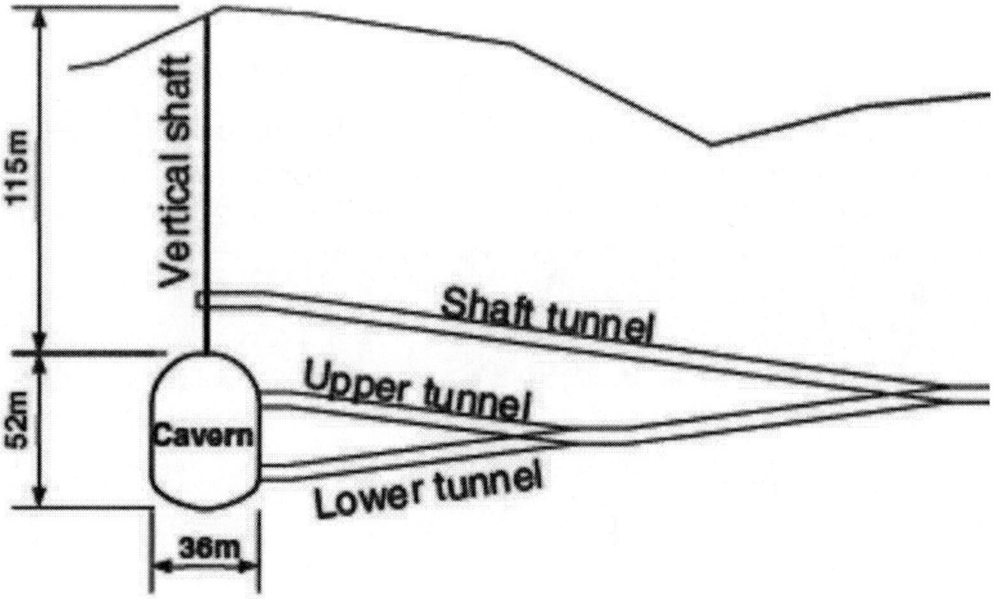

# 499130 Solar Thermochemical $CO_2$-Splitting Using Redox Cycles of Cr-Doped Mn-Based Perovskites

**Abstract:**

Finding novel strategies to enable the transition away from a fossil fuel-based energy economy to the efficient use of renewable energy resources is a global challenge of modern society.The solar-to-fuel technology is a promising approach to transform one of the most abundant renewable sources into fuels, while mitigating $CO_2$ emissions.[1] Solar-driven thermochemical cycles operate on redox processes of oxides in which $H_2O$ and $CO_2$ are converted to syngas ($H_2$ and CO). In this technology, solar energy is collected in concentrated solar power plants and stored chemically in synthetic fuels, through a 2-step thermochemical cycle[1,2]. Currently, the benchmark material used for the solar-to-fuel process is $CeO_2$. Despite the promise of $CeO_2$, the high reduction temperature of 1500 °C to reach usable fuel production yields opposes challenges for reactor design, long-term microstructural degradation and cost. In addition, traditionally redox cycling is operated in temperature-swing regimes. Namely, oxidation is performed at 500 °C lower than reduction, which poses high energy penalties for the process. Alternatively, thermochemical redox cycles could be operated under isothermal conditions benefiting from faster kinetics and long-term materials stability.[1] Recent advances proved that perovskite-type oxides could operate at significantly decreased temperature.[2,3,4] Namely, by utilizing solid solutions of $La_{0.6}Sr_{0.4}Cr_{0.8}Co_{0.2}O_3$ thermal reduction could be carried out at a 300 °C lowered temperature, while producing the same level of CO when compared to state-of-the-art $CeO_2$.[5] However, such materials exhibited slower kinetics than $CeO_2$ and further research is needed in order to find more efficient perovskite materials based on abundant and inexpensive elements that could work in more kinetically favorable conditions.

Consequently, in this work we explore $La_{0.6}Sr_{0.4}Cr_xMn_{1-x}O_3$ perovskite compositions due to favorable thermodynamics of Cr- and Mn-based perovskites for $CO_2$ splitting in accordance to recent thermodynamic models.[4,5,6,7] $La_{0.6}Sr_{0.4}Cr_xMn_{1-x}O_3$ perovskite powders were synthesized through a modified Pechini method covering a compositional range of chromium from x=0 to 1. Although $O_2$ release decreased with increasing addition of Cr, results showed that $La_{0.6}Sr_{0.4}Cr_{0.85}Mn_{0.15}O_3$ outperformed $CeO_2$ and $La_{0.6}Sr_{0.4}MnO_3$ under isothermal redox cycling operation. Namely, $La_{0.6}Sr_{0.4}Cr_{0.85}Mn_{0.15}O_3$ yields 350 µmol/g, whereas $La_{0.6}Sr_{0.4}MnO_3$ 250 µmol/g for 1400 °C and $pCO_2$=0.5 atm. Conversely, it was observed that Cr-rich perovskites performed poorer than $La_{0.6}Sr_{0.4}MnO_3$ under temperature-swing operation.

These results illustrate the beneficial influence of Cr-doping on the fuel production for isothermal operation, confirming trends obtained from thermodynamic modeling by Bork et al..[7] Furthermore, results corroborated the high operational flexibility that easy-tunable perovskite materials offer. Accordingly, proper materials design based on thermodynamic models and further experimental confirmation are crucial for

determining the most efficient compositions at given working conditions.

1. Muhich, C. L., Ehrhart, B. D., Al-Shankiti, I., Ward, B. J., Musgrave, C. B., & Weimer, A. W. *Wiley Interdisciplinary Reviews: Energy and Environment*, 5(3), 261–287. (2016).

2. Kubicek, M., Bork, A. H.& Rupp, J. L. M. *J. Mater. Chem. A, in review* (2017).

3. McDaniel, A. H., Miller, E. C., Arifin, D., Ambrosini, A., Coker, E. N., O'Hayre, R.,et al. *Energy & Environmental Science*, 6(8), 2424. (2013).

4. Yang, C.-K., Yamazaki, Y., Aydin, A., & Haile, S. M. *Journal of Materials Chemistry A.*, 2, 13612–13623. (2014)

5. Bork, A. H., Kubicek, M., Struzik, M. & Rupp, J. L. M. *J. Mater. Chem. A***3,** 15546–15557 (2015).

6. Bork, A. H., Povoden-Karadeniz, E. & Rupp, J. L. M. *Adv. Energy Mater.***7,** 1601086 (2017).

7. Bork, A. H., Povoden-Karadeniz, E. & Rupp, J. L. M. *to be submitted.*

# 482636 Study on the Performance of $Fe_2O_3$-$MgO/Al_2O_3$ As the Oxygen Carrier in Chemical-Looping Hydrogen Generation

**Abstract:**

$Fe_2O_3$-$MgO/Al_2O_3$, prepared by the impregnation method, was characterized by BET, TPR, XRD, Raman and XPS methods. It was shown that $Mg^{2+}$ went into $Fe_2O_3$ crystal lattice and solid solution was formed in $Fe_2O_3$-$MgO/Al_2O_3$. As a result, oxygen content of $Fe_2O_3$-$MgO/Al_2O_3$ increased and the grain size of $Fe_2O_3$ decreased. Both were in favour of oxidation-reduction reactions. $Fe_2O_3$-$MgO/Al_2O_3$ was used as the oxygen carrier in Chemical-Looping Hydrogen Generation, which is an oxidation-reduction reaction. The evaluation performances indicated that the oxygen carrier was so active that methane was totally oxidized at 850°C and atmosphere, and the hydrogen yield was high correspondingly. Combination of characteristic results and evaluation results, it was revealed that solid solution of MgO and $Fe_2O_3$ enhanced the oxygen content and dispersion of $Fe_2O_3$, which is beneficial to the methane conversion and hydrogen yield in chemical-looping hydrogen generation.

# 489930 Chemical Looping Gasification with Red Mud As the Oxygen Carrier and Catalyst for Hydrogen-Rich Syngas Production

**Abstract:**

Chemical looping gasification (CLG) is an innovative, air-blown technology used to produce $H_2$-rich syngas, with a relatively high ratio of $H_2$:CO, where air does not directly contact fuel. A CLG system consists of two separate reactors: an air reactor (AR) and a fuel reactor (FR). The oxygen and heat required for fuel gasification is transported from the AR to the FR via a solid oxygen carrier (usually a transition metal oxide), reducing the energy demand required for oxygen separation. Meanwhile, the depleted oxygen carrier (OC) can catalyze in-situ water gas shift reaction, producing $H_2$-rich syngas ready for Fisher-Tropsch synthesis and eliminating the costly external water-gas shift reactor. The dual-function of OC makes it a key point for process intensification in this technology. Red mud, byproduct from alumina industry, has been proposed as a cost-effective OC material for the chemical looping process by UKy-CAER. This work further investigated the performance of red mud as an oxygen carrier and catalyst for the CLG process in a fluidized bed reactor. It was verified that the gasification rate of coal char was enhanced approximately by 1.5-3 times in the bed of red mud particles. Deactivation and agglomeration was not observed during 20 continuous cycles. The effect of OC:fuel mass ratios on the produced syngas composition was studied as well. As the ratio of OC:fuel decreased, syngas fraction increased due to the oxygen-lean condition. $H_2$:CO ratios in the range of 2.1-4.5 were obtained in the bed of reduced red mud, demonstrating the catalytic function of red mud in the water gas shift reaction.

# 491552 Updates on the Operation of Netl's 50-Kw Chemical Looping Combustion Test Facility

**Abstract:**

The United States Department of Energy's National Energy Technology Laboratory (NETL) is developing transformational fossil fuel combustion technologies that result in more efficient power cycles and/or substantial reductions in GHG emissions. One such technology is Chemical Looping Combustion (CLC), which has the potential to achieve lower cost of electricity than a conventional power plant with a conventional $CO_2$ absorption system. Although the ideas behind chemical looping have a long history, CLC has numerous technical issues that must be addressed to prepare this technology for a commercial debut. Two major areas are (1) development of an economic oxygen carrier and (2) solids handling. These areas are being studied in a small 50 $kW_{th}$ natural gas circulating test facility. This paper will describe the test facility and recent operating experience using iron-based oxygen carrier materials, which include raw hematite (a "commodity" oxygen carrier) and a manufactured copper and iron oxide oxygen carrier supported by alumina. Operating experience has indicated good methane conversions for the manufactured carrier but poorer performance for the hematite material.

# 493978 Assessment of Looping Combustion and Gasification of Carbon (CarboLoop) in a Twin Fluidized Bed Reactor

**Abstract:**

CarboLoop is a process for looping combustion of carbon with inherent separation of $CO_2$, developed by Salatino and Senneca (2009, 2010), which represents an alternative to chemical looping combustion (CLC) of solid carbon. Unlike other CLC processes, CarboLoop makes no use of oxygen carriers, but takes advantage of the inherent propensity of carbons to uptake oxygen at low temperatures, forming oxygenated surface complexes. Surface oxides are eventually released as CO and $CO_2$ by exposure to higher temperatures (Brown et al., 1992; Haynes, 2001). The CarboLoop concept requires the utilization of two reactors (most typically two interconnected fluidized beds): one acting as Oxidizer, where the carbon is kept in contact with air at relatively mild temperature (200-400°C) to foster oxygen chemisorption; the other acting as Desorber, operating at higher temperature (700-800°C) where desorption of the oxygenated C-O surface compounds is promoted. The ratio between CO and $CO_2$as desorption products depends on the nature of the carbon and is influenced by the desorption temperature. The original CarboLoop concept, aimed at chemical looping combustion of solid carbon, can be easily implemented into a chemical looping gasification (CLG) version, by proper definition of operational conditions of the desorption step.

An experimental campaign based on discontinuous experiments in a thermogravimetric analyzer (Salatino and Senneca, 2009, Salatino et al., 2010) has been directed to provide the proof-of-concept of the process. An extensive experimental campaign based on the combined use of TGA and DSC (Senneca et al., 2013) complemented by microstructural characterization by XPS (Levi et al., 2015) contributed to shed light on important and non trivial mechanistic and thermochemical features of the cyclic oxidation-desorption reactions, and on the chemical nature of surface oxides involved.

In this study the CarboLoop concept is tested under conditions that approach those of a realistic looping setup, consisting of a dual interconnected fluidized bed reactor. The "Twin Bed" test reactor (Coppola et al., 2016) has been purposely developed for the characterization of looping processes at the bench scale while preserving the time-temperature history that particles experience in a realistic looping conditions. It consists of two lab-scale bubbling beds of an inert bed, acting as thermal ballast, operated batchwise, connected to each other by a rapid solids transfer line. Carbon samples are fed to the system and undergo sequential steps of Oxidization and Desorption of pre-set duration by rapid transfer from one reactor to the other. The fuel tested is a bituminous coal char with size range of 400-1000μm. The Oxidizer was operated in air at different temperatures in the range 200-300°C with a holding time of 20 min. The desorption stage was carried out at 700-800°C with the same holding time of 20 min in $N_2$. The progress of char oxidation has been monitored following the profiles of CO and $CO_2$concentration at the exhaust of the Desorber over iterated cycles. The effect of multiple cycles on char oxidation/desorption propensity has been investigated. A systematic characterization of the influence of the operating conditions holding in the oxidation and in the desorption steps has been performed, and analyzed in the light of a simple semi-lumped kinetic scheme. Results are relevant to the development of a tentative process layout for the implementation of the

CarboLoop concept.

**References**

P. Salatino, O. Senneca, Preliminary Assessment of a Concept of Looping Combustion of Carbon, Ind. Eng. Chem. Res. 48 (2009) 102-109.

P. Salatino, O. Senneca, Int. Patent application WO2010/026259A2 (2010).

P. Salatino, O. Senneca, L. Cortese, CarboLoop: a novel concept of looping combustion of carbon, Proc. of the 1st International Conference on Chemical Looping 2010, Lyon (FR).

T.C. Brown, A.E. Lear, B.S. Haynes, Oxygen chemisorption on carbon, Proc. Combust. Inst. 24 (1992) 1199-1206.

B.S. Haynes, A turnover model for carbon reactivity I. development, Combust. Flame 126 (2001) 1421-1432.

O. Senneca, P. Salatino, L. Cortese, Assessment of the thermochemistry of oxygen chemisorption and surface oxide desorption during looping combustion of coal char, Proc. Comb. Inst. 34 (2013) 2787-2793.

G. Levi, O. Senneca, M. Causà, P. Salatino, P. Lacovig, S. Lizzit, Probing the chemical nature of surface oxides during coal char oxidation by high-resolution XPS, Carbon, 90 (2015) 181-196

A. Coppola, F. Scala, L. Gargiulo, P. Salatino, A twin-bed test reactor for characterization of calcium looping sorbents, Powder Technol. 2016 (Article in press) http://dx.doi.org/10.1016/j.powtec.2016.11.067.

# 494362 Super-Dry Reforming of $CH_4$

**Abstract:**

**Super-dry reforming of $CH_4$**

**L.C. Buelens**[1], **V.V. Galvita**[1,*], **H. Poelman**[1], **G.B. Marin**[1]

[1] *Laboratory for Chemical Technology, Ghent University, Technologiepark 914, 9052 Ghent, Belgium*

**Corresponding author: vladimir.galvita@UGent.be*

## 1. Introduction

During the past decade, environmental research targeting the mitigation of $CO_2$ emissions has been shifting from $CO_2$ storage towards $CO_2$ utilization. The conversion of $CO_2$ to CO could provide a cheap $C_1$ building block for the chemical industry. Development of processes for $CO_2$ conversion are necessary to fully exploit these properties. In this respect, the conversion of diluted $CO_2$ streams into CO by means of chemical looping is particularly interesting [1]. Among the chemical looping processes, super-dry reforming of $CH_4$ takes up a prominent position as it allows for intensified $CO_2$ utilization and can be driven by renewable energy sources such as biogas [2].

The super-dry reforming process (Figure 1) relies on the combination of a reforming catalyst, typically nickel based, with an oxygen carrier and $CO_2$ sorbent material, which are typically iron oxide and calcium oxide based. In the $CH_4$ oxidation step, a $CO_2$ rich mixture of $CH_4$ and $CO_2$ is dry reformed into syngas over the reforming catalyst. This syngas is fed to a mixed bed of oxygen carrier and $CO_2$ sorbent, where CO and $H_2$ are converted into $CO_2$ and $H_2O$ upon reduction of the oxygen carrier, while $CO_2$ is fixated by the $CO_2$ sorbent. In the $CO_2$ reduction step, the oxygen carrier and $CO_2$ sorbent materials are regenerated either by increasing temperature or isothermally by means of a sweep gas. A major advantage of this process is its potential to achieve a three times higher capacity for $CO_2$ conversion than conventional dry reforming of $CH_4$.

The oxygen carrier is the heart of the super-dry reforming process, periodically cycling between reduction by syngas and reoxidation by $CO_2$, thus allowing the conversion of $CO_2$ into CO. Indeed, the choice of oxygen carrier material is governed by the degree of reducibility of the oxygen carrier and its capacity to convert $CO_2$ into CO. The reforming catalyst aids in achieving deepest possible reduction of the oxygen carrier by first converting the fuel to syngas. The role of the $CO_2$ sorbent in super-dry reforming is threefold: (i) A tremendous decrease in coke formation on the nickel catalyst can be achieved because a higher $CO_2$:$CH_4$ feed ratio can be used without compromising the product purity.

(ii) The $CO_2$ sorbent also enhances the reducibility of the oxygen carrier by a continuous in situ removal of $CO_2$. Hence, $H_2O$ is the main product during the first step of super-dry reforming, even though 4 different processes (dry reforming, WGS reaction, oxygen carrier reduction, $CO_2$ removal) occur. (iii) In the $CO_2$ reduction step of the super-dry reforming process, decomposition of the $CO_2$ sorbent provides $CO_2$ for the oxygen carrier reoxidation, hence producing CO.

The aim of this work is to study the stability and activity of functional materials applied in super_dry reforming.

## 2. Experimental

As dry reforming catalyst, a nickel based material was prepared [3]. As oxygen carrier and $CO_2$ sorbent, an iron oxide based [4] and calcium oxide [2] or lithium orthosilicate based material were prepared. Nanomorphology and chemical information were obtained by materials characterization via STEM-EDX. For studying solid phase transitions, time-resolved in situ XRD experiments were performed in a Bruker-AXS D8 Discover apparatus with Cu K*f¿* radiation and a Vantec linear detector. For studying the gas phase product yields, a quartz tube microreactor (inner diameter of 10mm) was used. The reactor, equipped with K_type thermocouples contacting the reactor walls at the position of the catalyst bed, was heated by an electric furnace. EkviCalc [5], a software package containing an extensive database of chemical species, was used for performing thermodynamic calculations.

## 3. Results and discussion

By combining a $Ni/MgAl_2O_4$ dry-reforming catalyst, $Fe_2O_3/MgAl_2O_4$ oxygen carrier and $CaO/Al_2O_3$ $CO_2$ sorbent in the isothermal super-dry reforming process, a higher CO production yield was obtained than through conventional dry reforming. The increased yield of CO can be explained through the inherent separation of $H_2O$ from CO, hence avoiding equilibrium limitations in CO yield caused by the water-gas shift reaction. In fact, this two-step process owes its success to the application of Le Chatelierfs principle in both steps. In the $CH_4$ oxidation step, the reduction of iron oxide is improved by continuous removal of $CO_2$ in the form of $CaCO_3$. In the $CO_2$ reduction step, the introduction of an inert sweep gas initiates the decomposition of $CaCO_3$ upon which the oxygen carrier is reoxidized by $CO_2$. Here, the formation of CO provides a driving force towards the complete decomposition of $CaCO_3$. Overall, the theoretical yield of super-dry reforming goes up to four molecules of CO per $CH_4$ with a space-time yield of 7.5 $mmol_{CO}\ s^{-1}\ kg_{Fe}^{-1}$ at 1023 K. [2]

The effect of the choice of materials (reforming catalyst, oxygen carrier and $CO_2$ sorbent) and their cyclic stability are investigated. A major challenge in the design of the reforming catalyst is the mitigation of catalyst deactivation through carbon deposits. One promising strategy is the addition of iron to obtain a $Ni\text{-}Fe/MgAl_2O_4$ catalyst, in which iron assists in the oxidation and removal of carbon deposits from the catalyst surface [3]. The long-term application of an oxygen carrier is mostly limited by sintering and the undesired formation of inactive solid phases. To this end, $50Fe_2O_3/MgAl_2O_4$ seems to provide a good candidate in terms of material stability and activity based on cyclic redox experiments over 4 days time on stream. As for the $CO_2$ sorbents, CaO based and $Li_4SiO_4$ based materials are compared. One possible strategy for increasing the cyclic stability of the $CO_2$ sorbent is the application of a coating, such as $ZrO_2$ (Figure 2). The aim of doing so is to reduce the effect of crystallite growth due to sintering by providing a physical barrier between the different $CO_2$ sorbent particles. However, it needs to be addressed whether or not this coating has any adverse effect on the properties of the $CO_2$ sorbent material, such as a reduced accessibility of the core material for $CO_2$. Hereto, both experimental evidence and thermodynamic considerations are taken into account.

## 4. Summary

This work discusses the material properties, important for the real-life implementation of the super-dry reforming process. On the one hand, the properties of a structured $CO_2$ sorbent material, based on $Li_4SiO_4$ and $ZrO_2$, are investigated in view of its application. On the other hand, the effect of redox cycling on different $Fe_2O_3/MgAl_2O_4$ oxygen carrier materials is elaborated on.

**Acknowledgment**

This work was supported by the Long Term Structural Methusalem Funding of the Flemish Government, the Interuniversity Attraction Poles Programme, IAP7/5, Belgian State _ Belgian Science Policy and the Fund for Scientific Research Flanders (FWO; project G004613N). L.C. Buelens acknowledges financial support from the Institute for the Promotion of Innovation through Science and Technology in Flanders (IWT Vlaanderen). The authors also thank Prof. Christophe Detavernier (Department of Solid State Sciences, Ghent University) for access to in situ XRD equipment.

**References**

[1] V.V. Galvita, H. Poelman, C. Detavernier, G.B. Marin, Appl. Catal. B-Environ. 164 (2015) 184-191.

[2] L.C. Buelens, V.V. Galvita, H. Poelman, C. Detavernier, G.B. Marin, Science 354 (2016) 449-452.

[3] S.A. Theofanidis, V.V. Galvita, H. Poelman, G.B. Marin, ACS Catal. 5 (2015) 3028_3039.

[4] N.V.R.A. Dharanipragada, L.C. Buelens, H. Poelman, E. De Grave, V.V. Galvita, G.B. Marin, J. Mater. Chem. A. 3 (2015) 16251_16262.

[5] B. Nolang, EkviCalc and Ekvibase, version 4.30 (2013) Svensk Energi Data: Balinge, Sweden.

Figure SEQ Figure * ARABIC 1 Schematics of the super-dry reforming process

Figure SEQ Figure * ARABIC 2 STEM-EDX characterization of $Li_4SiO_4$-$ZrO_2$ $CO_2$ sorbent material. (A) EDX elemental mapping of Zr and Si, clearly showing the presence of a silica based particle, coated by zirconia. (B) Corresponding HAADF-STEM image. Scale bar represents 100 nm.

# 498083 Efficiency Evaluation of Combined Cycle Power Plants Integrated with Fixed-Bed Chemical-Looping Combustion Reactors

**Abstract:**

$CO_2$ produced from the combustion of fossil fuels for power generation is the main greenhouse gas contributing to climate change. The mitigation of $CO_2$ emissions in power generation can be achieved by increase in renewable industry penetration, improvement of power plant efficiency, and integration of carbon capture and sequestration (CCS) technologies. Specifically, chemical-looping combustion (CLC) is a promising CCS technology for power generation with low energy penalty and inherent $CO_2$ separation.[1] CLC involves the use of a metal oxide as an oxygen carrier which transports the oxygen from the air to the fuel, and avoids direct contact between fuel and air. Commonly, CLC is realized as two interconnected reactors, a Reducer and an Oxidizer. In the Reducer, fuel is oxidized by the metal oxide into combustion products ($CO_2$, $H_2O$), and pure $CO_2$ can be captured after the stream of combustion products is condensed. In the Oxidizer, the reduced metal oxide is reacted with air. Other CLC reactor designs include fixed bed reactors[2] and reverse flow fixed bed reactors,[3] which provide process intensification and reactor modularization advantages compared to traditional fluidized beds.

The focus of this work is on combined cycle (CC) power plants integrated with fixed-bed CLC reactors, which consist of a fixed bed reactors island, an open cycle air gas turbine and a water/steam cycle. This plant configuration is of interest due to its high efficiency, low installation cost, and faster commercialization time.[4,5] We explore the feasibility of fossil-fueled power plants with CLC on the basis of "virtual" power plant prototypes, as shown in Figure 1. Prototype plant models were first validated against static, full load data from an existing natural-gas-fueled supercrictical power plant with an efficiency of 57.9%, and then the conventional combustion chamber was replaced by a high-pressure fixed-bed CLC island. Dynamic optimization of high-pressure fixed-bed reactors was performed to improve the utilization and efficiency of the gas turbine of the power plant.[5] The main factors impacting the integrated CLC-CC power plants efficiency were explored, including the CLC reactor operating strategy, pressure ratio, gas turbine inlet temperature and temperature of ambient air. This analysis showed an optimal efficiency of up to 49% being feasible for the integrated CLC-CC plant. This was accomplished by utilizing all the reactor exhaust streams to improve the Brayton and Rankine cycles efficiency. Dynamic simulation of the integrated CLC-CC plant showed that the integrated CLC-CC power plant generates a relatively flat power output and CC is only slightly affected by the discontinuous batch nature of the operation of the fixed-bed CLC reactors. Plant simulation under part-load and transient load operation was also studied and a plant level control architecture was devised for efficiency optimization. The developed simulation and optimization framework enabled in-depth studies of stability, feasibility and optimization of power cycles with CLC technologies embedded.

Figure 1: Simplified diagram of the CC integrated with CLC. (CPR: Compressor; G: Power generator; GT: Gas turbine; SH: Superheater; RH: Reheater; EVA: Evaporator; ECO: Economizer; HP: High-pressure turbine; LP: Low-pressure turbine; CON: Condenser; BFP: Boiler feed pump; PH: Preheater; OX: Oxidation; RED: Reduction; HR: Heat removal).

**References**

1. Herzog, H. J. New technologies could reduce carbon dioxide emissions to the atmosphere while still allowing the use of fossil fuels. Environ. Sci. Technol. (2001).

2. Han, L. & Bollas, G. M. Dynamic Optimization of Fixed Bed Chemical-Looping Processes. Energy 112C, 1107–1119 (2016).

3. Han, L. & Bollas, G. M. Chemical-looping combustion in a reverse-flow fixed bed reactor. Energy 102, 669-681 (2016).

4. Kehlhofer, R., Hannemann, F., Rukes, B. & Stirnimann, F. Combined-Cycle Gas & Steam Turbine Power Plants. (PennWell, 2009).

5. Chen, C., Han, L. & Bollas, G. M. Dynamic Simulation of Fixed-Bed Chemical-Looping Combustion Reactors Integrated in Combined Cycle Power Plants. Energy Technol. 4, 1209–1220 (2016).

6. Zhou, Z., Han, L. & Bollas, G. M. Overview of Chemical-Looping Reduction in Fixed Bed and Fluidized Bed Reactors Focused on Oxygen Carrier Utilization and Reactor Efficiency. Aerosol Air Qual. Res. 14, 559–571 (2014).

**Acknowledgements**

This material is based upon work supported by the National Science Foundation under Grant No. 1054718. CC gratefully acknowledges support by the GE Graduate Fellowship for Innovation. This work was partially sponsored by the United Technologies Corporation Institute for Advanced Systems Engineering (UTC-IASE) of the University of Connecticut. Any opinions expressed herein are those of the authors and do not represent those of the sponsor.

# 500396 Fate of Sulfur in Coal-Direct Chemical Looping Systems

**Abstract:**

The fate of sulfur in the coal-direct chemical looping system (CDCL) was investigated in the sub-pilot reactor with a heat-traced gas analysis system. The CDCL process achieves 96.5% carbon capture efficiency with a 26.8% increase in cost of electricity compared to a conventional pulverized coal power plant without $CO_2$ capture. The sulfur balance was successfully closed during the injection of high sulfur coal. More than 69% of sulfur from coal exit the reactor as $SO_2$ and $H_2S$ in the reducer flue gas while less than 5% exit as $SO_2$ in the combustor spent air. The remaining sulfur was retained in coal ash as inorganics. The finding suggests an acid gas removal system targeting both $H_2S$ and $SO_2$ is required to meet the recommended quality of $CO_2$ stream for sequestration and transportation. Using the determined ratio of $SO_2$ and $H_2S$, a properly designed Claus plant can potentially enable the recovery of elemental sulfur. On the other hand, the combustor spent air was found to comply with the US EPA sulfur emission regulation and can be released to the atmosphere without a costly acid removal system. The relationship between the sulfur and carbon capture efficiencies was established experimentally and was found to be proportional to each other throughout the experiment at a ratio of 0.8 below 93% of carbon capture efficiency and near 1 above 93%. This was attributed to the delayed release of organic sulfur during incomplete char gasification in the reducer. The finding reaffirms the effectiveness of the counter-current moving bed design in minimizing the amount of carbon and sulfur emission in the combustor spent air with an average carbon and sulfur capture efficiency of 96.5 and 95%. Sulfur deposition on the iron based oxygen carriers did not affect the system performance and complete removal of deposited sulfur was observed during oxidation in a thermogravimetric analyzer. The findings demonstrate the robustness of the coal-direct chemical looping system to handle high sulfur coal without complicated acid gas cleaning scheme or severe performance penalties. A 250 $kW_{th}$ pilot scale coal-direct chemical looping reactor has been constructed by Babcock and Wilcox Power Generation Group in Barberton, OH that is undergoing unit testing with both low and high sulfur coal in 2017.

# 498543 Production of Chemicals from the Thermal Cracking of Castor Oil, Castor Oil Methyl Ester and Ricinoleic Acid

**Abstract:**

This work evaluates the experimental results of three different processes to obtain heptaldehyde, undecylenic acid and methyl undecylenate through castor oil, castor oil methyl ester, and ricinoleic acid. Heptaldehyde and undecylenic acid are important chemicals on food, packaging, ink, solvent and polymers industries. Methyl undecylenate is commonly used as a Flavoring agent and also an important route to obtain undecylenic acid. Methyl undecylenate was obtained by the thermal cracking of castor oil methyl ester. In order to obtain heptaldehyde, methyl undecylenate, and undecylenic acid, the thermal cracking was performed using commercial castor oil, castor oil methyl ester (obtained by castor oil transesterification reaction) and commercial ricinoleic acid, respectively. Thermal cracking temperatures were 530, 545, 560 and 575 °C with 400 g/h mass flow on the feed. All materials were blended with distilled water, considering the same proportion (75 % material, 25 % water). Temperature influence on bio-oil yields was observed, likewise different behaviors to each material. Iodine index, acid index, and density were determined to all bio-oil samples and their content were analyzed by GC-FID. The best condition to obtain higher yields of each chemical were pointed. With regard to the castor oil, higher yields of heptaldehyde and undecylenic acid were found on experiments at 530 °C. The same temperatures resulted in higher yields concerning the ricinoleic acid. The cracking of methyl ricinoleate results on higher yields of heptaldehyde at 530 °C and methyl undecylenate at 560 °C.

# NEEM COATED UREA- A FISCAL APPROACH FOR FARMERS

*Sara Ahsan, Ali Ayub*

## Abstract

Soil fertility is determined by three major elements namely Nitrogen, Phosphorus & Potassium (N, P, K) of which nitrogen plays a very important role. For this reason, Urea (Containing 46% of N) consumption is very high all over the world. Generally, the plants are able to recover a fraction of Nitrogen present in Urea and this fraction is known as Nitrogen Use Efficiency (NUE). This NUE stands anywhere between 30 to 50%. Thus, two third of Urea Nitrogen escapes from the soil in the form of nitrogen, ammonia and nitrates etc. and is not used by the plants. Engro Fertilizers, Pakistan, researched and experimented use of "Azadirachta indica" commonly known as neem oil as nitrification inhibiting agent. Based upon the results of extensive laboratory trials, coating urea with neem helps in slow release of nitrogen, nourishing the crops for a longer period. The process reduces pollution of groundwater. There is an increase in crop yield and also serves as an efficient natural insecticide. It also serves as an anti-caking agent and will reduce the amount of urea required by a plant by ~10-20%.

This paper explicates Engro's experience and research on neem coated urea, methods of preparing an effective neem oil emulsion, quantities of raw materials to be used per unit of urea, its uses and results obtained after extensive laboratory and commercial trials.

## INTRODUCTION

Snam Progetti divides Urea plant into following major sections i.e.

1. Urea synthesis and NH3, CO2 recovery at high pressure.
2. Urea purification and NH3, CO2 recovery at medium and low pressures.
3. Urea concentration.
4. Urea prilling.
5. Waste water treatment.

Snamprogetti ammonia stripping process is characterized by a urea synthesis loop operating at about 160 Kg/cm2 (abs); with ammonia to carbon dioxide molar ratio at urea reactor inlet of 3.3 to 3.6. This allows a high CO2 conversion into urea in the reactor itself. There are two kinds of chemical reactions at the same time in the urea reactor:

$$\mathbf{2NH_3 + CO_2 \leftrightarrow NH_2\text{-}COO\text{-}NH_4 + 32560\ kcal/kmol\ of\ carbamate\ (at\ 1\ atm;\ 25°C)}$$

$$\mathbf{NH_2\text{-}COO\text{-}NH_4 \leftrightarrow NH_2\text{-}CO\text{-}NH_2 + H_2O - 4{,}200\ kcal/kmol\ of\ urea\ (at\ 1\ atm;\ 25°C)}$$

First reaction is very fast & exothermic and the second one is weakly endothermic and occurs in the liquid phase at low speed. Downstream the urea synthesis, the decomposition and relevant recovery of unconverted chemical reagents occurs. It is carried out in three subsequent steps: High Pressure Decomposition in H.P. stripper; Medium pressure Decomposition in M.P. Decomposer and, finally, Low Pressure Decomposition in Low Pressure Decomposer.

The decomposition reaction is the reverse reaction of the first one above showed, viz.:

$$\mathbf{NH_2\text{-}COO\text{-}NH_4 \leftrightarrow 2NH_3 + CO_2 - Heat,}$$

And, as can be inferred from the equation, it is promoted by reducing pressure and/or adding heat.

The urea reactor effluent solution enters the stripper, under slightly lower pressure than the urea reactor, where a fair part of the unconverted carbamate is decomposed in NH3 and CO2. The overall yield of the H.P. synthesis loop referred to CO2 is as high as 83-85% (on molar basis). Downstream of the stripper residual carbamate and ammonia are recovered in two recycle stages operating at 18.5 kg/cm2 (abs) (namely M.P. section) and 4.8 kg/cm2 (abs) (namely L.P. section) respectively.

Ammonia and carbon dioxide vapors from the stripper top, after mixing with the carbonate recycle solution from M.P. section, are condensed, at the same pressure level of the stripper itself, in the 1st and 2nd High Pressure carbamate condensers, thus producing the L.M.P and the L.P. steam which are used in downstream sections. After separating the inert gases which are passed to M.P. section, the carbamate solution is finally recycled to the reactor bottom by means of a liquid/liquid ejector, which exploits H.P. ammonia feed to reactors as motive fluid.

Downstream of the two recovery stages urea is concentrated to 99.75% via two vacuum concentrator stages. The solution leaving the L.P. decomposer bottom with about 70% wt. urea, is sent first to the tube of vacuum pre concentrator and then pumped to first vacuum concentrator both operating at a pressure of 0.34 kg/cm2 (abs). The mixed phase coming out from the process side of the first vacuum concentrator enters the first gas-liquid vacuum separator from where vapors are again extracted by the first vacuum system while the urea melt (~ 95 wt%), by gravity & differential pressure flow, enters the bottom of second vacuum concentrator operating at a pressure of 0.03 kg/cm2 (abs). The extracted urea melt (~99.75% pure) from the second concentrator is sent to prilling tower. Droplets of molten urea from the prilling bucket fall along the natural draught prilling tower and solidify and cool while encountering a countercurrent air flow.

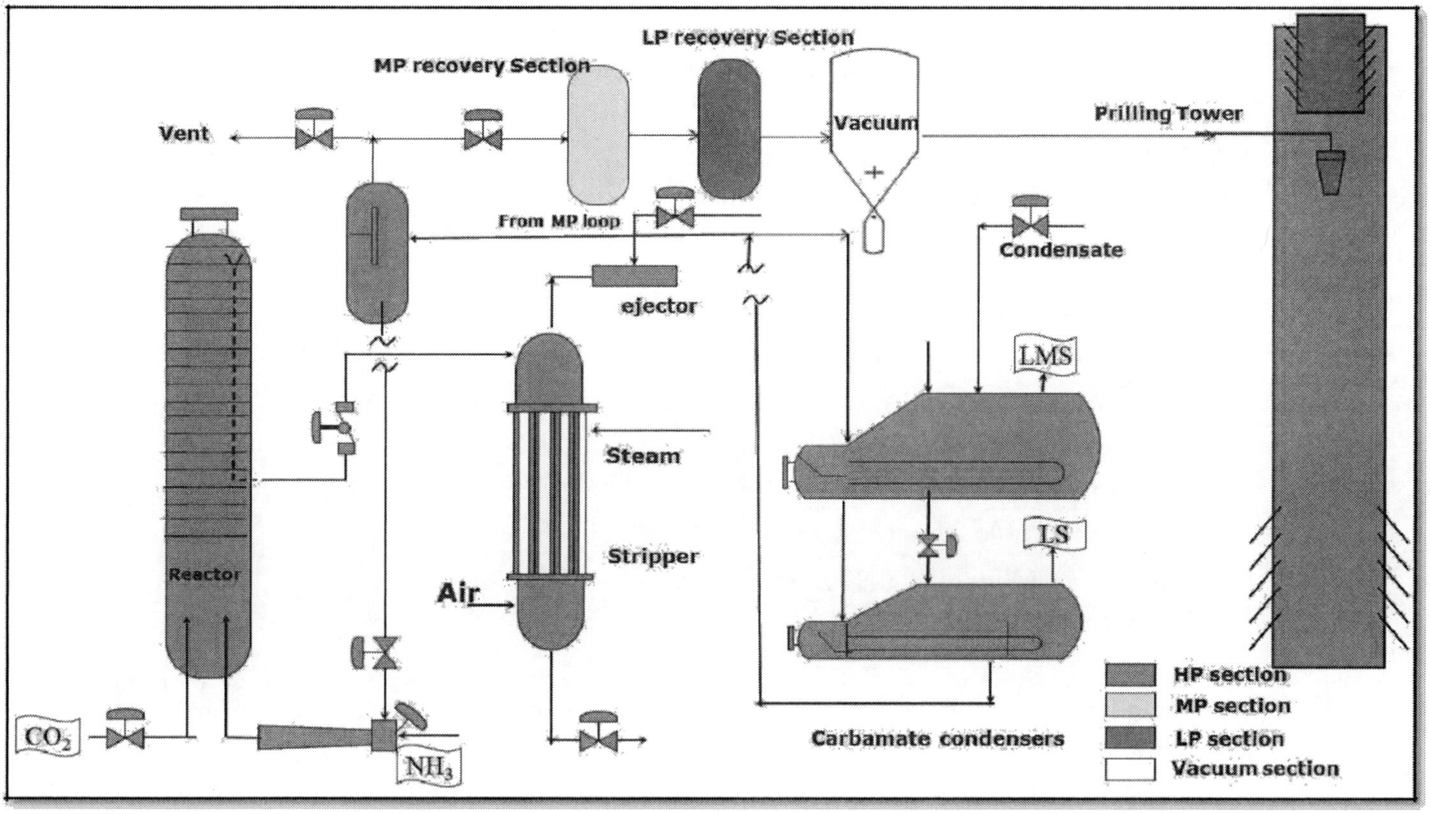


**Figure 1. Process Description:** Picture elucidating a short process description of Snam's urea plant.

## NEEM COATING ON UREA PRODUCT- AN EXPERIENCE AT ENGRO

*Purpose of coating Neem on Urea Product:*

With 46% N content, Urea is the most popular Nitrogenous fertilizer around the world. When Urea is applied to soil, it is first hydrolyzed into Ammonical form (Ammonium Ion i.e. NH4+) and then to nitrite (NO2-), followed by to nitrate (NO3-) forms by the process called nitrification. Nitrification process is necessary for making nitrogen available to plants. However, if the process of nitrification is too rapid; nitrogen will escape to atmosphere and plants will not be able to recover it from Urea efficiently. Generally, the plants are able to recover a fraction of Nitrogen present in Urea and this fraction is known as Nitrogen Use Efficiency (NUE). This NUE stands anywhere between 30 to 50%. Thus, two third of Urea Nitrogen escapes from the soil and is not used by the plants. Hence this lead us to regulate the urea hydrolysis and nitrification. This regulation is done by some chemical/natural agents called Nitrification inhibitors. However, most of the nitrification inhibitors are costly chemicals (examples nitrapyrin, dicyandiamide and ammonium thiosulphate) which are not economically feasible.

It has been scientifically established that Neem oil serves as an effective nitrification inhibitor whose benefits when coated on urea area mentioned below:

- Neem coating leads to more gradual release of urea, helping plants gain more nutrients and resulting in higher yields.

- Lower underground water contamination due to leaching of urea.
- Neem serves as a natural insecticide and an anticaking agent

***Properties of Neem Oil:***

The properties of neem oil used are as follows:

1. Neem Oil purity = 98.5%+
2. Color= Yellowish brown clear liquid
3. Moisture = 0.5% Max.
4. Specific gravity = 0.9 to 0.94 at 30 oC
5. Iodine value = 65 to 80
6. Azadirachtin Content = min 150 ppm
7. Saponification Value = 180 to 205
8. Up saponi cable matter = < 2 %
9. Density of neem oil = $0.92 g/cm^3$
10. Viscosity of neem oil = 35CSt

As per standard limits of PSQA, Azadirachtin content in final product must be greater than 50 ppm corresponding to 250ppm of neem oil coat. Engro started its coating setting a standard in Pakistan of 500ppm of neem oil coat. So, 0.5kg of neem oil is required per ton of urea. Cost of neem oil is 2.21$ per kg.

## Process employed for coating of Prilled Urea

***Process Description:***

Coating of neem oil on urea product requires preparation of solution of neem oil in water. Neem oil is not soluble in water so it requires an emulsification agent. On Laboratory scale, tests were done on the use of Lab grade 25% Ammonical water as an emulsifying agent. The results were favorable for the use of Ammonical water as an agent. It was then decided to use it in our emulsion preparation method.

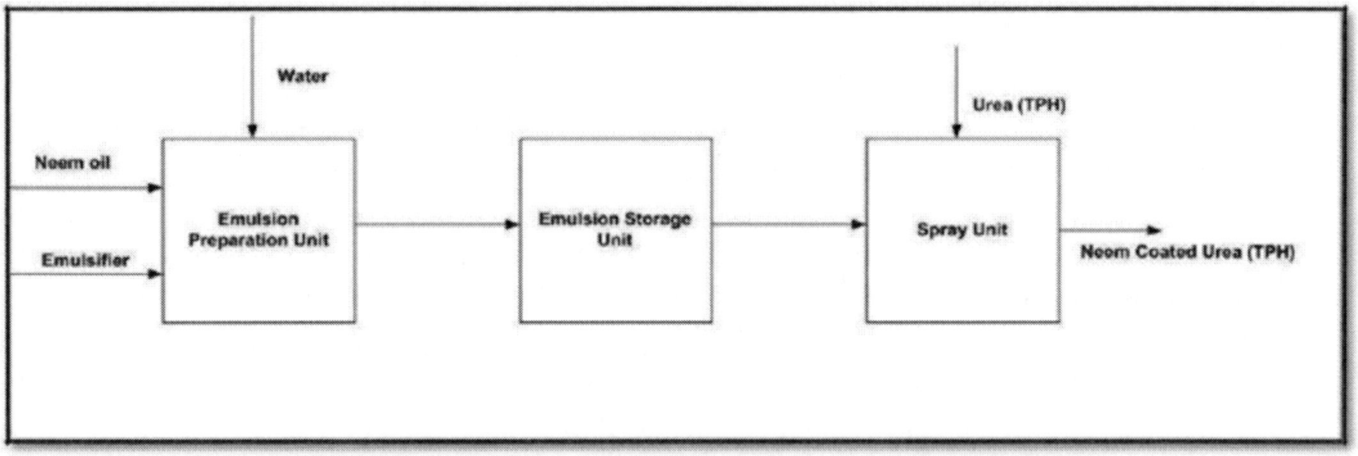


**Figure 2: Block Diagram of emulsion preparation process**

Below two images depicts neem solution change in color before and after addition of emulsifying agent.

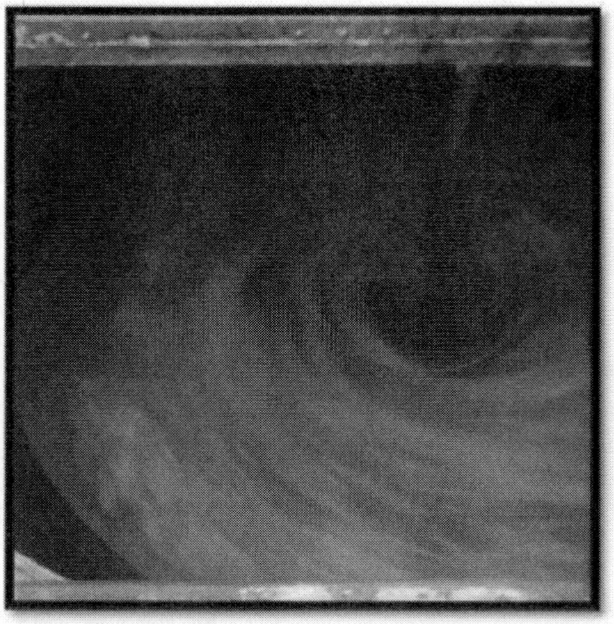

**Figure 3:**Neem Oil & water (Before addition of Ammonical solution)

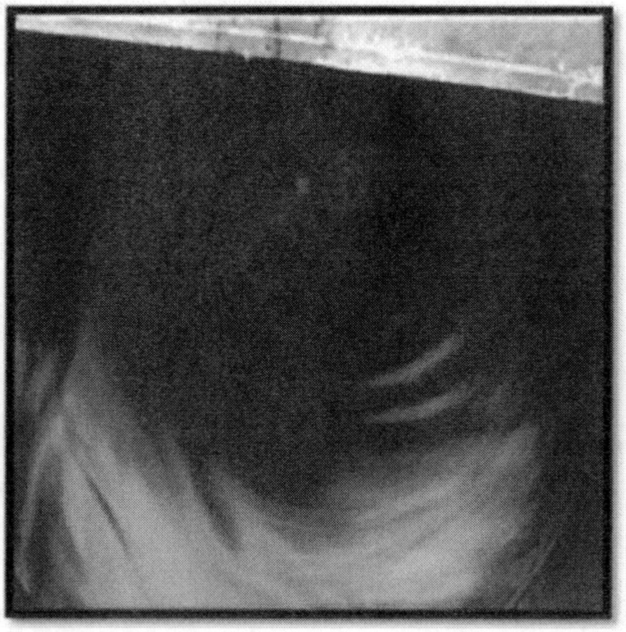

**Figure 4:**Neem emulsion (After addition of Ammonical solution)

Design capacity of Engro's EnVen plant is 3835 MTPD which is equivalent to 160TPH of product over the belt. For coating 500 ppm of neem oil over this product, first and the foremost step was to add accurate quantities of raw materials and emulsifying agent. For the aforementioned urea production, total neem emulsion required was 3835kg, with pure neem oil amount of 1917.5kg (50% of total emulsion). 0.1grams of 25% Ammonical solution is required per 50grams of pure neem oil used and the remaining being water.

Engro's Laboratory researched and registered its test for determining amount of Azadirachtin in the coated product. Below is the Process flow diagram of the facility used during coating activity

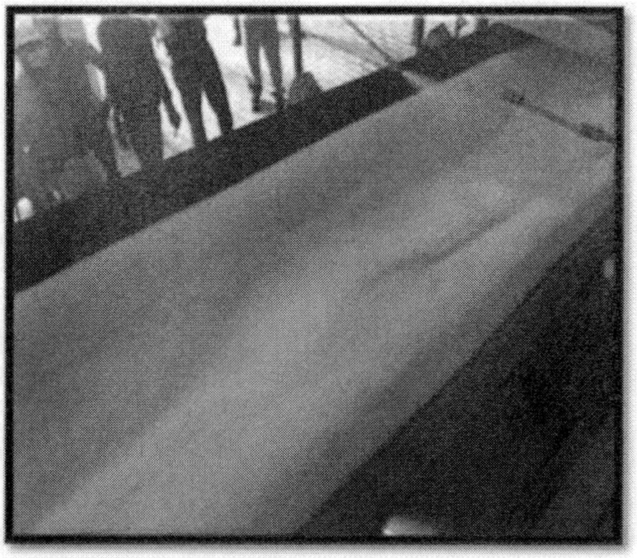

**Figure 5:** Coated Urea on Conveyer Belt

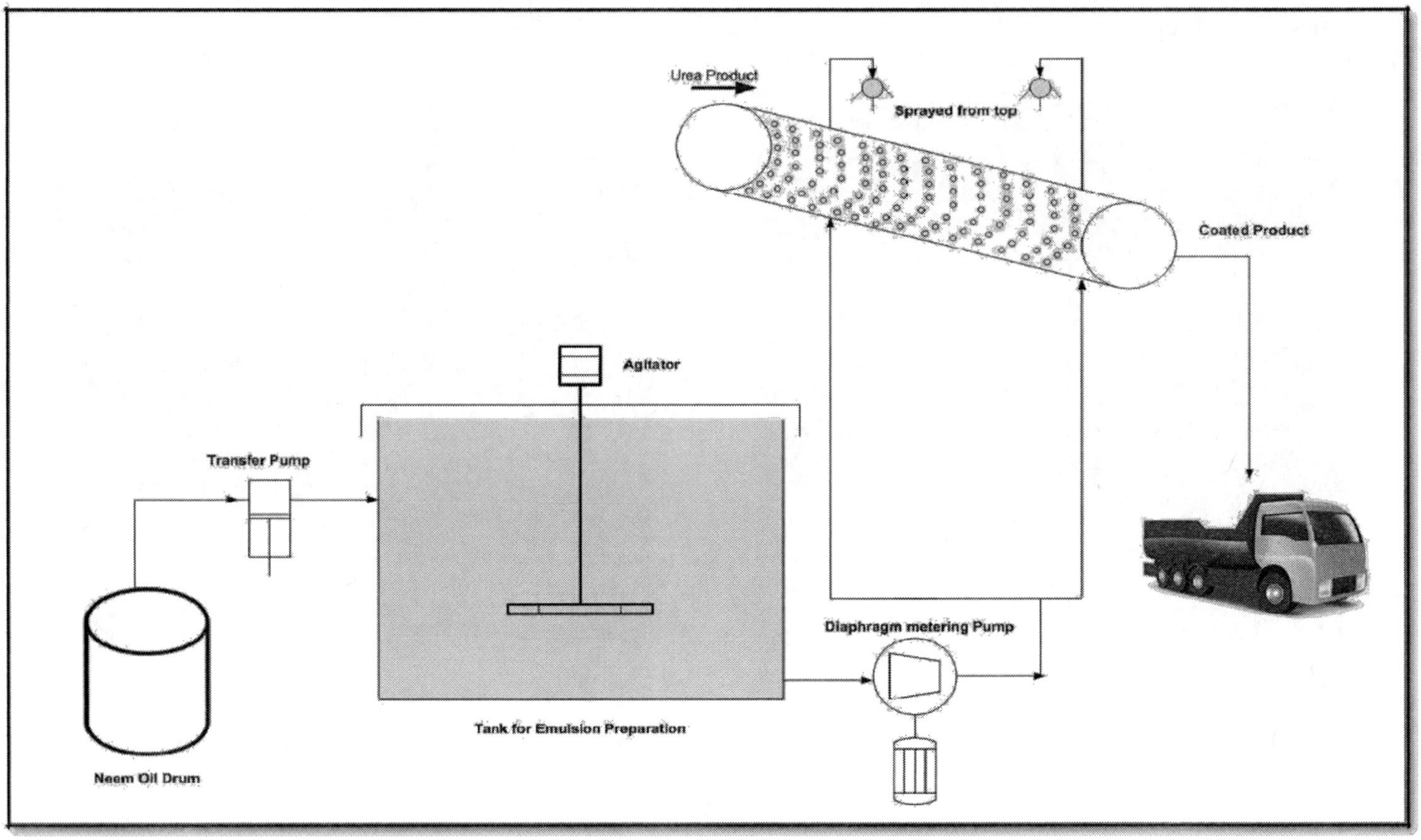


**Figure 6: Block Diagram of emulsion preparation process**

Before stating coating, stroke testing of the diaphragm metering pump was done. This testing was made basis for selection of the pump strokes during neem coating. The stroke testing results are as mentioned below:

**Table 1.** Pump stroke testing results

| Strokes | Flow (Litre/min) | Flow (Litre/hr) |
|---|---|---|
| 10 | 0.5 | 30 |
| 20 | 0.980 | 58.8 |
| 30 | 1.37 | 82.2 |
| 40 | 1.87 | 112.2 |
| 50 | 2.35 | 141 |
| 60 | 2.6 | 156 |
| 70 | 3.15 | 189 |
| 80 | 3.48 | 208.8 |
| 90 | 3.9 | 234 |
| 100 | 4.3 | 258 |

For a coat of 500 ppm corresponding strokes needs to be adjusted at 70%. To determine efficaciousness of the spraying unit, we set the strokes primarily at 70% and collected a sample then adjusted them to 50% and collected another sample. These samples reported Azadirachtin content in them to be 99.5ppm and 90ppm. Minimum limit of the active agent was 50ppm, so, strokes were further reduced to 45% and coating as started. Following are the results obtained during coating.

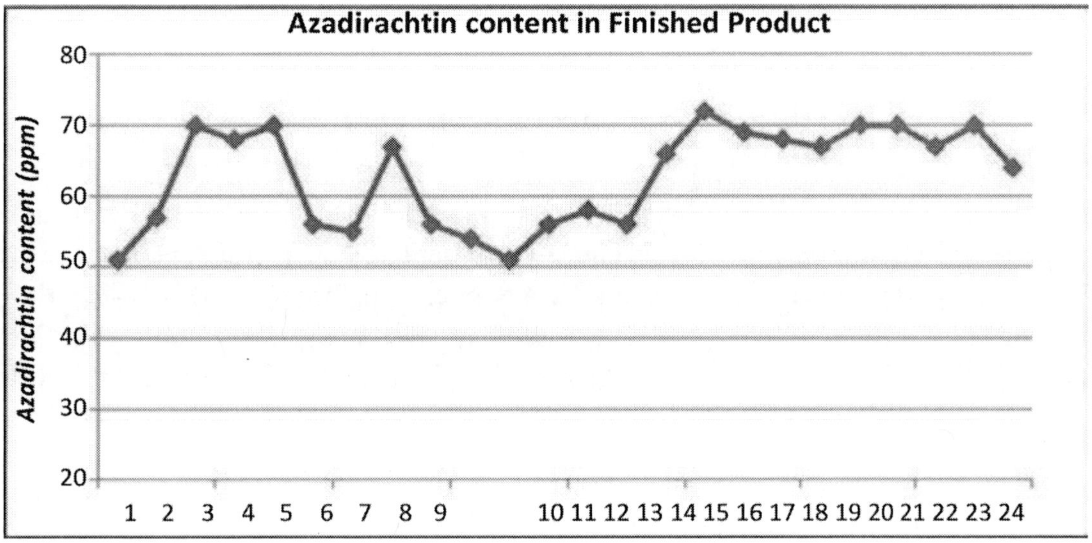


**Graphical representation 1**: Azadirachtin content in Finished Product

***Required Equipment***

- One transfer Pump for transferring neem oil into the tank where emulsion is to be prepared (Capacity of used pump: $4m^3/hr$)
- A tank of capacity depending on the design capacity of plant (Tons per hour of product passing through belt conveyer) (Capacity of used Tank: $4.82m^3$)
- Agitator for mixing and emulsion preparation (Rpm of Used agitator: 70)
- Neem emulsion storage tank (Capacity of used Tank: 4.82 $m^3$)
- Gear Pump for transferring neem emulsion to nozzles (Capacity of gear pump used: 220LPH)
- Spray nozzles for spraying neem emulsion on urea over the belt conveyer. (Capacity of each nozzle: 1.6LPM)

## Neem in Pakistani Market:

Neem story is very different in Pakistan. Its oil extraction is being done in the country but at a small scale. Neem seed and oil are not traded in Pakistani commodity market. As this tree seed and leaves have toxic properties, so it is not fed to the animals, hence each year all its fruit and leaves are wasted. Consequent result of less production of oil is high prices. After lots of struggles, a pharmaceutical company agreed on selling at 9$ per kg of neem oil. But purchasing at this rate was neither economically feasible nor gives a positive message to the market. Hence, if we want to start neem coating in Pakistan permanently, either its market needs to be improved or else oil should be imported from India or any other country where prices of neem oil are low.

In case we are successful in marketing this Neem coated Urea, it will start a total new industry in Pakistan with lots of parties setting up efficient Neem oil extraction plants and lively hood to thousands of people.

## Discussion and Benefits of neem coating on Urea

The decrease in the local urea sales whereby increasing the sale of imported urea led local fertilizer producers to slash their prices. Consequently, to deplete our urea inventory price was lowered by PKR 70. It is important to mention here that government, in Fiscal Year 2016-17 (FY17) budget, reduced urea prices by PKR 400/bag to PKR 1,400/bag. But this price cutting didn't result in the expected sales. So, Engro decided of neem coating on Urea Product, hoping to attract the customers with its uses. Indian government decision of coating 100% produced urea further strengthened our decision. For test run purposes, 10,000 MeT of urea was coated and sent to selected warehouses for marketing.

Engro Fertilizers Limited started first ever coating in Pakistan and homogenized the procedures for production of Neem Coated Urea. To have a uniform and consistent coating of Neem emulsion on urea prills and maintaining the contents of Azadirachtin in final product, Laboratory results and tests were performed after every two hours.

Summarizing the benefits of neem coating on urea product as under:

- Increase in price of coated urea per 50kg bag
- Decrease in nitrification and consequent increase in the yield of crops thus strengthening agricultural business
- Reduction in raw material consumption due to decrease in the amount of fertilizer required by plants
- Reduction in environmental pollution of ground water due to leaching of nitrates and gaseous emissions
- Lively hood to thousands of people
- Profit to the farmers due to reduction in the amount of fertilizers required per crop by approximately 30%
- Neem serves as a natural insecticide

# 496747 Technoeconomic Assessment of Two-Step Process for Upgrading and Utilizing Used Railroad Ties in an Integrated Biorefinery

**Abstract:**

In the USA, over 21 million wood railroad ties are produced per year to replace the approximately 20 million used ties that have provided excellent primary service in railroad track over 40 years. Due to the presence of carcinogenic preservatives, these used ties will primarily be either landfilled or incinerated as they do not meet U.S. Environmental Protection Agency (U.S. EPA) stringent requirements. Consequently, a two-step pyrolysis process was developed to upgrade used ties by first extracting wood preservatives (Step One) and converting upgraded used ties into biofuels, bioproducts, power, and heat (Step Two). This study evaluates the techno-economic viability of the proposed two-step process embedded in integrated pyrolysis biorefinery with liquid fuel, power and recovered preservatives as product streams.

The technoeconomic analysis assumed 2.33 million metric tons of used railroad ties are available yearly according to statistics by the Railroad tie association. This analysis considers 3 plants spread across the U.S. to minimize transportation cost with 2358 tons/day throughput each. The rail road ties biorefinery consists of eight areas including preprocessing (A100), thermal desorption (A200) and fast pyrolysis and vapor quenching (A300), hydroprocessing (A400), hydrogen generation (A500), power generation (A600), cooling and other utilities (A700) and waste management (A800). The minimum selling prices as well as sensitivity analyses were performed to determine the impact of thermal desorption conditions on the overall economic.

# 496556 Valorisation of Three-Phase Olive Mill Wastewater with the Addition of High Protein Co-Substrates

**Abstract:**

### Introduction

Nowadays national and international legislation attempt to enforce reductions in greenhouse gas emissions, while minimization of dependency to fossil fuel gathers all-around agreement. At the same time optimization of waste management strategies towards zero waste generation is a major challenge for modern society. Olive Mill effluents (OME) is the effluent generated during the production of olive oil and accounts for the 60% of the overall weight of the olive fruit. OME is rich in carbon and phenols while its nitrogen content together with the pH are very low rendering common treatment methods ineffective to tackle it. A valorization process that can be used for the management of OME is Anaerobic Digestion (AD). A key attribute of energy efficiency schemes has been the utilization of negative value substrates for biogas generation in anaerobic digesters. Anaerobic digestion (AD) is a biological method for the treatment and valorization of organic material. Its product can be recovered and utilized for heat and/or electricity generation. The efficiency of the process is affected by the operational conditions selected for the digesters and it depends highly on the type of the organic substrate used as feedstock.

Common organic substrates for AD systems include municipal solid waste, agroindustial byproducts and waste including slaughterhouse and dairy industry waste. Several types of AD feedstock are rich in fat, oil and grease (FOG). This is a feature that is highly desirable, as fat has the highest biomethane potential of all natural macromolecules. However, high fat content can inhibit the digesters through flotation, accumulation, pH reduction, and a number of intracellular inhibition processes. Furthermore, agroindustial wastewater, including olive mill effluents (OME), is characterized by high concentration of phenols, low nitrogen availability and low pH. All of the above are known to inhibit the digestion process especially in overloaded or unstable systems.

Aim of this work was to investigate the effect of external addition of nitrogen into digesters operating with OME as single substrate. The co-substrates evaluated were low-value byproducts of animal and plant origin, namely bone and meat meal, blood meal, feather meal, mixture of feather and blood meal, fish meal, soy been meal and maize gluten.

### Material and methods

OME was collected from a three-phase olive mill. The inoculum used was collected from four, six-liter digesters operating under steady state conditions with cattle manure as substrate and an OLR of approximately 3kg/VSm3-d. Protein-rich substrates were collected from a pet food manufacturer. All substrates were stored in a freezer at -20°C after collection until further utilization.

In this work 250mL Duran bottles were used as batch digesters and for assessing the biomethane potential and biodegradability of the different substrates. In order to evaluate the batch process and generate comparable results two loading rates (10 and 20 $kgVS/m^3$) were evaluated. The I/S ratio was set at 2 and 1 respectively. In the second stage of the experiment, four identical stainless steel reactors were used. The reactors were heated through the recirculation of hot water within the water jacket, feeding and removal of the substrates was taking place manually, while the mixing of the reactors was achieved with the application of immersed propellers and direct current mixing motors.

## Results and discussion

### Mono digestion of the substrates

The highest biogas production of the substrates under mono-digestion conditions was obtained using bone and meat meal with a value of 511 $mLCH_4/gVS_{added}$. This was approximately 16% higher compared to the production of biogas using blood meal and feather meal mixture with a production of 427 $mLCH_4/gVS_{added}$. Related to the plant-derived substrates the specific methane production of gluten and soy bean meal amounted to 495 $mLCH_4/gVS_{added}$ and 460$mLCH_4/gVS_{added}$ respectively.

### Co-digestion of OME with high nitrogen substrates

In the second stage of this study OME was co-digested with high nitrogen substrates. The highest biogas production was attained using a combination of OME and fishmeal with a yield of 482 $mLCH4/gVS_{added}$ corresponding to the 77% of the ultimate production for this mixture. Digestion of the other mixtures gave also very positive results with the mixture of OME & blood meal and of OME & soybean meal offering yields of 454 and 446 $mLCH4/gVS_{added}$ corresponding to the 74 and 72% of the ultimate production for the two mixtures respectively. The lowest biogas production registered was the one obtained from a mixture containing OME & bone and meat meal with a yield of 418 $mLCH4/gVS_{added}$, equal to the 68% of the theoretical value.

### Validation of the results in CSTR systems

In order to validate the results of the previous experimental stages, the biogas yield using four substrate mixtures composed of olive mill wastewater (OMWW) and protein-rich substrates were assessed in semi-continuous digesters. Feeding was provided daily throughout the week including the weekends. The four substrates chosen for analysis in the semi continuous systems were:

(a) bone and meat meal as it is the co-substrate with the highest market availability and the lowest price.

(b) blood meal; this substrate was selected due to its negative effects recorded when high loading rate was used in the batch digestion.

(c) fishmeal; this substrate offered the highest bio-methane production of all substrate mixtures assessed in the batch digestion stage.

(d) The last of the substrates assessed was gluten. This was the only plant-derived protein-rich substrate assessed; its selection was based on its higher protein content over the second candidate substrate, namely soybean meal.

In this experimentation stage the highest biogas production was achieved by the mixture containing blood meal with an average production equal to 449 mLCH4/gVS added, followed by biogas production from bone and meat meal with 443 mLCH4/gVS added. The lowest production was achieved when using the mixture containing FM with an average yield of 416 mLCH4/gVS added. Our results indicate that the type of protein added is insignificant; in all cases the yields of the four different mixtures were within 7% of each other ranging between 449 and 416 mLCH4/gVS added. In all cases VFA concentrations were very low and no inhibition was recorded in any of the reactors treating the different substrates.

## Conclusion

OMWW, a heavily polluted effluent can be valorized through the anaerobic digestion process into biogas and digestate. The addition of nitrogen-rich substrates offers the opportunity to achieve a steady process with high bioconversion efficiency. Furthermore, the type of the co-substrate seems to be insignificant as no substantial differences in biogas yield were registered when fish, animal or plant derived proteins were used as nitrogen-carrying additives. This work offers a framework for the possibility to valorize the OMWW in treatment plants in the vicinity of the olive mills without the requirement of manure addition as carrier substrate. This development would minimize the risk of bacterial contamination from transportation of manure near olive mills. According to our results one ton of nitrogen-rich substrates can be used for the treatment of more than $50m^3$ of OMWW while at the end of the process the recovered effluent has significantly better characteristics as soil improver when compared to raw OMWW.

# 492445 Evaluation and Comparison ofÂ the Scale-up Potential for Autotropic andÂ Heterotrophic Algal Oil Production

**Abstract:**

Significant research has been completed over the past decade to develop technologies to facilitate the production of algal lipids for use as feedstock for next generation renewable fuel facilities. The interest for microalgae is motivated by the substantial increase in yield per land area compared to oilseed crops reducing the pressure to supplant food crops with energy crops. However, the challenges to produce oil from algae at scales approaching crop oil production facilities have been great. One of those barriers is the need to get photons from sunlight or artificial sources to the growing algae as an energy source. It has recently been proposed that growing the algae heterotropically might overcome this scale-up barrier. The transition to heterotrophic halts the photosynthesis process, and requires an organic carbon source to provide energy, as the heterotrophic strain of microalgae is not able to produce energy compared to the autotrophic which produces an energy source though photosynthesis. Additionally, the transition has been proven to increase the lipid content of the microalgae by replacing the chlorophyll cells produced during photosynthesis. If the strain of microalgae used is non-light dependent it will negate the requirement of industrial scale, clear photo bioreactors or open ponds. Additional chemicals and catalysts will be required for the growth and development of the microalgae strains for the growth media.

In this study, a group of senior chemical engineering students developed preliminary designs and then performed an economic analysis for algal oil production utilizing both autotropic and heterotrophic growth strategies. Both proposed systems utilizes the strain of microalgae *Chlorella Vulgaris* as the raw material. *Chlorella Vulgaris* has a lipid content ranging between 15-35 wt%.

The study involved the development of Input/output diagrams, block flow diagrams, and process flow diagrams for compete processes starting from lab algae inoculation through to final lipid oil product purification. Preliminary equipment sizing was performed. The capital costs were estimated using a factored broad cost estimation method based on vendor quotes for the largest equipment and cost chart data for the rest of the equipment. Revenues and operating costs for a 20 year project life were also estimated.

The heterotrophic option will yield a higher lipid content which will generate a larger amount of oil when compared to the autotrophic strain but will require a higher quantity of inputs. The production throughput was comparable to a small scale oilseed production facility. Neither option is currently economically competitive with a commercial soybean oil processing facility. However, the heterotrophic option had significant advantages over the autotropic design. The key barriers to commercialization were identified so that future R&D activities can focus in these areas.

MOMENTIVE™
inventing possibilities

# Process Developments in Continuous Silicone-Polyether Copolymers

- Akshay Kundan
- Chauncey Rinard
- Steven Bahr



10/30/2017

# BACKGROUND ON SILICONE-POLYETHER COPOLYMERS (SPC)



MOMENTIVE™

# Urethane Additives

**POTENTIAL APPLICATIONS, KEY PRODUCTS & TYPICAL BENEFITS**

- Improve reaction mix compatibility
- Determine reaction rate and productivity
- Provide bulk foam stabilization
- Regulate cells size and openness
- Make smooth surface of molded foam
- Improve properties like fire, comfort
- Improve dimensional stability of low-density PU
- Help CASE systems for easy pouring and faster cure
- Make PU leather smooth and abrasion-resistant
- Enhance foam physical properties and fire behavior

**MOLDED FOAMS**
- Automotive seating
- Furniture
- Pillows and mattresses

**SLAB FOAMS**
- Furniture
- Bedding
- Technical foams
- Laminated foams

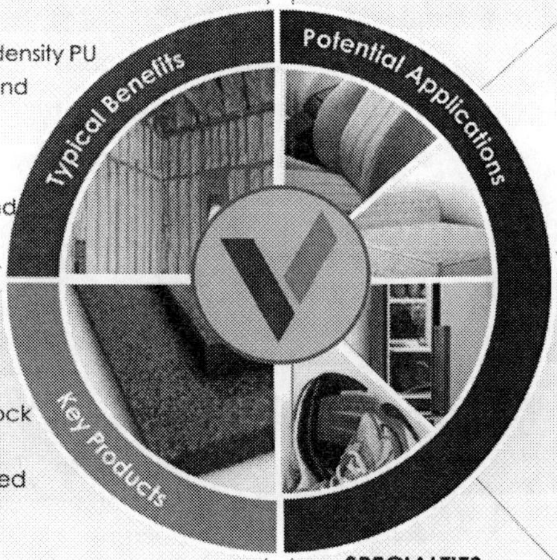


- Niax* silicones and catalysts for slabstock foams (conventional, high resilience and visco-elastic)
- Geolite* process modifiers for slabstock foams
- Niax silicones and catalysts for molded foams based on TDI, MDI and TM systems
- Niax silicones, catalyst and special additives for rigid foams
- Niax additives for specialty PU applications
- Fomrez† special PU catalysts

**RIGID FOAMS**
- Fridge, cool store, water heater, display units
- Refrigerated trucks, reefer containers
- Construction panels, SIPs
- Insulation panels, boards
- Gap filling
- Insulated pipe & tanks
- Mining/grouting
- Decorative, structural, packaging, floral

**SPECIALTIES**
- Carpet backing
- Electronics
- Footwear
- Artificial turf
- Artificial leather

3

MOMENTIVE

*Niax and Geolite are trademarks of Momentive Performance Materials Inc.
†Fomrez is a trademark of Chemtura Corporation, used with permission by Galata Chemicals LLC.

# Momentive Urethane Additives Portfolio

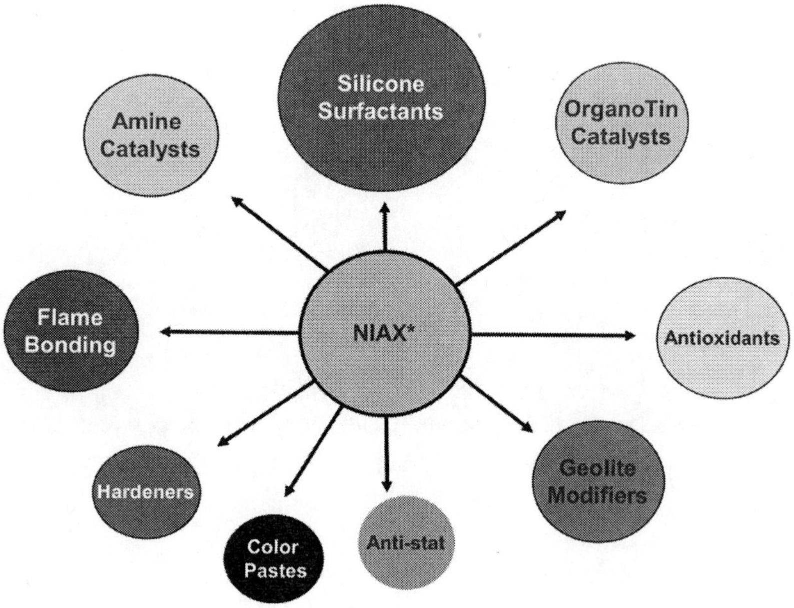


4 

MOMENTIVE

## Why Silicone Surfactants? Emulsification and Stabilization

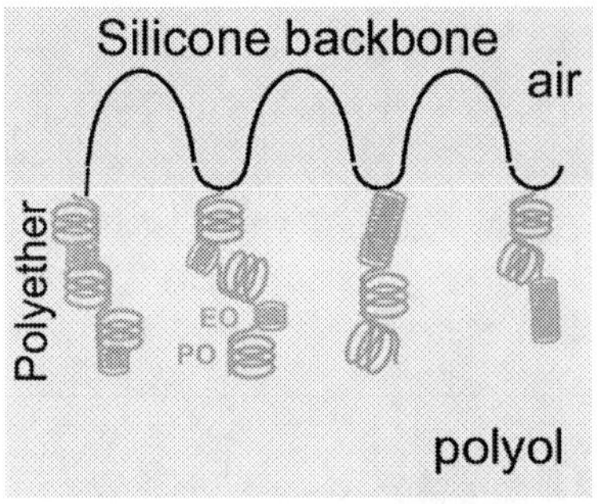


| Surfactant<br>1% Surfactant in a 3000 MW Triol | γ<br>(dynes/cm) |
|---|---|
| None (pure polyol) | 35 |
| Nonylphenolethoxylate | 32 |
| Sulfonated Fatty Ester | 31 |
| Silicone/Polyether | 25* |
| Fluorocarbon/Polyether | 24 |

*The only product that will stabilize a urethane foam. Surface Tension reduction is a necessary but not sufficient requirement for foam stabilization

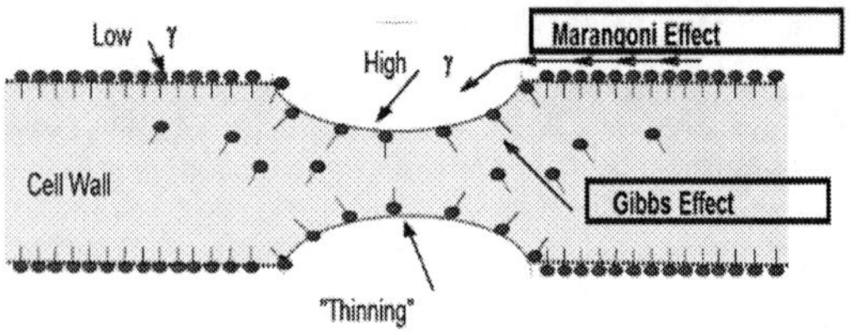


**Gibbs-Marangoni**: Thinning causes lower local concentration of surfactant which raises γ causing a γ gradient in the film. This gradient is a restorative force drawing more surfactant to lower the γ. The surface is said to become "elastic".

**Ewers-Sutherland**: Similar concept except this approach suggests that as the surfactant mono-layer moves to reduce the local γ gradient it drags with it sufficient bulk material to help heal the rupture.



MOMENTIVE

## Copolymer Chemistry

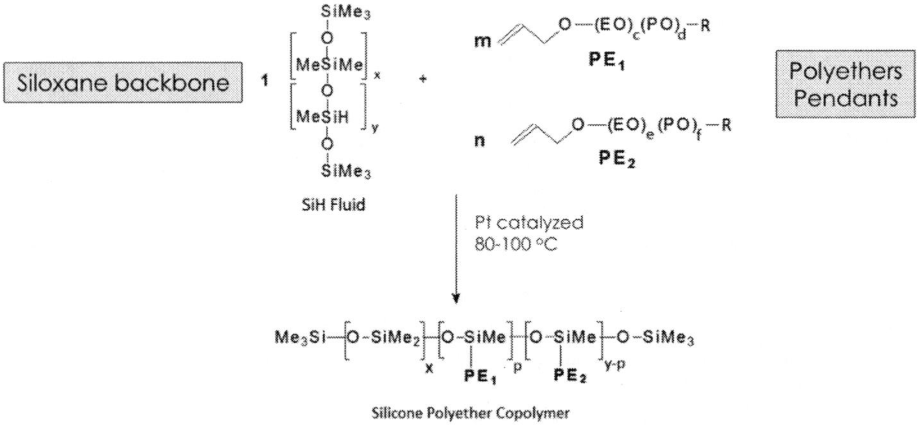


- The reaction is governed by compatibility, reaction selectivity and randomization.
- Potential Technologies in MPM:
  - Batch process
  - Continuous Copolymer Unit (CCU)
  - Static mixing plug flow reactors
  - Multistage Mixer (MSM)
  - Patented Technologies: CCU, Static mixing plug flow reactors, MSM.



MOMENTIVE

# IP REVIEW



MOMENTIVE™

## Innovation – Continuous SPC Manufacturing Technology

- Detailed IP search of available patents for producing silicone copolymers continuously.
- Process development boom came in early 2000s.

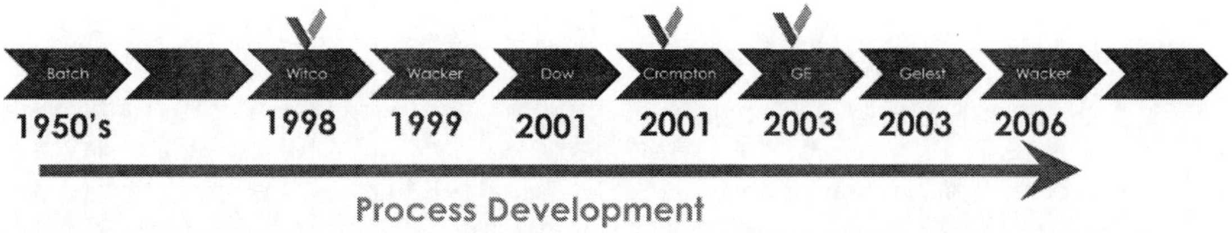


| Year filed | Company | Patent # | Process |
|---|---|---|---|
| **1998** | **Witco Corporation** | **US 5986022** | **CSTR and PFR (CCU)** |
| 1999 | Wacker Chemie Gmbh | US6350824 B1 | Pre-mixed and fed into the reaction zone |
| **2001** | **Dow Corning Toray Silicone** | **US 6,410,772 B2** | **CSTR, tubular reactor and PFR** |
| 2001 | Crompton | US20020099137 A1 | Static Mixing PFR in series or parallel |
| **2003** | **General Electric Company** | **US 6,897,280 B2** | **Multi-stage blade – mixed plug flow reactor (MSM)** |
| 2003 | Gelest Inc | US 20040176561 A1 | PFR and a tubular reactor packed with a solid, acidic polymerization catalyst |
| 2005 | Wacker Chemie Ag | US7145028 B2 | Integrated loop tube reactor |

8 

MOMENTIVE™

# CONTINUOUS COPOLYMER UNIT (CCU)

PATENT NO.: 5986022
YEAR APPROVED: 1999
ASSIGNEE: WITCO CORPORATION



MOMENTIVE™

## CCU Design

- 2 CSTR's and 1 PFR in series
- Reaction has to complete at least 80-85% before entering PFR to avoid phase separation.
- Properties slightly different from batch products.
- CSTR Limitation: Not all of any of the reactants can be consumed completely.

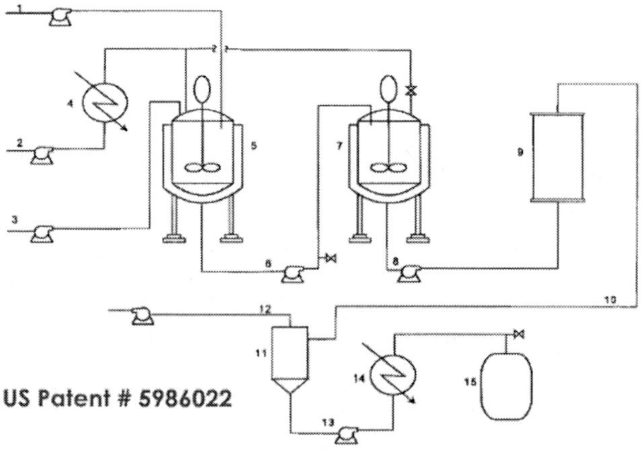

**US Patent # 5986022**

- PFR Limitation: Immiscible SiH fluid and olefinic compound will phase separate very rapidly and reaction slows down or ceases completely.
- Benefits from a batch: Low Operating Cost (Conversion Cost Equivalent to LCC), Reduced Footprint, Reduced Waste, Solvent Reduction, Capacity.
- Cons: Versatility

**Has its advantages over batch but versatility is missing.**

10  MOMENTIVE™

# MULTI-STAGE MIXER PATENT (MSM)

PATENT NO.: US 6,897,280 B2
YEAR APPROVED: 2005
ASSIGNEE: GENERAL ELECTRIC COMPANY



MOMENTIVE

## MSM Patent Overview

- Evolution of CCU Technology.
- ID 50 mm, L – 0.6 m, 8 stages of 4-8 min residence time, tubular jacketed reactor. Offset channels to prevent back mixing.
- Eddies and vortices of sufficient intensity in multi-stage blades shear biphasic liquid mixture and disperse one phase into another for rapid reaction.
- Last stage of reactor is plug flow. 75-90% of reaction is finished before entering this region.
- Front end equipment required
  Flow meters, buffer, catalyst addition systems, stripping unit
- Back end equipment required
  Product stripper, product cooler, product filters and blend equipment (tanks or mixer)

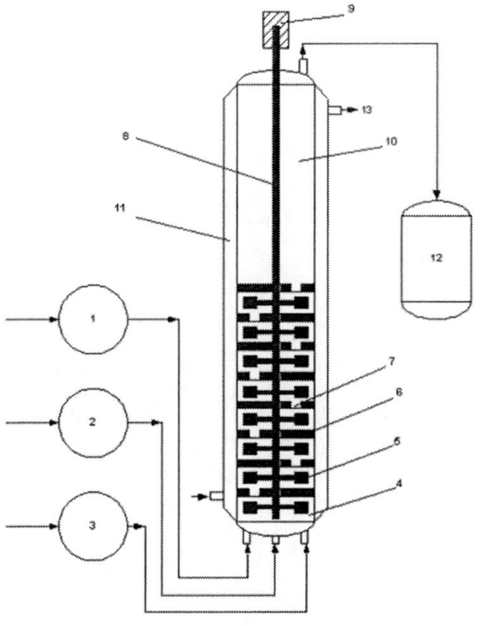


US 6,897,280 B2 (May 24, 2005)

2nd Generation CCU




MOMENTIVE

## Batch/CCU/MSM

### Comparison of processes for different SPC (US Patent# US 6,897,280 )

| Raw material | Process | SiH (cc H2/g) | Appearance | Viscosity (cst) | Foam Rise (cm) | Flow Air (porosity) | Cell Structure |
|---|---|---|---|---|---|---|---|
| Siloxane A + $PE_1$ + $PE_2$ + Buffer + Pt catalyzed | Batch | ND | Clear | 2349 | 38.0 | 5.2 | 7 |
| | CCU | **0.4** | **Hazy** | 2461 | **36.8** | 4.2 | **10** |
| | MSM | ND | Clear | 2482 | 38.6 | 6.3 | 7 |
| Siloxane B + $PE_3$ + $PE_4$+ Buffer + Pt catalyzed | Batch | ND | Clear | 2771 | 40.5 | 2.2 | 6 |
| | CCU | **0.1** | **Hazy** | 2569 | **Total Foam Collapse** | | |
| | MSM | ND | Clear | 2626 | 39.2 | 7.0 | 7 |
| Siloxane B + $PE_5$ + $PE_4$ + Buffer + Pt catalyzed + DPG | Batch | ND | Clear | 1003 | 37.9 | 6.6 | 7 |
| | CCU | ND | Clear | 942 | 37.1 | 6.1 | **8** |
| | MSM | ND | Clear | 1036 | 37.9 | 5.6 | 7 |
| Siloxane B + $PE_6$ + $PE_4$ + Buffer + Pt catalyzed | Batch | ND | Clear | 1794 | 40.9 | 5.1 | 7 |
| | CCU | ND | Clear | 1851 | 39.0 | 5.9 | 7 |
| | MSM | ND | Clear | 1833 | 40.7 | 5.1 | 7 |

**MSM produced similar to batch grade. CCU is slightly different.**




MOMENTIVE

# DOW PATENT

PATENT NO.: US 6,410,772 B2
YEAR APPROVED: 2002
ASSIGNEE: DOW CORNING TORAY SILICONE CO., LTD.



MOMENTIVE

## Dow Patent Overview

- ID: 50 mm, L = 1 m, 33 disks
- Blade or pin stirrers severely impact plug flow conditions.
- 2 or more disks maintain plug flow conditions.
- Rotation of disks facilitates mixing of various components & improves efficiency while maintaining plug flow conditions.
- Reactor capacity significantly improved by adding a pre-mixer.
- Static reactor downstream (25 mm ID, L - 3m, 1.5L ) solves the feed rate & elongation of tubular reactor issues.
- Loading of SiH fluid to PE is about 1.
- Residence time of 4 min for feed rate of 50kg/hr achieved.
- **Limitations:** Single PE product.

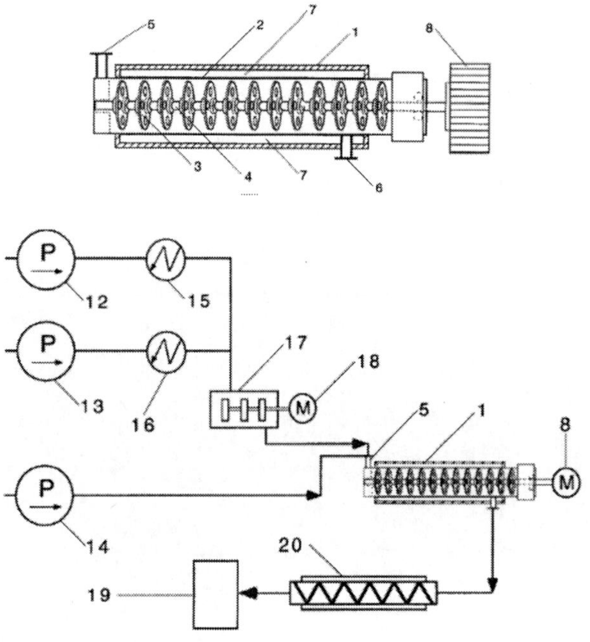


**US Patent # 6,410,772 B2**





## Summary of Comparisons

| | Batch | CCU | MSM |
|---|---|---|---|
| Reactor | Single tank | At Least One 100-300 Gallon Tank Plug Flow Reactor | Tubular reactor with multiple stages of stirring blades |
| Pros/Cons | Versatility | Low Operating Cost (Conversion Cost Equivalent to LCC)<br>Reduced Footprint<br>Reduced Waste<br>Solvent Reduction<br>Capacity | Low Operating Cost (Conversion Cost Lower than LCC)<br>Versatility<br>Small Footprint<br>Reduced Waste<br>Solvent Reduction<br>Capacity |
| Capability | Terminal/Pendant Copolymers<br>Single/Multiple Polyethers<br>Wide Range of MW's/Viscosities | Terminal/Pendant Copolymers<br>Single Polyether<br>Lower End MW's/ Viscosities | Terminal/Pendant Copolymers<br>Single/Multiple Polyethers<br>Wide Range of MW's/Viscosities |
| Product grade | | Different than batch | Similar to batch |

**Process Development key to achieving good quality SPC**

# Thank You.

**DISCLAIMER: THE MATERIALS, PRODUCTS AND SERVICES OF Momentive Performance Materials Inc. and its subsidiaries and affiliates DOING BUSINESS IN LOCAL JURISDICTIONS (collectively "SUPPLIERS") ARE SOLD BY THE RESPECTIVE LEGAL ENTITY OF THE SUPPLIER SUBJECT TO SUPPLIERS' STANDARD CONDITIONS OF SALE, WHICH ARE INCLUDED IN THE APPLICABLE DISTRIBUTOR OR OTHER SALES AGREEMENT, PRINTED ON THE BACK OF ORDER ACKNOWLEDGMENTS AND INVOICES, AND AVAILABLE UPON REQUEST. ALTHOUGH ANY INFORMATION, RECOMMENDATIONS, OR ADVICE CONTAINED HEREIN IS GIVEN IN GOOD FAITH, SUPPLIERS MAKE NO WARRANTY OR GUARANTEE, EXPRESS OR IMPLIED, (i) THAT THE RESULTS DESCRIBED HEREIN WILL BE OBTAINED UNDER END-USE CONDITIONS, OR (ii) AS TO THE EFFECTIVENESS OR SAFETY OF ANY DESIGN INCORPORATING SUPPLIERS' PRODUCTS, MATERIALS, SERVICES, RECOMMENDATIONS OR ADVICE. AFOREMENTIONED EXCLUSIONS OR LIMITATION OF LIABILITY ARE NOT APPLICABLE TO THE EXTENT THAT THE END-USE CONDITIONS AND/OR INCORPORATION CONDITIONS CORRESPOND TO THE RECOMMENDED CONDITIONS OF USE AND/OR OF INCORPORATION AS DESCRIBED BY SUPPLIER IN ITS PRODUCT DATA SHEET AND/OR PRODUCT SPECIFICATIONS. EXCEPT AS PROVIDED IN SUPPLIERS' STANDARD CONDITIONS OF SALE, SUPPLIERS AND THEIR REPRESENTATIVES SHALL IN NO EVENT BE RESPONSIBLE FOR ANY LOSS RESULTING FROM ANY USE OF ITS MATERIALS, PRODUCTS OR SERVICES DESCRIBED HEREIN.** Each user bears full responsibility for making its own determination as to the suitability of Suppliers' materials, services, recommendations, or advice for its own particular use. Each user must identify and perform all tests and analyses necessary to assure that its finished parts incorporating Suppliers' products, materials, or services will be safe and suitable for use under end-use conditions. Nothing in this or any other document, nor any oral recommendation or advice, shall be deemed to alter, vary, supersede, or waive any provision of Suppliers' Standard Conditions of Sale or this Disclaimer, unless any such modification is specifically agreed to in a writing signed by Suppliers. No statement contained herein concerning a possible or suggested use of any material, product, service or design is intended, or should be construed, to grant any license under any patent or other intellectual property right of Suppliers or any of its subsidiaries or affiliates covering such use or design, or as a recommendation for the use of such material, product, service or design in the infringement of any patent or other intellectual property right.



 

MOMENTIVE

# 497659 Esterification of Carboxylic Acids Present in the BIO-OIL Produced By Thermal Cracking of Triacylglycerols

**Abstract:**

The production of alternative fuels from renewable sources has been object of several studies in recent years. A promising alternative is the conversion of biomass to biofuels by thermal cracking. Through this process, it is possible to obtain a liquid product called bio-oil, with physicochemical properties similar to the petroleum crude oil. However, bio-oil might contain a high amount of carboxylic acids that results on a high acid index ranging from 110 mg KOH/g of sample to 207 mg KOH/g of sample, whereas the maximum required by the current legislation (EN 14104) is 0.5 mg KOH/g of sample. In this work, the bio-oil was produced from commercial soybean oil at 525°C and a mass flow rate of 350 g/h. The objective of this study is to perform esterification reactions in order to reduce the acidity index of the bio-oil. These reactions present a methodology of easy operation, low cost and may reduce the bio-oil corrosivity. For this, alcoholic reagents, ethanol and methanol, were used for the esterification reactions. The influence of different temperatures, reaction times, catalysts and, stoichiometric ratios between the reagents on the reduction of the acid number and improvement of the bio-oil quality were evaluated. Good results were obtained with the esterification reaction using ethanol, sulfuric acid, bio-oil to ethanol ratio of 10:5, temperature of 60°C and reaction time of 90 minutes. The acid index of the bio-oil sample decreased from 111.33 mg KOH/g of sample to 34.64 mg KOH/g of sample, resulting in a reduction of approximately 70%.

# 506291 Flex-to-Stretch Electronics

**Abstract:**

Stretchable electronic devices maintain electrical connections when subjected to stress or strain and are therefore useful for enabling electronics with new mechanical properties. However, these devices often require power or electrical signals from non-stretchable components. Connecting flexible and stretchable substrates is a possible solution that maintains the electrical connection within a medium that can withstand deformation and prolonged use. The ability to create a robust electrical connection between these mechanically disparate components may enable new types of hybrid devices. In this study we present a basic method to fabricate such new hybrid devices, utilizing liquid metals that form fantastic stretchable interconnects because they are both conductive and fluidic. We characterized the mechanical and electrical properties of these hybrid devices. While we focus on silicon elastomers and liquid metals, there are numerous other methods of fabricating stretchable devices that share many properties with the devices in this work.

# 506292 Systematic Analysis of Cloud Point and Crystallization in Fatty Acid Ethyl Ester Biodiesel Mixtures

**Abstract:**

Saturated biodiesel components, fatty acid methyl esters (FAMEs) or fatty acid ethyl esters (FAEEs), crystallize at high temperatures, which limits their commercial application. FAEEs are particularly advantageous biodiesel components because they can be renewably sourced and have lower cloud points (CP) than FAMEs with the same carbon-chain length. This research systematically evaluates FAEE cloud point and crystallization data to aid in the development of FAEE biodiesel blends with practical cold flow properties. The effects of cold-flow improving additives dimethyl azelate (DMA) and triacetin (TA) on FAMEs and FAEEs are also analyzed. Neat FAEE CP increases with carbon-chain length and decreases with greater degree of unsaturation, and saturated/saturated FAEE mixtures show partial cocrystallization. Additives lowered the CP, but did so via different mechanisms. Thermodynamic ideal solution modeling accurately predicts binary FAEE CP with and without additives within 2% of experimental values.

# Feasibility Study of Ionic Liquid Desalination Design

Material Physics and Applications Division, Los Alamos National Laboratory
PI: Zachary Cosenza RLM: George Scott Goff

LA-UR 17-26682

## Introduction

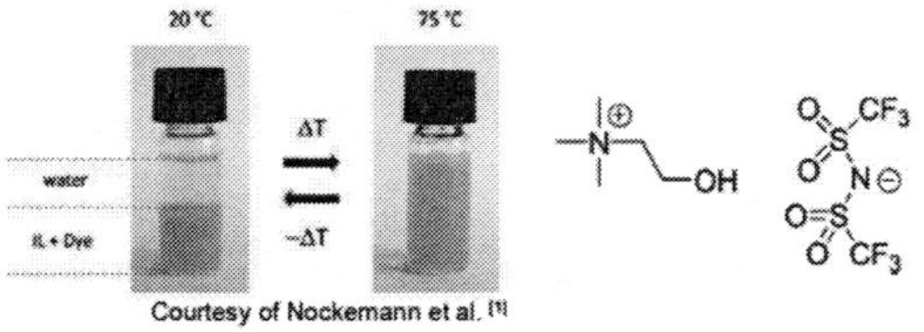


Courtesy of Nockemann et al. [1]

- Ionic liquids (ILs) are a class of room temperature salts in liquid form.
- Water has been shown to be soluble in ILs with T dependence, while the IL remains insoluble.
- Temperature may be used to drive water from a saline into the IL and back out at modest T.
- Current desalination methods (reverse osmosis and multiple-effect evaporation) run at high T or P
- Proposal is to **develop liquid-liquid extraction system using ILs** and saline.

## Baseline System

Continuous Process uses Octanoic Acid as solvent

Batch Process (N = 1) uses [Choline][TF2N] as solvent

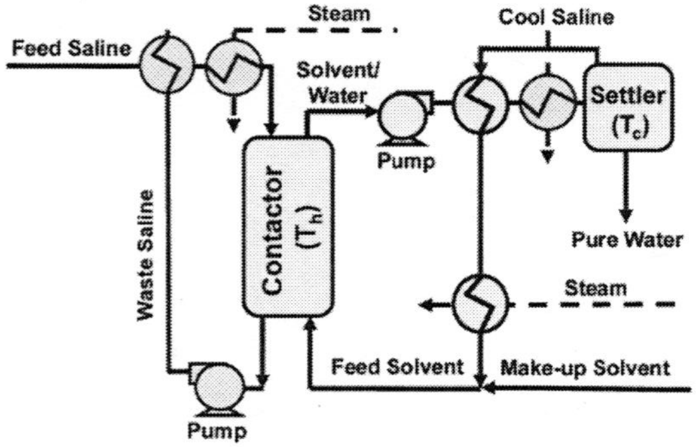


### Major Process Parameters

**Temperature** – Separation efficiency is driven by the difference in contactor ($T_h$) and settler ($T_c$) temperatures.

**Solvent** – The process efficiency is limited by equilibrium, and in turn by the solvent used.

**Saline Input** – It has been shown [3] that equilibrium is altered by saline concentrations, with higher salt concentrations reducing the driving force.

## Methods

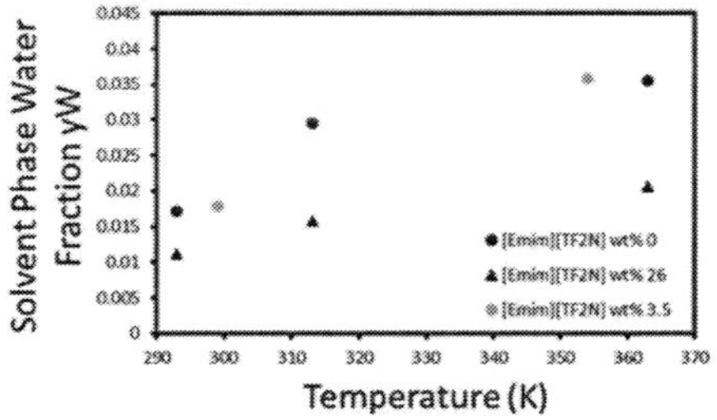


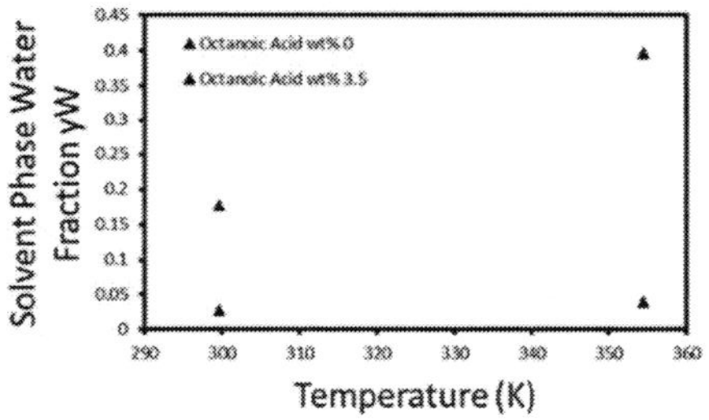


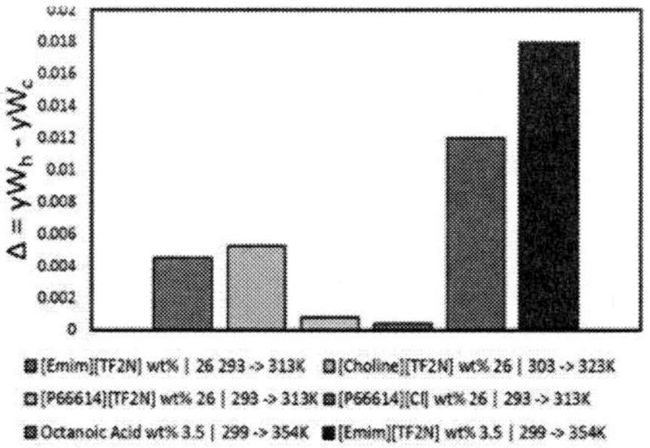


## Analysis

***Operating Line (OL):***

$$\frac{V_m}{F} = \mathcal{N} * \iota_m + \frac{V_1}{F} - \mathcal{N} * \iota_1$$

**Boundary Conditions:**

(i) Set $\mathcal{N}$ (slope)
(ii) Set inlet water conc.
(iii) Pinch Point solution
(iv) Set inlet solvent conc.

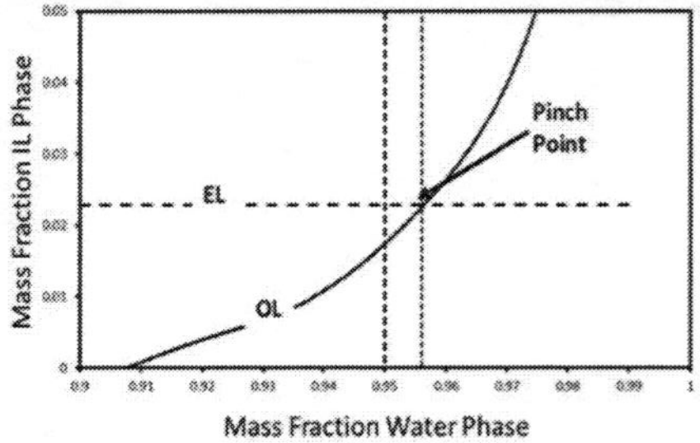


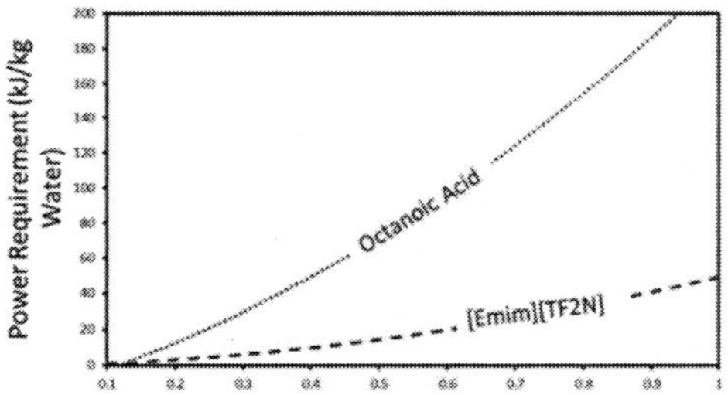


**Energy Requirements OA and Emim**, 353K with $L$ = 3.0 kg/s 3.5 wt% water baseline

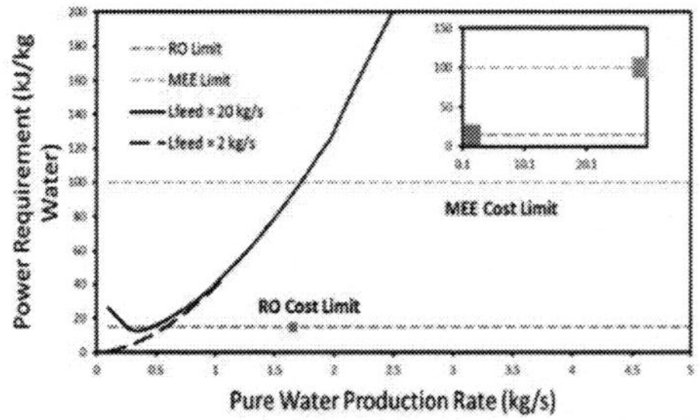


**Cost Curve w/ Parameter Variation,** RO ■ [14, 23] and MEE ■ [14, 24] based on pure water production (353K) and 3.5 wt% water

## Conclusions

- Like the conclusion reached by Sanap et al [2], the continuous process is highly pinched at N = 1 at all upper process temperatures.
- Compared to the current literature work (Octanoic Acid as a solvent), [Emim][TF2N] has higher extractive and cost efficiency.
- If coupled with waste heat the marginal energy cost is close to that of RO, but far from MEE.
- Data on droplet size, mixing times, mass and heat transfer coefficients, and fixed costs would help scale-up process beyond basic thermodynamics.

## References

[1] Nockmann, *J. Phys. Chem.* Ed 113. 1429-1437 (2009)
[2] D.B. Sanap et al, *Desalination* Ed. 357 150-162 (2015)
[3] Bajpayee, Anurag Directional Solvent Extraction Desalination, MIT 2012, *MIT Libraries*
[4] Zhong Yujiang, *Environ. Sci. Tech.* Ed. 50 1039-1045 (2016)

## Acknowledgements

Los Alamos NATIONAL LABORATORY

# 506320 Preparation and Characterization of Shape Memory Assisted Self-Healing Coatings

**Abstract:**

Shape memory assisted self-healing (SMASH) coatings have the ability heal nearly completely after damage with the application of heat, making them useful for corrosion resistance in steel structures (1). This summerâ€™s project sought to expand upon existing research to develop an efficiently healing and transparent SMASH coating that could be used in technological applications, such as on cell phone or computer screens (2) . The materials tested were comprised of electrospun poly(Îµ-caprolactone) (PCL) fiber mats coated in epoxy with differing glass transition temperatures (Tg). The healing efficiency of the samples were tested by damaging each one with a razorblade and healing them over varying times and heats in an oven. The transparency of each sample was observed visually before and after heating, a technique which will be refined later in the research process. Once the most ideal combination is found, further experimentation will be conducted to determine whether the ideal epoxy/PCL combination is also one that can withstand multiple damage-healing cycles to the same spot without becoming opaque. References (1) Nejad, H. B. Thermally/Mechanically Responsive Polymeric Composites with Shape Memory/Self-Healing Properties. Syracuse University 2015, 79-95. (2) Luo, X. Thermally Responsive Polymer Systems for Self-Healing, Reversible Adhesion and Shape Memory Applications. Syracuse University, ProQuest Dissertations Publishing 2010. 3437600.

# 506372 Defluoridation of Ground Water Using Impregnated Aluminum

**Abstract:**

Groundwater is an important source of drinking water and it is estimated that more than 200 million people worldwide are drinking groundwater with fluoride concentrations greater than the WHO guideline value of 1.5 mg/L. The majority of these cases occur in the developing world. fluoride has long been found to have a beneficial effect on dental health, although when present in drinking water at concentrations much above the WHO guideline value (and national drinking water limit for most countries) of 1.5 mg/L, long term use can result in development of dental fluorosis or at its worst, crippling skeletal fluorosis. The effects on health depend on a number of factors including dose from other sources of fluoride (toothpastes, food), age, gender and general health status.

# 506536 Ordering pH-Responsive Polymer-Grafted Nanoparticles in Flow Coating Process

**Abstract:**

Poly(acrylic acid) (PAA)-grafted silica nanoparticles with different grafting densities and brush lengths were synthesized via surface-initiated ATRP, and their deposition patterns on silicon substrates were investigated by systematically varying the speed of moving substrates, pH values, and free polymer poly(vinylpyrrolidone) (PVP) concentrations in aqueous solution. Brush-brush interactions are tuned by varying the PAA chain conformations with pH and also with the addition of free PVP. Different deposition patterns were observed for nanoparticles at low grafting density and long brush length at a velocity of 0.4 mm/min and a PVP concentration of 0.24 mg/ml. At pH 11, nanoparticles assembled into well-regulated lines parallel to the moving direction of the blade. However, at pH 9, nanoparticles showed stronger repulsion, which led to Rayleigh-instability, and a dot pattern formation. At pH 3, particle attractions get strong due the hydrogen bonding between PAA and PVP, hence Bénard cell patterns were observed. The mechanisms of hydrogen bonding between two different polymers and also physical bridging within particles are demonstrated to organize the grafted nanoparticles in a flow-coating process. Our findings will enable the easy fabrication of patterned particle domains for functional polymer films.

# 500922 *Porphyridium cruentum* Grown on Swine Waste has Minimal Changes to its Fatty Acid Composition

**Abstract:**

**Title:**

*Porphyridium cruentum* Grown on Swine Waste Has Minimal Changes to Its Fatty Acid Composition

**Authors:**

**Humeyra B. Ulusoy-Erol**[1], **Mariana Lara Menegazzo**[2], Emily Gottberg[1], Jessica Vaden[1], Maryam Asgharpour[1], Christa N. Hestekin[1], and Jamie A. Hestekin[1]

[1] Ralph E. Martin Department of Chemical Engineering, University of Arkansas, Fayetteville,

AR, United States.

[2] Department of Engineering, Federal University of Grande Dourados, Dourados, MS, Brazil.

Phone number: +1 479 575 3492; Fax number: +1 479 575 7926

**Abstract**

Environmental problems like eutrophication due to wastewater are becoming more significant because of the dramatic increase in the population. Microalgae are considered one of the best alternatives for wastewater treatment processes due to their significant uptake mechanism for nitrogen and phosphorus as well as the ability to grow in different water types. *Porphyridium (P.) cruentum* which is a red algae and an important source of omega-3 and omega-6 fatty acids can be grown in different types of water. *P. cruentum* grown on swine waste were found to have no statistical differences in terms of the biomass productivity and fatty acid methyl ester (FAME) composition compared to a control media. In fact, *P. cruentum* yielded a higher percentage of saturated fatty acids compared to unsaturated fatty acids when grown on swine waste indicating that it would also have potential to be used as a biofuel.

# Granulation of Anammox Sludge in Upflow Reactors

Chong-Jian Tang*, Peng Liu, Cheng-Shan Duan, Cheng Yu, Yu-Xia Song, Li-Yuan Chai

[a]Department of Environmental Engineering, School of Metallurgy and Environment, Central South University, Changsha 410083, China;
[b]Chinese National Engineering Research Center for Control and Treatment of Heavy metal Pollution (CNERC–CTHP), Changsha 410083, China
(*E–mail: chjtang@csu.edu.cn)

**Abstract**
Nitrogen pollution has caused severe eutrophication in water bodies and becomes a key environmental problem around the world. The efficient and cost-effective removal of nitrogen from ammonium-rich wastewater has attracted much attention. Anammox (ANerobic AMMonium OXidation) process is a promising biological nitrogen removal technology for treatment of ammonium–rich wastewater, which could directly oxidize ammonium to dinitrogen gas under anoxic condition. It also contains high value in scientific significance as it contributes more than 50% of the global nitrogen cycle. However, a great challenge for Anammox process is that the chemolithoautotrophic Anammox bacteria exhibit a slow initial growth rate, resulting in poor enrichment of Anammox biomass and poor stability of reactors. Granulation of the Anammox sludge has shown great advantages to effectively enhance the biomass enrichment. In this study, we reviewed the characteristics of Anammox bacteria, the formation, structure and flotation of Anammox granules under high hydraulic loadings. Full–scale applications of granular Anammox process were also presented. The major points are 1) Anammox bacteria secrete a large amount of extracellular polymeric substances (EPS) and contain plentiful hydrophobic functional groups that facilitate the biomass granulation; 2) granulation enhances the sludge settling property and retention time, which contributes to the extremely high nitrogen removal rate (NRR) of 77 kg $m^{-3}$ $d^{-1}$ of Anammox upflow reactors; 3) flotation of Anammox granules frequently occurs under nitrogen removal rate higher than 10 kg $m^{-3}$ $d^{-1}$, which raises as a big problem for further improvement of nitrogen removal rate and stable operation of upflow reactors; the over-production of extracellular polymeric substances (EPS) (especially proteins) under high hydraulic conditions has been considered to be a key reason for Anammox granules' flotation. This review sheds light on the enhanced nitrogen removal from wastewaters by Anammox granules.
**Keywords**
Anammox; biological nitrogen removal; sludge granulation; settling property; flotation; extracellular polymeric substances

## INTRODUCTION

**Formation of granular sludge is considered as an effective means to overcome the negative effects caused by various environmental conditions and diverse inhibitors during the Anammox process (Hulshoff Pol et al., 2004). Anammox bacteria can secrete extracellularpolymeric substances (EPS), and tend to aggregate together to form granules even under high hydraulic shear stress (Tang et al., 2011). The aggregated granularsludge with a larger diameter and higher density was prone to settling inside reactors according to Stokes equation (Lu et al., 2012). Further, the granular sludge features a stable ecological structure where anaerobic bacteria mainly distributes inside the granules, and aerobic bacteria locates in the outside layer of granules (Gonzalez-Gil et al., 2015; Speth et al., 2016). For the case of Anammox granules (Vazquez–Padin et al., 2010; Speth et al., 2016), the aerobic ammonium–oxidizing bacteria (AOB) (i.e., nitrifying bacteria) in the outer layer can consume the low level dissolved oxygen in reactor. This shell prevents the inhibitors in aqueous solution from penetrating into the inner part of granules with Anammox bacteria, effectively abating the negative influences caused by oxygen and inhibitors.**

**Therefore, the granulation of Anammox biomass plays a key role in stable and high–rate**

operation of Anammox process. However, the formation of Anammox granules and other related informationand mechanism arestill not well addressed. The present review gives an overview on the characteristics of Anammox bacteria and Anammox granules. Particularly, the formation and flotation of Anammox granules, and the related mechanisms are comprehensively addressed. The full-scale application of granularAnammox process is also discussed in this paper.

## FORMATION OF ANAMMOX GRANULES

### Effect of different inocula

Anammox granulationthatoccursin upflow reactors under high loading rate was ascribed to the natureof the Anammox bacteria. Inocula are an important factor thataffects the fast granulationof Anammox biomass leading to different startup performance of Anammox reactors. As tabulated in Table 1, thestartupperformance seeded with differentinoculaexperienced fourphases (i.e., cell lysis, lag, activity elevation, and stationary phase). The startup time, especially the cell lysis duration time, was significantly reduced by seeding flocculentnitrifying sludge and denitrifying sludge as compared to the anaerobicgranularsludge (Tang et al., 2013). However, thenitrogen removal performance for reactorseeded with anaerobic granular sludge was 2–3 times higherthanflocculentsludges (see Table 1). This higher nitrogen removal rate was ascribed to the easier formation of Anammox from anaerobic granular sludge. As reported by Tang et al. (2010), granularsludge was a good carrier for Anammox bacterial growth providingnuclei for granulation.Thus, the Anammox bacteria grow on the surface of the anaerobic granularsludge resulting in the direct transition from anaerobic granular sludge to Anammox granules (Tang et al., 2011). The Anammox granules were gradually enriched inside the upflow reactors and finally fully–filled the whole reactor (as shown in Fig. 1). Due to the challenging and long startupperiodfor Anammox granulationfromthe flocculentdenitrifying sludge and nitrifying sludge (Tang et al., 2013), researchers have selected the granularsludge (anaerobicor aerobic) as inoculumto seed Anammox reactors for improved nitrogen removal performance.

**Table 1.** Comparison of startup performance with different inocula (Tang et al., 2013)

| inoculum | cell lysis phase(d) | lag phase (d) | activity elevation phase (d) | stationary phase (d) | nitrogen removal rate (kg $m^{-3}$ $d^{-1}$) |
|---|---|---|---|---|---|
| anaerobic granular sludge | 1~28 | 29~57 | 58~98 | 99~122 | 2.09 |
| denitrifying sludge | 1~3 | 4~42 | 43~79 | 80~86 | 1.03 |
| nitrifying sludge | 1~3 | 4~21 | 22~73 | 74~80 | 0.71 |

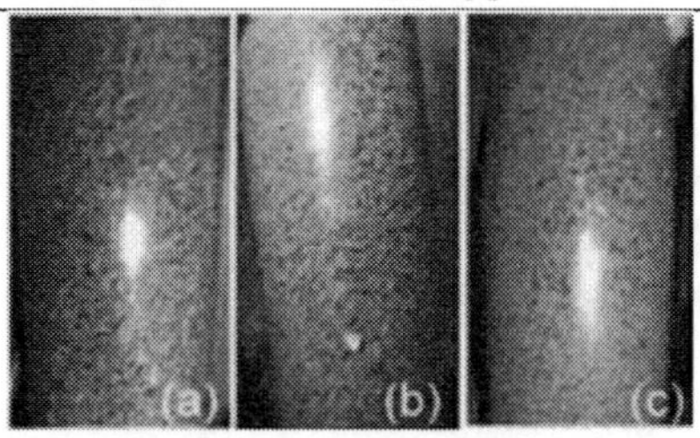


Figure 1. Transition of Anammox granules from anaerobic granular sludge (a: Day 10; b: Day 150; and c: Day 400). (Tang et al., 2011)

### Formation process of Anammox granules

The structureof Anammox granules was divided into four levels from micro-scale to macro-scale, (i.e., single cell, microbial cell cluster, subunit and granule) (Lu et al., 2012). The single cell underTEMobservation has been illustrated above, whereas the microbial cell cluster was among 200 μm in diameter(Fig. 2d). The neighboring microbial cell clusters were connected by the filamentous bacteria–extracellularpolymeric substances (EPS) bonds(Lu et al., 2012). Inside the microbial cell clusters, microbial cells were assumed to be packed together by EPS. From the scanning electron

**microscopy (SEM) picture, we can identify that considerable cocci aggregate together (the white arrow), showing the microbial cells cluster. The subunit was visually observable. Fig. 2a shows that an Anammox granule is consisted of several subunits, as indicated by the white arrows. The subunits ranged about 1 mm, which could be separated from the granule by external pressure. The force cohering these subunits to form a large scale (2–5 mm) granule might be from ESP bonding (Lu et al., 2012).**

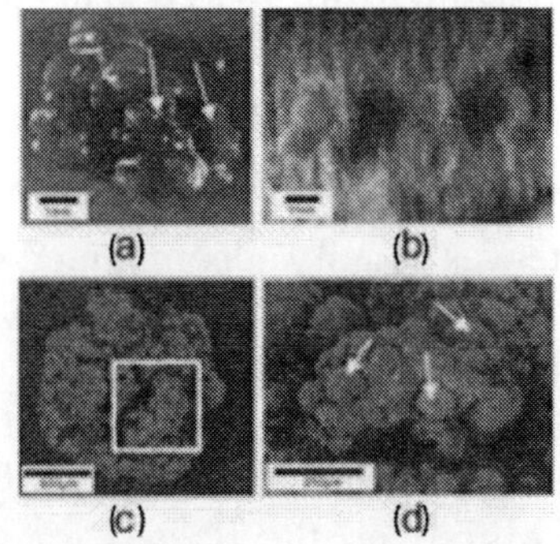

**Figure 2.** The structure of Anammox granule. (a) single granule; (b) subunits separated from a granule; (c) SEM picture of a granule (the white rectangle indicates a subunit); and (d) the microbial cells clusters (the white arrows) in a subunit (Lu et al., 2012).

**In conclusion, the Anammox cells firstly gathertogether to form a microbial cell cluster, followed by the clusters aggregation through the filamentous bacteria–EPS bonds to form a subunit. Then, several subunits coheres each other under the extensive mixing condition in high hydraulically–loaded upflow reactors, where collision among the subunits easily occurs by the aiding of EPS. Finally, granules are formed inside the upflow reactors.**

## EXTRACELLULAR POLYMERIC SUBSTANCES

**Extracellular polymeric substances (EPS) are considered to be a key reason for biomass granulation (Hulshoff Pol et al., 2004; Liu et al., 2009). They are secreted by bacteria as sticky materials constituting proteins, polysaccharides, humic acids, and lipids that assisted cell adhesion. Therefore, EPS should be helpful to initiate the aerobic as well as anaerobic granulation process (Hulshoff Pol et al., 2004; Liu et al., 2009; Hou et al., 2015). Accumulation of the secreted EPS improves the biological adhesion and microbial aggregation, since EPS could bridge bacterial cells and otherpartides into an aggregate (Hou et al., 2015).**

**The comparison of EPS content in different biological granules is listed in Table 2. The autotrophic Anammox granules contained a high EPS content as compared to the heterotrophic granules (Tang et al., 2011), which is beyond our previous expectation. As reported, the autotrophic bacteria usually secreted low EPS contentbecause they were using inorganic carbonas carbonsource instead of organic compounds (Tang et al., 2011). However, the Anammox bacteria secreted a large quantity of EPS under high hydraulic loadings (Tang et al., 2011). Therefore, extensive granulation of Anammox biomass inside upflow reactors occurs with shorthydraulic retention time.**

**Table 2.** Extracellular polymeric substances (EPS) content in different microbial granules.

| granular sludge | EPS (mg gVSS$^{-1}$) | | | reference |
|---|---|---|---|---|
| | proteins | polysaccharides | PN/PS | |
| Anammox granules | 164.4±9.3 | 71.8±2.3 | 2.29 | Tang et al., 2011 |
| aerobic granules | 40 | 16 | 2.5 | Zheng and Yu, 2007 |
| phenol–degrading granules | 240±13 | 61.0±9.4 | 3.93 | Adav et al., 2008 |
| anaerobic granules | 42.7±37.8 | 17.3±6.8 | 2.5 | Wu et al., 2009 |
| hydrogen–producing granules | 70.9±4.5 | 115.6±5.2 | 0.6 | Mu and Yu, 2006 |
| denitrifying granules | N.A. | N.A. | 2.2 | Franco et al., 2006 |

N.A.: not available; PS: polysaccharides; PN: proteins.

**Still, the roles between polysaccharide and protein in biomass granulation are controversial. Hou et al. (2015) concluded that protein contained high content of hydrophobic amino acids and loose structure to fully expose inner hydrophobic groups which promote Anammox sludge aggregation. In contrast, several researchers showed that polysaccharide plays a more significant role in granulation (Ni et al., 2010). The polysaccharide possesses a number of negatively charged functional groups such as carboxyl and hydroxyl which could function as a bridging interaction to easily form granules.**

## FLOTATION OF ANAMMOX GRANULES AND CONTROL STRATEGIES

After granulation of Anammox biomass was realized, flotation of Anammox granules could occur. The increase of nitrogen removal rate was usually accompanied with increased hydraulic loading rate and nitrogen gas production, resulting in extensive mixing between gas bubbles and granules. Thus, the nitrogen gas bubbles produced by Anammox bacteria were trapped inside the granules (Chen et al., 2010). As a result, the density of the granules decreased resulting in the sludge flotation and even sludge washout. In this situation, reactor performance eventually deteriorated during the continuous elevation of nitrogen loading rate. Moreover, the floated granules assembled continuously in the settler and blocked up the effluent pipe, resulting in malfunction of the reactor operation. Therefore, the flotation of Anammox granules is a significant problem that affects the high–rate and stable operation of Anammox reactors (Chen et al., 2010).

The flotation of Anammox granules has attracted a great deal of attentions in recent years. It seems that flotation of granules is inevitable during the operation of high–rate Anammox reactors, yet the mechanism of flotation still remains unclear. Most of researchers insisted that the physical property of Anammox is the key reason. For example, Chen et al. (2010) reported that flotation occurred due to the blocking of gas tunnels and thus ineffective exhaust of nitrogen gas to the outside. Tang et al. (2009) applied force analysis to evaluate the bubble column inside UASB reactor, and confirmed that gas accumulation led to the whole granules flotation. Our recent study compared the physical, chemical, and microbial properties of settled and floated granules. It was found that with the increase of nitrogen removal rate, EPS secretion, especially protein content, of Anammox granules increased, thus imposing the adhesion and bridging ability to form big granules from small aggregates. Gas bubbles were formed between the joint of aggregates consequently (Fig. 3). Furthermore, the gas tunnels from inner part to the surface of granules were easily buried under the high EPS secretion (Chen et al., 2010), resulting in the direct adhesion of hydrophobic bubbles on granular surface (as shown in Fig. 3b). Due to the decrease in granules density, flotation of Anammox granules inevitably occurred in upflow reactors and was further enhanced with increasing nitrogen removal rates and gas production rates. Subsequently, deterioration of reactor performance prevailed.

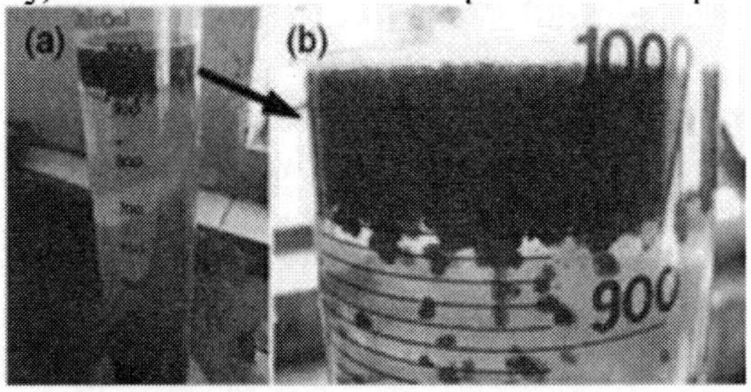


**Figure 3.** Pictures of Anammox granular flotation (a and b). The red Anammox granules tend to be floated after adsorption of dinitrogen gas bubbles produced by Anammox bacteria. As we can see from (b) gas bubbles are easily found in granules surface.

## FULL–SCALE APPLICATION OF ANAMMOX GRANULES IN WASTEWATER TREATMENT

Anammox process has been widely used for nitrogen removal from ammonium–rich wastewater.

The development of granule–based Anammox reactors with NRR higher than 70 kg $m^{-3}$ $d^{-1}$ further increased the application significance of Anammox process for treatment of real wastewaters including sludge liquor, landfill leachate, monosodium glutamate (MSG) wastewater and pharmaceutical effluents (Lackner et al., 2014). However, due to the slow growth rate of the Anammox bacteria, the application of full–scale Anammox–based process required an extremely long period, mainly due to the slow startup course (van der Star et al., 2007; Lackner et al., 2014). For the first full-scale Anammox reactor (70 $m^3$) located in Rotterdam, Netherland (NL), the startup lasted 3.5 years (van der Star et al., 2007). The first full-scale Anammox reactor (60 $m^3$) implemented in China for treatment of monosodium glutamate wastewater (built by Zhejiang University) spent about 4 years for enrichment of Anammox biomass. In order to accelerate the startup, an upscaling strategy of "lab–pilot–full–scale" Anammox biomass enrichment has been proposed.

Once the Anammox reaction initiates inside reactors, the startup performance would be significantly accelerated (van der Star et al., 2007). Therefore, after successful implementation of the first full-scale reactor, the Anammox biomass has been largely enriched, resulting in the widespread installations of full–scale Anammox-based processes. By the year 2014, there were more than 100 full–scale Anammox-based processes around the world, among which more than 50% are sequencing batch reactors (SBR), 88% being operated as single–stage systems, and 75% for side–stream treatment of municipal wastewater (Lackner et al., 2014). Although the granular systems only consisted 20% of all full-scale applications, the nitrogen removal performance was substantially higher (Lackner et al., 2014). For example, the nitrogen removal rate of the first full–scale Anammox reactor (designed by Paques) with granular sludge reached 9.5 kg $m^{-3}$ $d^{-1}$, while the average nitrogen loading for full–scale SBR (without granular sludge) is about 1.7 kg $m^{-3}$ $d^{-1}$ (Lackner et al., 2014). Since 2006, Paques has designed granular reactors as one-stage implementations, with the majority of their systems applied for industrial wastewater treatments. This shift from two- to one-stage installations was mainly driven by the lower investment costs (Lackner et al., 2014).

Most of the full-scale Anammox processes were applied to treat reject water which contains the reliable low COD (chemical oxygen demand)/N ratio, especially in Europe (van der Star et al., 2007; Lackner et al., 2014). For example, the first full–scale reactor (70$m^3$) in Rotterdam, NL (van der Star et al., 2007) and the largest full–scale reactor (22000 $m^3$) in Blue Plains, US (Lackner et al., 2014) were all used to treat reject water from wastewater treatment plant. The Anammox process was then expanded to treat landfill leachate, slaughterhouse effluents, potato processing wastewater, distillery effluent, and even metal processing wastewater (Lackner et al., 2014).

Speth et al. (2016) investigated the Anammox biomass (Fig. 4) in a full–scale partial nitritation and Anammox (PNA) process (600 $m^3$) in Olburgen, NL. They discovered 23 near–complete draft genomes that represent the majority of the microbial community, among which 19 have no close relatives being previously cultivated or sequenced and 6 belong to bacterial phyla without any cultivated members. A metabolism diagram (Fig. 4c) of the granules in the full-scale reactor has been proposed. As shown in Fig 9c, organic carbon (C–org) was oxidized to carbon dioxide ($CO_2$) on the granule surface by autotrophs. Also, on the granule surface, $NH_4^+$ was oxidized to nitrite ($NO_2^-$) that was either reduced to nitric oxide (NO) or further oxidized to nitrate ($NO_3^-$). The $NO_3^-$ formed in the granules was reduced again in the anaerobic core, either with C–org or molecular hydrogen ($H_2$) as electron donor. $H_2$ can be formed through fermentation of organic carbon by CHB2 and CFX1 (two hydrogenase–encoding organisms). This cyclic feeding resulted in additional organic carbon removal from the system. NO formed from $NO_2^-$combined with $NH_4^+$and converted to dinitrogen gas by Anammox.

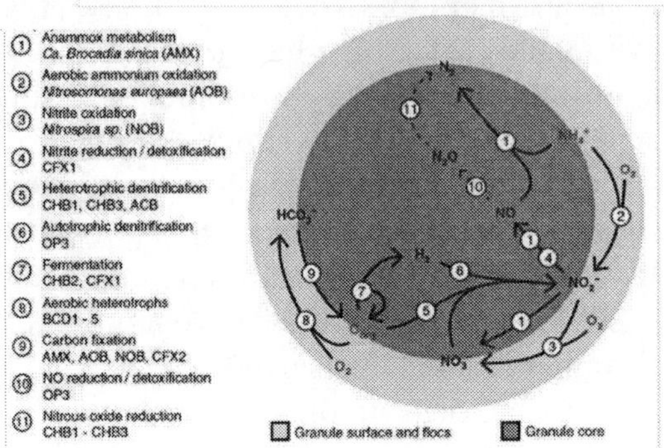


**Figure 4.** Schematic diagram of nitrogen conversions in the Olburgen PNA granules (Speth et al., 2016).

## CONCLUSIONS

**Anammox process has been considered to be the most sustainable nitrogen removal technology for ammonium–rich wastewater treatment due to the high nitrogen removal performance, low operational cost, and sludge production. However, due to the extremely low slow growth rate of the chemolithoautotrophic Anammox bacteria, the enrichment of Anammox biomass is quite difficult, and the process is easily inhibited by environmental conditions, which significantly restricts the application of this environmental–friendly process. Granulation of Anammox biomass provides an effective method for stable and high-rate operation of Anammox upflow reactors. Anammox bacteria could secrete large amount of extracellular polymeric substances and many hydrophobic functional groups, which significantly contributes to Anammox biomass granulation The Anammox granules feature a stable ecological structure with AOB on the surface and Anammox bacteria in the inner section of granules, which also improve the stability of Anammox reactors. Unfortunately, flotation of Anammox granules occurs inevitable under high shear stress and high nitrogen loading conditions. Floated granules assemble continuously in the settler and block up the effluent pipe, resulting in malfunction of the reactor operation. However, the mechanism behind the flotation still remains unclear, and effective control methods are highly desirable during the subsequent investigations. Since the granular Anammox process possesses much higher nitrogen removal performance, more installations of full–scale granular Anammox upflow reactor for treatment of ammonium-rich wastewater should be applied in future**

## REFERENCES


1. Hulshoff Pol LW, de Castro Lopes SI, Lettinga G, Lens PN (2004) Anaerobic sludge granulation. Water Res 38: 1376–1389.
2. Lu HF, Zheng P, Ji QX, Zhang HT, Ji JY, Wang L, Ding S, Chen TT, Zhang JQ, Tang CJ, Chen JW (2012) The structure, density and settlability of anammox granular sludge in high–rate reactors. Bioresource Technol 123: 312–317.
3. Lackner S, Gilbert EM, Vlaeminck SE, Joss A, Horn H, van Loosdrecht MCM (2014) Full–scale partial nitritation/anammox experiences: An application survey. Water Res 55: 292–303.
4. Speth DMichiel R, in't Zandt H, Simon GC, Bas ED, Jetten MSM (2016) Genome–based microbial ecology of anammox granules in a full–scale wastewater treatment system. Nature Comm 7:11172.
5. Tang CJ, Zheng P, Chai LY, Min XB (2013) Characterization and quantification of anammox start–up in UASB reactors seeded with conventional activated sludge. Int Biodeter Biodegr 82: 141–148.
6. Tang CJ, Zheng P, Qaisar M (2009) The shear force amendments on the slugging behavior of upflow Anammox granular sludge bed reactor. Sep Purif Technol 69: 262-268.
7. Tang CJ, Zheng P, Wang CH, Mohmmod Q, Zhang JQ, Chen XG, Zhang L, Chen JW (2011). Performance of high–loaded ANAMMOX UASB reactors containing granular sludge. Water Res 45: 135–144.
8. Van der Star WRL, Abma WR, Bolmmers D, Mulder JW, Tokutomi T, Strous M, Picioreanu C, van Loosdrecht MCM (2007) Startup of reactors for anoxic ammonium oxidation: Experiences from the first full–scale anammox

reactor in Rotterdam. Water Res 41:4149–4163.
9. Ni BJ, Chen YP, Liu SY, Fang F, Xie WM, Yu HQ (2009) Modeling a granule–based anaerobic ammonium oxidizing (ANAMMOX) process. Biotechnol Bioeng 103: 490–499.
10. Liu XW, Sheng GP, Yu HQ (2009) Physicochemical characteristics of microbial granules. Biotechnol Adv 27: 1061–1070.
11. Hou XL, Liu ST, Zhang ZT (2015) Role of extracellular polymeric substance in determining the high aggregation ability of anammox sludge. Water Res 75:51–62.
12. Chen JW, Ji QX, Zheng P, Chen TT, Wang CH, Mahmood Q (2010) Floatation and control of granular sludge in a high–rate anammox reactor. Water Res 44: 3321–3328.
13. Ni BJ, Hu BL, Fang F, Xie WM, Kartal B, Liu XW, Sheng GP, Jetten M, Zheng P, Yu HQ (2010) Microbial and physicochemical characteristics of compact anaerobic ammonium–oxidizing granules in an upflow anaerobic sludge blanket reactor. Appl Environ Microbiol 76: 2652–2656.
14. Tang CJ, Zheng P, Zhang L, Chen JW, Mohmmod Q, XG Chen, Hu BL, Wang CH, Yu Y (2010c) Enrichment features of anammox consortia from methanogenic granules loaded with high organic and methanol content. Chemosphere 79: 613–619.
15. Vazquez–Padin J, Mosquera–Corral A, Campos JL, Mendez R, Revsbech NP (2010) Microbial community distribution and activity dynamics of granular biomass in a CANON reactor. Water Res 44:4359–4370.
16. Gonzalez-Gil G, Sougrat R, Behzad AR, Lens PNL, Saikaly PE (2015) Microbial community composition and ultrastructure of granules from a full-scale Anammox reactor. Microb Ecol 70:118–131.
17. Franco A, Roca E, Lema JM (2006) Granulation in high–load denitrifying upflow sludge bed pulsed reactors. Water Res 40: 871–880.
18. Adav SS, Lee DJ, Tay JH (2008) Extracellular polymeric substances and structural stability of aerobic granule. Water Res 42: 1644–1650.
19. Mu Y, Yu HQ (2006) Biological hydrogen production in a UASB reactor with granules. I. Physicochemical characteristics of hydrogen-producing granules. Biotechnol. Bioeng. 94, 980-987.
20. Wu, J., Zhou, H.M., Li, H.Z., Zhang, P.C., Jiang, J.(2009) Impacts of hydrodynamic shear force on nucleation of flocculent sludge in anaerobic reactor. Water Res. 43, 3029-3036.
21. Zheng, Y.M., Yu, H.Q.(2007) Determination of the pore size distribution and porosity of aerobic granules using size-exclusion chromatography. Water Res. 41, 39-46.
22. Franco, A., Roca, E., Lema, J.M.(2006) Granulation in high-load denitrifying upflow sludge bed pulsed reactors. Water Res. 40, 871-880.

# 490741 Using a Bio-Derived Solvent to Cast Polysulfone Ultrafiltration Membranes


**Abstract:**

Organic solvents derived from petroleum sources, such as N-methyl-2-pyrrolidone (NMP) and dimethylacetamide (DMAc), have been traditionally used to fabricate polymeric membranes. These solvents have a negative impact on the environment and human health since most of them are volatile and hazardous; therefore, using renewable solvents derived from biomass is of great interest in order to make membrane fabrication sustainable. In this study, methyl-5-(dimethylamino)-2-methyl-5-oxopentanoate (Rhodiasolvâ"‡ PolarClean) was used because it is a bio-derived, biodegradable, nonflammable and nonvolatile solvent. Polysulfone was chosen as the polymer to fabricate membranes due to its thermal stability, strong mechanical strength, good chemical resistance, and antifouling properties. The thermodynamics aspects of the polysulfone/PolarClean/water system were investigated. From cloud point curves and theoretical predictions, PolarClean showed the potential to be a better solvent for polysulfone, and dope solutions could be prepared at a reasonable temperature of 65 â„ƒ.

Polysulfone (PSf) membranes prepared with PolarClean were also investigated in terms of their morphology, porosity, water permeability, protein rejection and compared with the membranes prepared with traditional solvents. The pores of polysulfone/PolarClean membranes were sponge-like, which led to low water flux and high solute rejection, while those of polysulfone /NMP or polysulfone /DMAc membranes were finger-like macrovoids that led to high water flux and low solute rejection. The water flux of polysulfone/PolarClean membranes was 26.2 LMH, which was lower than polysulfone/DMAc membranes (146.9 LMH) and polysulfone/NMP membranes (48.9 LMH). However, BSA rejection rate using polysulfone/PolarClean membranes reached 90%, which was similar to polysulfone/NMP membranes (90%) and higher than polysulfone/DMAc membranes (87%). The difference of water flux and the solute rejection rate data likely corresponded to their morphologies. In conclusion, bio-derived solvents should be investigated further and may become promising replacements to traditional solvents.

# Reducing Specific Energy Consumption of Seawater Desalination: Staged RO or RO-PRO?

Mingheng Li

**Department of Chemical and Materials Engineering**
**California State Polytechnic University, Pomona**

October 30, 2017



## Outline

1. Motivation
2. Methodology
3. Results & Discussion
4. Conclusions

# Motivation



## Global Grand Challenges in Energy and Water

[1] Professor Richard Smalley, Nobel Laureate

# Membrane-Based Processes

- Reverse Osmosis (RO)

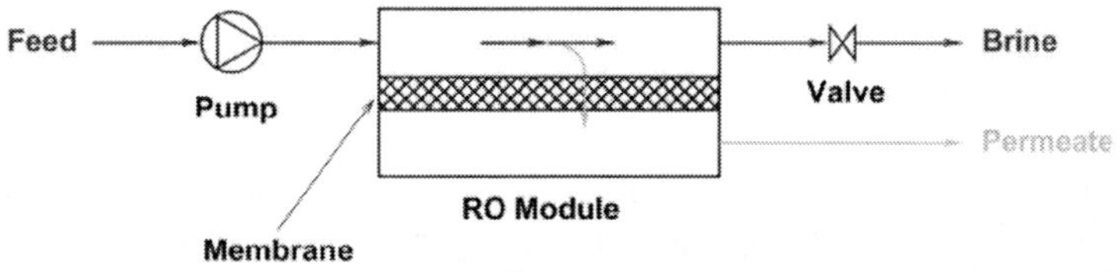


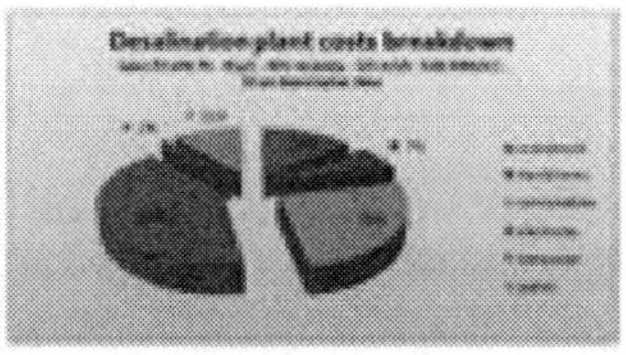

Pressure Retarded Osmosis (PRO)

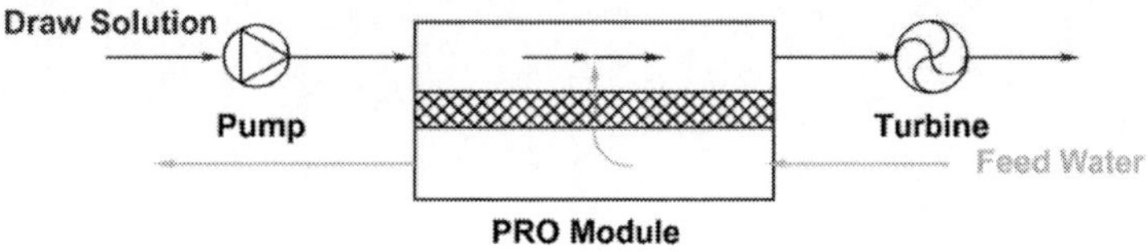


[1]https://www.lenntech.com/processes/desalination/energy/general/desalination-costs.htm



# Energy-Efficient Desalination Processes

- Hybrid RO-PRO with Energy Recovery Device (ERD)

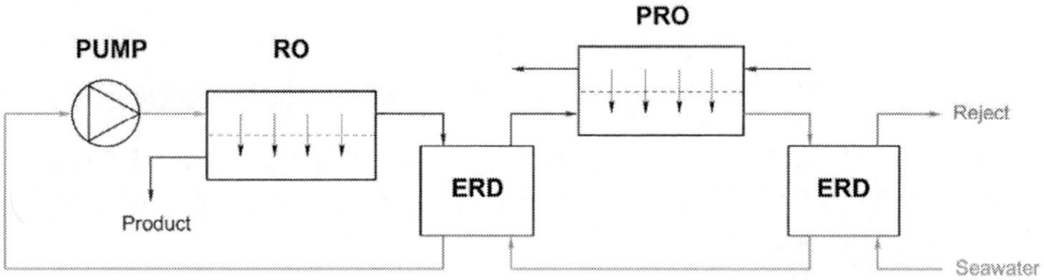


Multistage RO with ERD

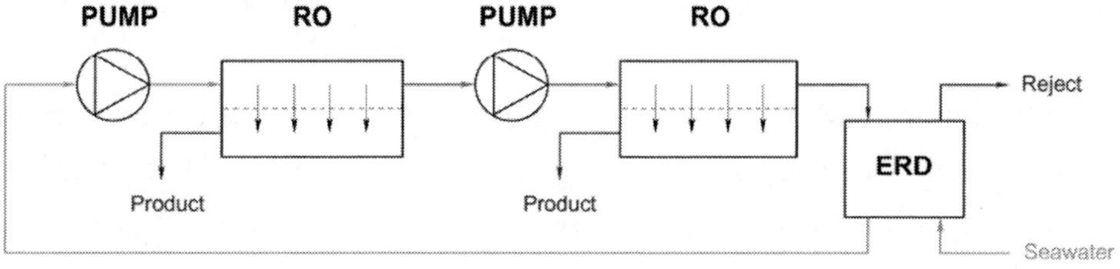

# Methodology



## Characteristic Equation of RO and PRO

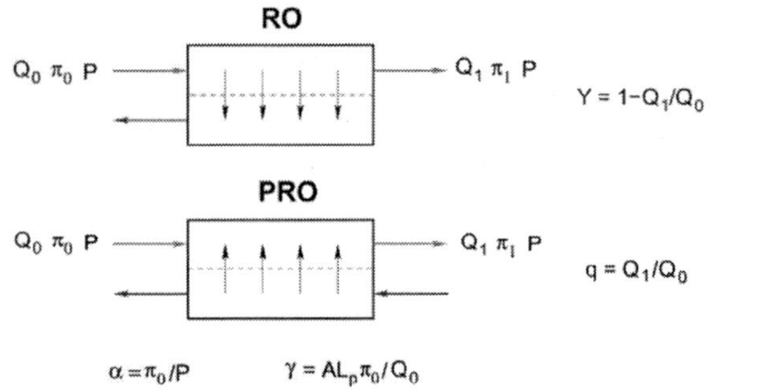


Assumptions:

- Negligible pressure drop
  Negligible concentration polarization
  Perfect transverse mixing of salt

$$\gamma \quad = \quad \alpha\left[1-q+\alpha \ln \frac{\alpha-1}{\alpha-q}\right]$$

| | | |
|---|---|---|
| $q$: | Dimensionless flow rate ($Y = 1 - q$) | $\rightarrow$ Performance parmater |
| $\alpha^{-1}$: | Dimensionless applied pressure | $\rightarrow$ Operating parameter |
| $\gamma$: | Dimensionless membrane area | $\rightarrow$ Design parameter |

[1] Li, IECR, 2010; AIChE J., 2015

## Model Accounting for PRO Efficiency

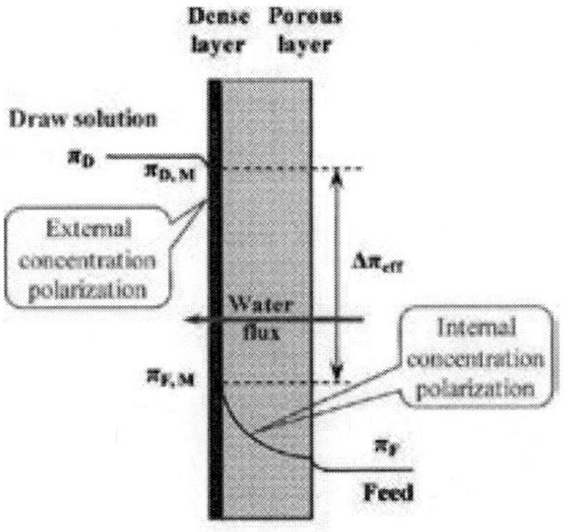


**PRO mode**

- Detrimental Effects in PRO
  - External Concentration Polarization (ECP)
  - Internal Concentration Polarization (ICP)
  - Reverse Salt Leakage

Equation accounting for PRO efficiency

$$\gamma = \alpha\left[1 - q^{ideal} + \alpha \ln \frac{\alpha - 1}{\alpha - q^{ideal}}\right]$$

$$q^{real} - 1 = \eta_{PRO}(q^{ideal} - 1)$$

[1]Wang, Ong, and Chung, IECR, 2010

[2]Li, AIChE J., 2017



## Optimization Model for RO-RO

$$\min_{\alpha_j, Y_j, \gamma_j} NSEC = \frac{1}{\eta_{pump} Y_{total}}\left[\frac{Y_1}{\alpha_1} + \frac{1}{\alpha_2} - \frac{\eta_{ERD}(1 - Y_2)}{\alpha_2}\right]$$

$$s.t.$$

$$0 = \gamma_j - \alpha_j\left[Y_j + \alpha_j \ln \frac{1 - \alpha_j}{1 - Y_j - \alpha_j}\right], \quad j = 1,2$$

$$0 = \gamma_{total} - \left[\gamma_1 + \gamma_2(1 - Y_1)^2\right]$$

$$0 = [1 - (1 - Y_1)(1 - Y_2)] - Y_{total}$$

$$0 \leq \gamma_j, \quad j = 1,2$$

$$0 \leq \alpha_j, \quad j = 1,2$$

$$0 \leq 1 - \alpha_j, \quad j = 1,2$$

$$0 \leq Y_j, \quad j = 1,2$$

$$0 \leq 1 - Y_j, \quad j = 1,2$$

$$0 \leq \alpha_1/(1 - Y_1) - \alpha_2$$

[1]Li, Desalination, 2017

## Optimization Model for RO-PRO

$$\min_{\alpha_j,\gamma_j,q_2} NSEC = \frac{1}{\eta_{pump} Y_1}\left[\frac{1}{\alpha_1} - \frac{\eta_{ERD}(1-Y_1)}{\alpha_1} - \frac{\eta_{ERD}\eta_{PRO}(q_2-1)}{\alpha_2}\right]$$

$$0 = \gamma_1 - \alpha_1\left[Y_1 + \alpha_1 \ln \frac{1-\alpha_1}{1-Y_1-\alpha_1}\right]$$

$$0 = \gamma_2 - \alpha_2\left[1 - q_2 + \alpha_2 \ln \frac{\alpha_2 - 1}{\alpha_2 - q_2}\right]$$

$$0 = \gamma_{total} - \left[\gamma_1 + \gamma_2(1-Y_1)^2\right]$$

$$0 \leq \gamma_j, \quad j = 1,2$$

$$0 \leq \alpha_1$$

$$0 \leq 1 - \alpha_1$$

$$0 \leq \alpha_2 - 1$$

$$0 \leq q_2 - 1$$

[1] Li, Desalination, 2017



# Results & Discussion

# RO-RO vs RO-PRO in NSEC

- Y = 30%

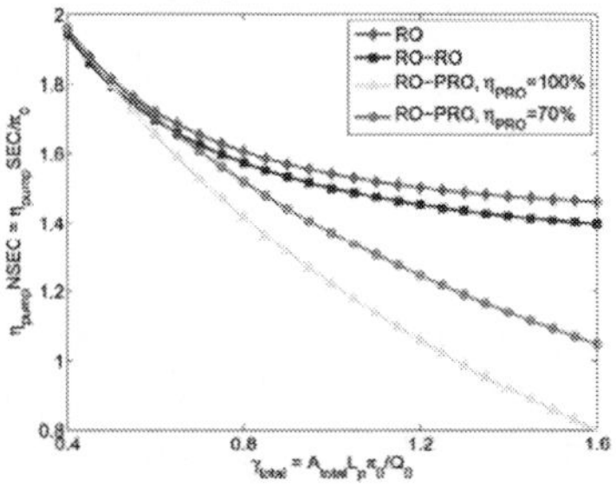


Y = 40%

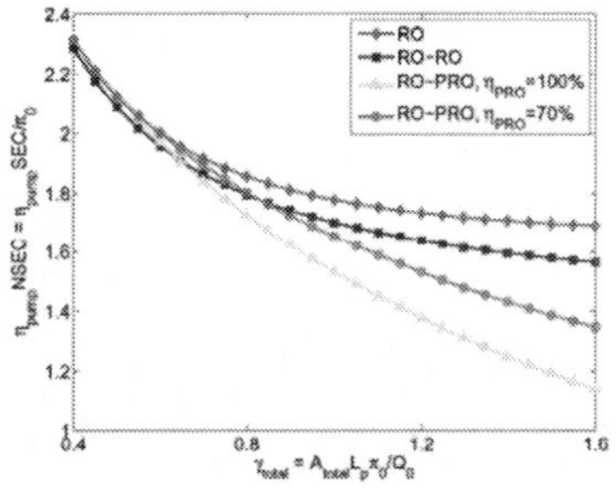


Y = 50%

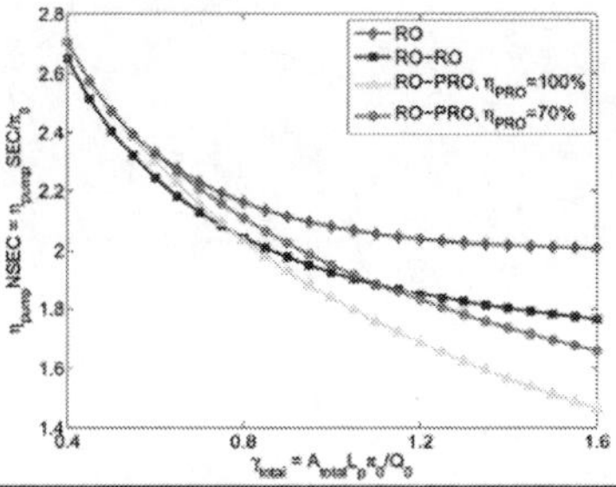


Y = 60%

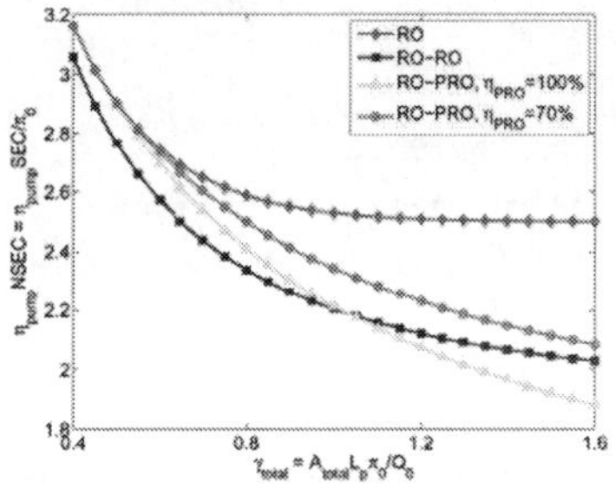


[1] Li, Desalination, 2017



# RO-RO vs RO-PRO in Membrane Area Allocation

- Y = 30%

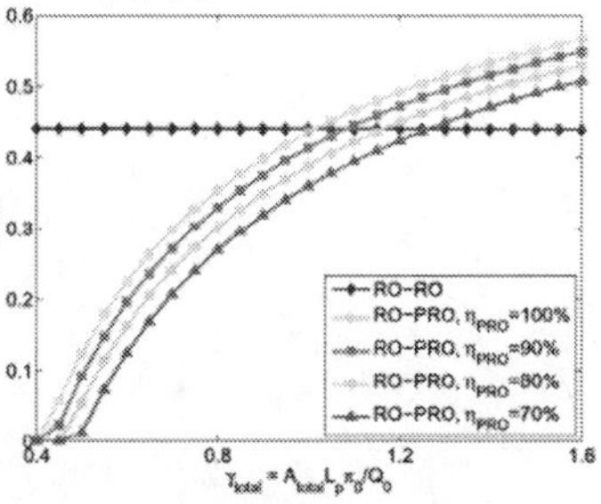


Y = 40%

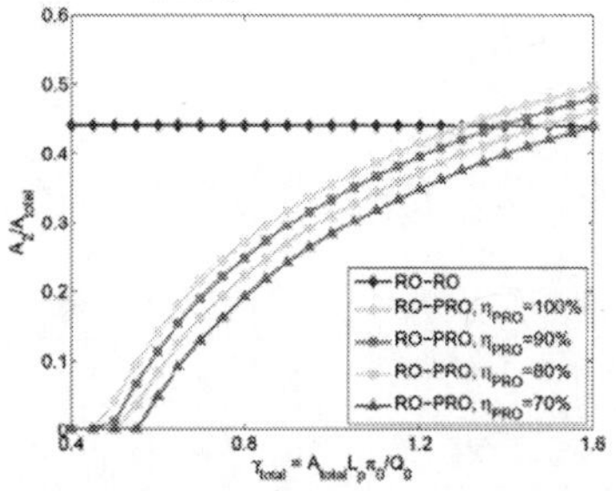


Y = 50%

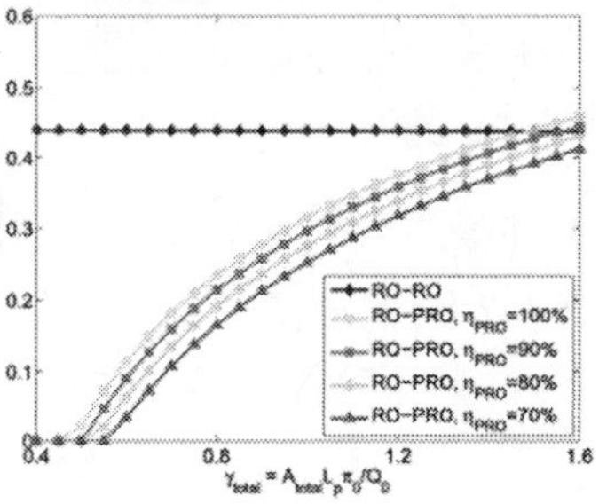


Y = 60%

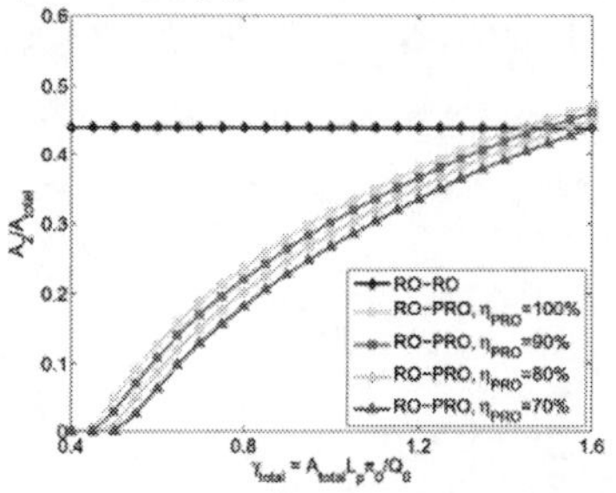

## Hydraulic vs Osmotic Energy Recovery in RO-PRO

- Y = 30%

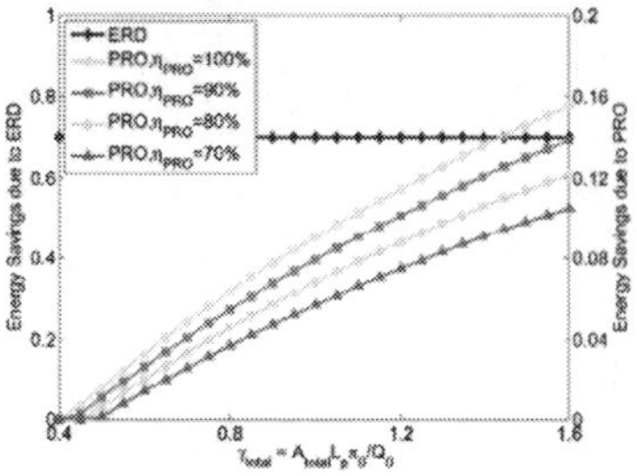


Y = 40%

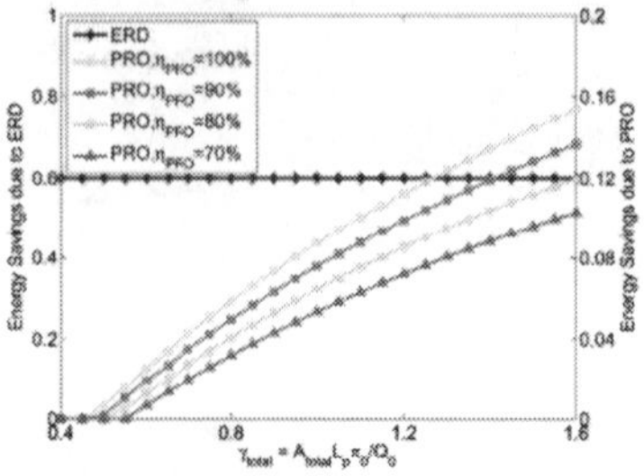


Y = 50%

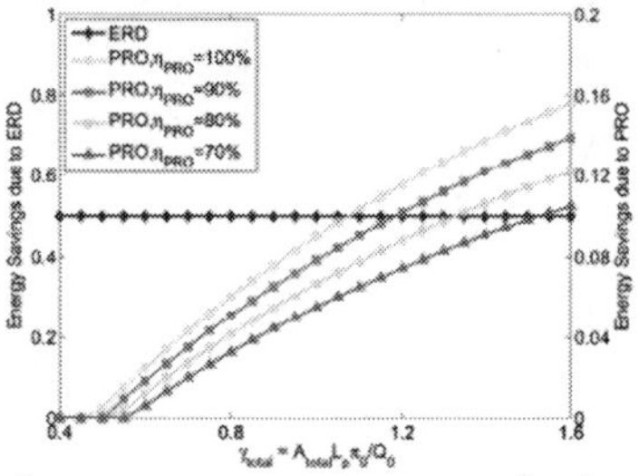


Y = 60%

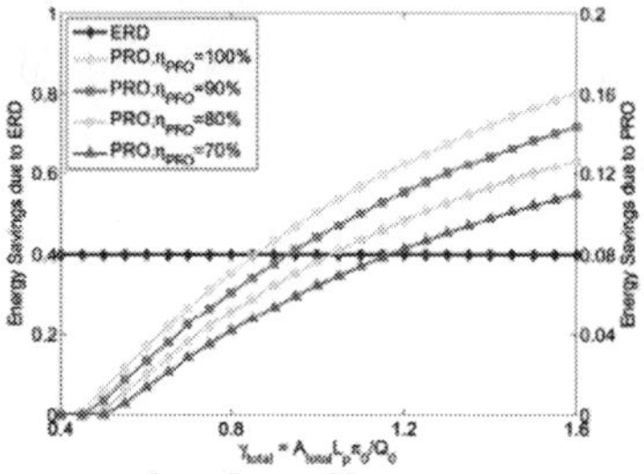


Osmotic energy recovery is by no means close to hydraulic energy recovery.



## Stage-Level Volume Gain Ratio in PRO

- Y = 30%

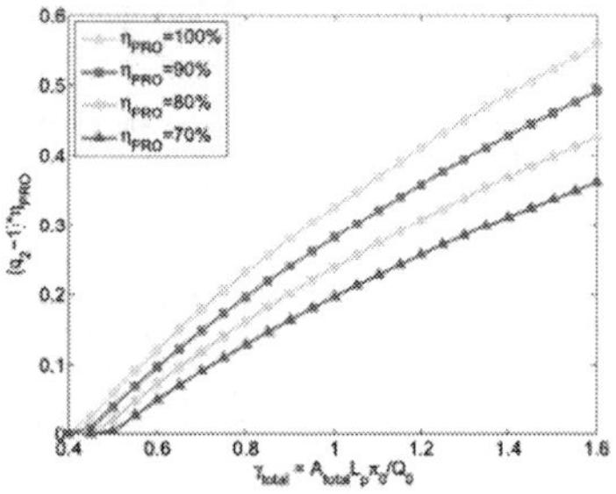


Y = 40%

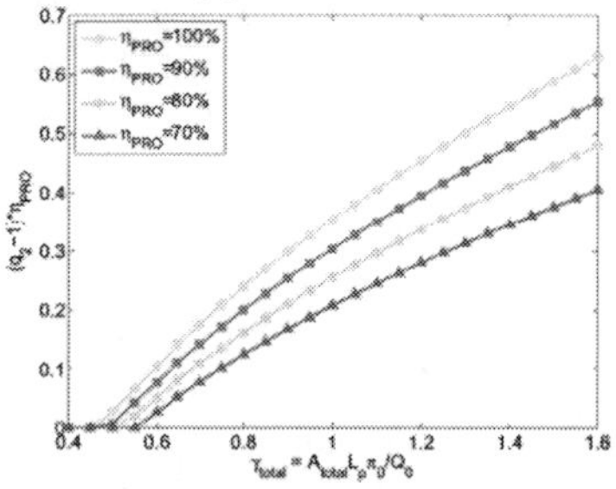


Y = 50%

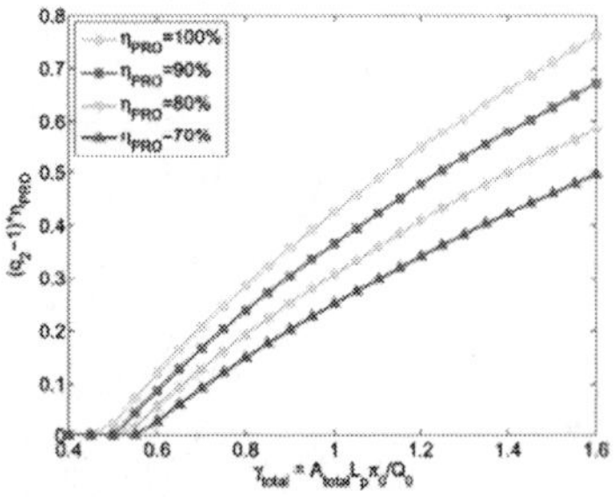


Y = 60%

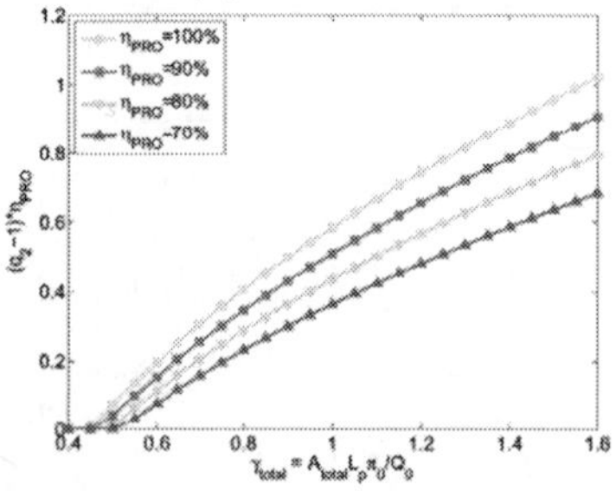

## System-Level Volume Gain Ratio in RO-PRO

- $\eta_{PRO} = 100\%$

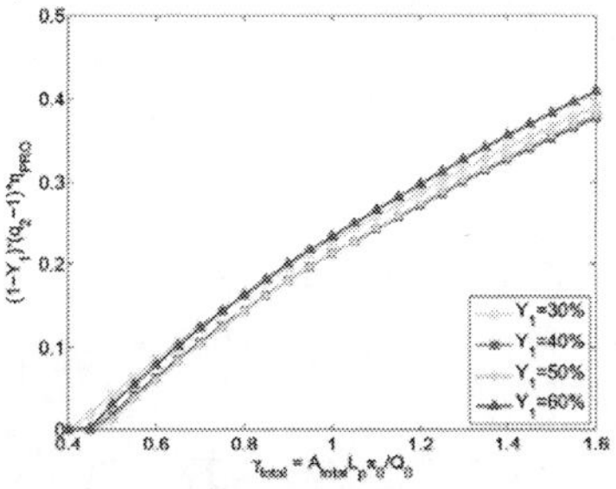


$\eta_{PRO} = 90\%$

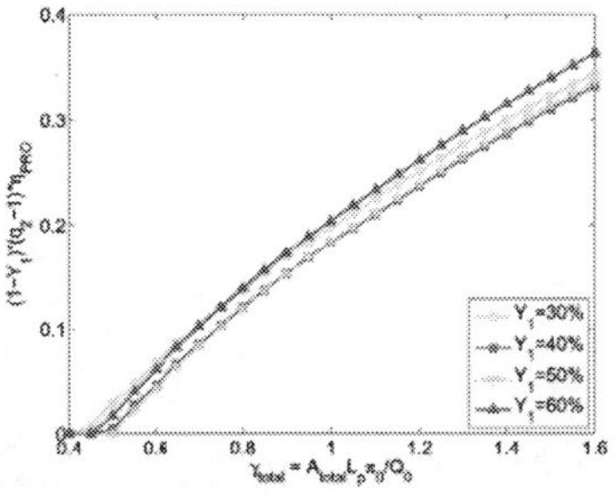


$\eta_{PRO} = 80\%$

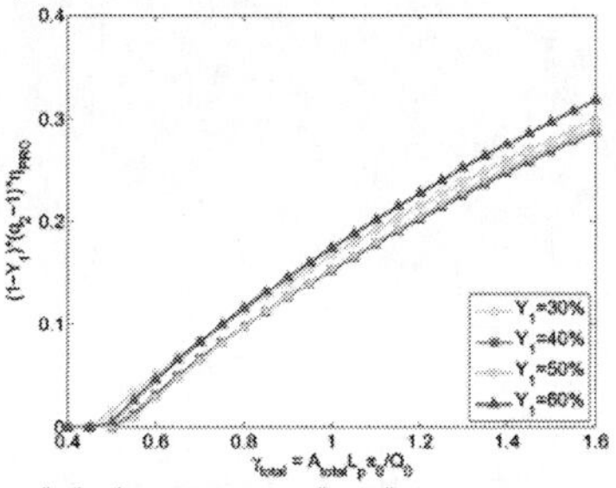


$\eta_{PRO} = 70\%$

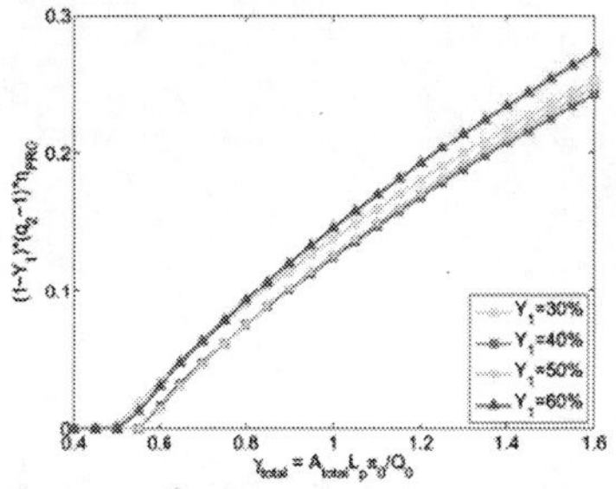


Volume gain is more or less independent of water recovery.



## Optimal Hydraulic to Osmotic Pressure Ratio in PRO

- Y = 30%

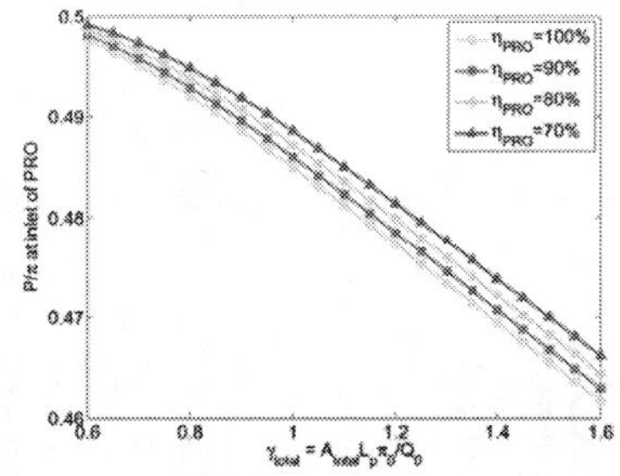


Y = 40%

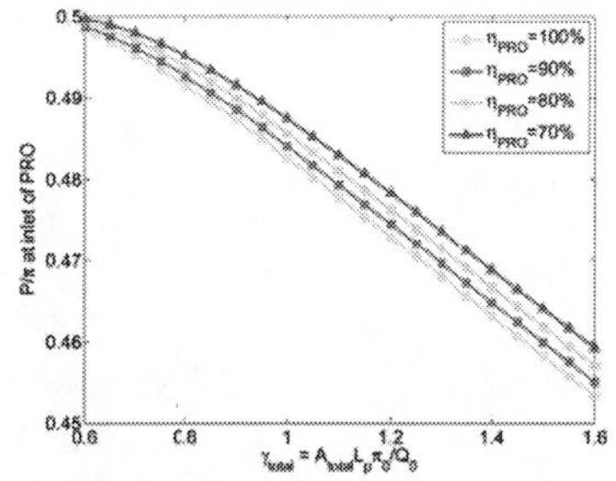


Y = 50%

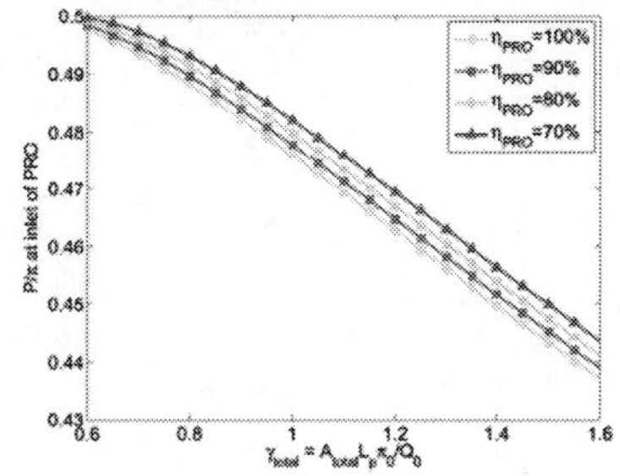


Y = 60%

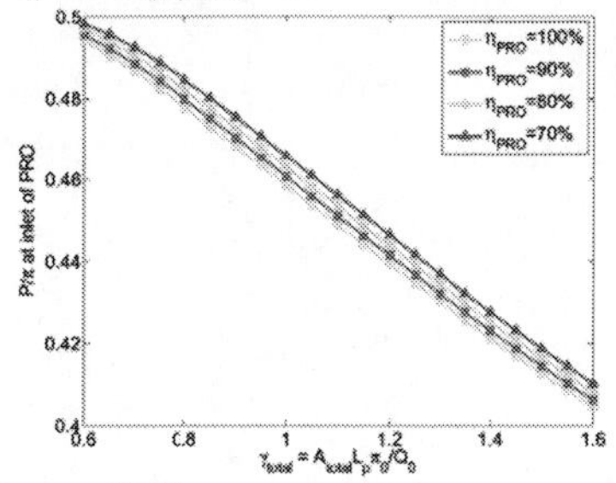


$P/\pi$ is always less than 0.5.

## RO-RO vs RO-PRO in NSEC $\eta_{ERD} = 80\%$, $\eta_{pump} = 80\%$

- Y = 30%

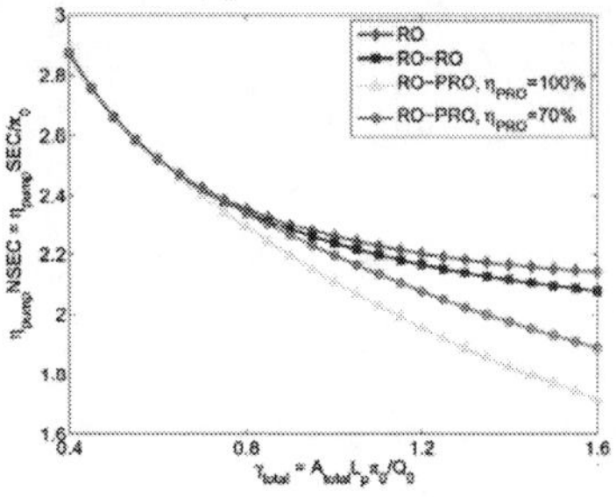


Y = 40%

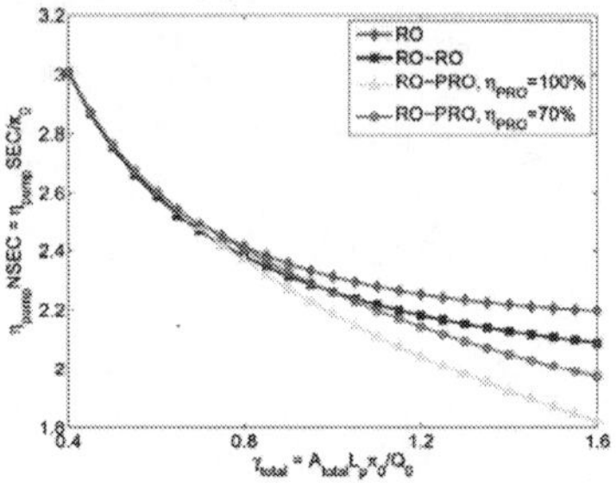


Y = 50%

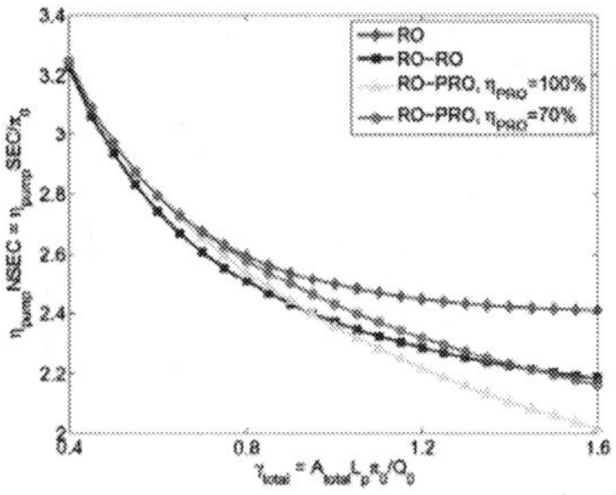


Y = 60%

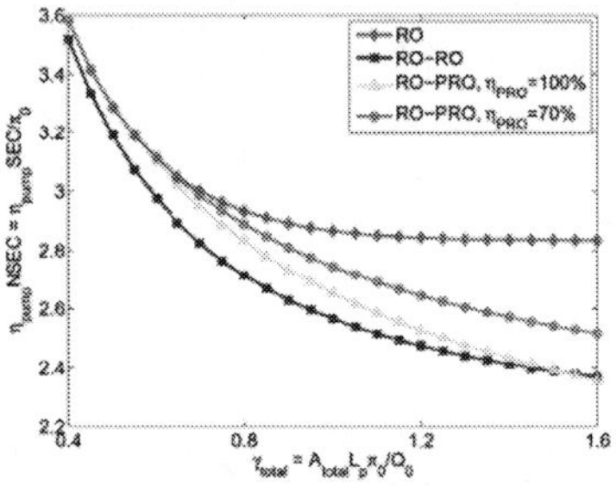




# Conclusions

## Conclusions

- RO-PRO excels when recovery is low (e.g., 30%), membrane area is abundant, and detrimental effects in PRO are not significant.
- RO-RO outperforms at higher recoveries (e.g., 60%) even though high-salinity brine is beneficial for PRO.
- Suppressing detrimental effects in PRO are beneficial for RO-PRO.
- Development of cheaper and higher performance PRO membranes will allow a large $\gamma_{total}$ to be implemented in plant design, which favors RO-PRO.

# 494376 Novel Thin Film Composite Nanofiltartion Membrane Prepared with Ethylenediamine Functionalized β-Cyclodextrin for Water Treatment

**Abstract:**

In this work, a novel amine monomer — ethylenediamine functionalized *β*-cyclodextrin (*β*-CD-EDA), was synthesized and employed for the preparation of thin film composite nanofiltration (NF) membrane via the interfacial polymerization with trimesoyl chloride (TMC), to incorporate the intrinsic CD cavities into the polymer network of the selective layer. The chemical structure of the synthesized *β*-CD-EDA was confirmed by HMRS, FTIR, TGA XPS and $^1$H NMR. The chemical properties, surface morphology and the separation performance of TFC membranes prepared with β-CD (TFC_β-CD) and β-CD-EDA (TFC_*β*-CD-EDA) as aqueous monomers were investigated and compared. The effects of the interfacial polymerization time and *β*-CD-EDA concentration on the membrane properties were investigated studied as well. The separation performances of TFC_β-CD-EDA membrane were further evaluated with feed solutions containing different solutes. It was found that TFC_*β*-CD-EDA membranes showed very high rejections to sodium sulfate, magnesium sulfate, sodium citrate, sodium gluconate and glucose. The anti-fouling performances of the TFC_β-CD-EDA membrane were also studied with bovine serum albumin (BSA), humic acid (HA), sodium alginate (SA) as simulated pollutants of protein, natural organic matter, and hydrocarbon in the natural water source, respectively, and the results showed that the membrane possessed excellent fouling resistance to these common foulants.

# 501885 Concentration of Titania Waste Acid By Direct Contact Membrane Distillation: Influence of Ferrous Sulfate and Polyacrylamide on Flux and Rejection

**Abstract:**

Titanium dioxide or titania ($TiO_2$) is an important compound widely used in many fields. One of the commercial processes to produce $TiO_2$ is the â€œsulphateâ€ process, in which for every ton of $TiO_2$ produced, around 6~8 tons of sulfuric acid (~20wt.%) will be generated. In our previous work, the concentration of the simulated acid by direct contact membrane distillation (DCMD) has been investigated preliminarily, an effort within the overall framework to identify new processes for reducing waste charge in $TiO_2$ industry. The current research is an extension with the emphases being influence on flux and rejection of DCMD by components brought along from $TiO_2$ generation: 1) inorganic salts; 2) organic compounds. At present stage, work regarding $FeSO_4$, a major inorganic salt, has been carried out. The experimental results indicate that crystallization of $FeSO_4$ evolved during the concentration. $FeSO_4$ crystals with different polymorphs settled on membrane surface and were also observed inside membrane structure. The rejection coefficient of the membrane in DCMDã€€did not correlate closely with $FeSO_4$ precipitation on and in the membrane structure. Additionally, higher feed flow rate changed obviously the surface fouling morphology and membrane flux variation. In the upcoming work, polyacrylamide (PAM) will be investigated in terms of membrane wetting. PAM is organic compound used in the step for flocculation settling of metatitanic acid ($H_2TiO_3$). The majority of PAM settles down during the process. How the residual PAM influence the MD will be investigated.

# 487614 Molecular Interaction Between Acidic Sppsu and Basic Hpei Polymers and Its Effects on Membrane Formation for Ultrafiltration

**Abstract:**

This study aims to investigate the fundamentals of fabricating flat-sheet ultrafiltration (UF) membranes from acid/base polymer complexes via non-solvent induced phase inversion and to explore their potential for ultrafiltration. The concept of acid/base polymer blends with the aid of electrostatic interaction has been employed to fabricate membranes for fuel cell and pervaporation, but not for ultrafiltration. Basic hyperbranched polyethyleneimine (HPEI) polymers and acidic sulfonated polyphenylenesulfone (sPPSU) polymer are blended together as casting solutions. It is found that the molecular interaction between them is greatly influenced by the HPEI amount, changing from ionic crosslinking with chain entanglement to polymer aggregation. As a result, the phase inversion mechanism and polymer precipitation rate during membrane formation as well as membrane morphology vary significantly as a function of HPEI content. With an increase in HPEI concentration, membrane morphology transforms from an initially fully sponge-like structure to a finger-like macrovoid cross-section, then back to a totally macrovoid-free structure again. In addition, the as-prepared UF membranes demonstrate great variations in swelling ratio, surface charge, roughness, hydrophilicity, mechanical strength and UF performance. At the optimal HPEI concentration of 0.3 wt%, the as-cast UF membranes exhibit the highest mechanical stability and swelling resistance with an optimal water flux of 724 LMH/bar and a reduced molecular weight cut-off of 86.4 kDa.

## 490524 Economic Assessment and Scale-up of an Eco-Friendly Continuous Bioleaching System for Recovery of Rare Earth Elements from End-of-Life Materials

**Abstract:**

We have previously developed an eco-friendly strategy to recover rare earth elements (REE) from recyclable materials that relies on microbial production of organic acids to solubilize and release metals from a solid matrix. Microbially mediated rare earth recovery processes offer an economical and green alternative that can be applied to low-value feedstocks for extraction of the valuable metals. This presentation will focus on our efforts to assess the economic benefits of this approach, seek ways to reduce costs, and to scale the process.

We have shown previously that REE could be leached from recyclable materials using microbially produced organic acids under laboratory bench-scale conditions (Reed, et. al., Hydrometallurgy (2016), 166:34-40). We have also conducted techno-economic analyses for bioleaching REE from various feedstocks (Thompson, *et. al.*, in Review). These analyses demonstrated that the profitability of this process was sensitive to the leach pulp density which directly correlated to the size (cost) of the batch bioreactor necessary to produce organic acids. Therefore, to provide organic acids in quantities and at rates useful for commercial bioleaching, we developed a continuous bioreactor process. Furthermore, for low grade feedstocks leaching approaches with low capital and operating costs we examined ways to minimize material processing such as heap leaching.

Gluconic acid has been shown to be effective for bioleaching of rare earth elements (REE) from recyclable materials and the *Gluconobacter oxydans (NRRL B58)* strain is an industrially proven bacterium for efficient production of gluconic acid from glucose. Conditions for producing an optimal lixiviant (exhibiting the lowest pH) with *Gluconobacter* were determined in a batch reactor system while varying temperature, stirring speed, aeration, glucose concentration, and other media components. At incubation conditions a lixiviant solution was produced with pH 2.14 and the measured gluconic acid concentration was 233 mM.

With this lixiviant we were able to leach 49% and 11% of the total REE from two industrially relevant feedstocks, fluidized catalytically cracking (FCC) catalyst and retorted phosphor powders, respectively, that were contacted with the lixiviant for one day at a 1.5% solid/liquid (m/v) ratio. Increased densities had lower leaching percentages but higher total REE recovery for the same lixiviant. Initially we applied lixiviant to feedstocks in shaking tubes or flasks; however by leaching in a stirred reactor we were able to increase REE recovery efficiency by 13-16%. Furthermore, under the optimal conditions *Gluconobacter* was able to sustain continuous organic acid production for over 100 hours at dilution rates ranging from 0.05 $hr^{-1}$ to 0.38 $hr^{-1}$. The effluent produced at the lowest dilution rate (with pH 2.20, 173 mM gluconic acid) performed similarly to the lixiviant produced in the batch system with respect to REE leaching from the feedstocks.

We also assembled and tested a simulated bench-scale heap leaching system. Heap leaching can be advantageous

compared to reactor leaching because not only are a tank and associated plumbing obviated but agitation of the solid-liquid mixture and associated energy costs are not required. We tested the system with FCC catalyst and 5× recirculation of the lixiviant, and found that REE recovery was almost as efficient to the batch system using a similar density and contact time.

Together our results demonstrated that microorganisms that produce organic acids can induce effective leaching of REE from solid waste materials, and we have been able to scale the process (1000-fold) for continuous production of organic acid, and improved leaching scale (250-fold) to further decrease potential leaching costs.

# 499820 One-Pot Production of Furans from Industrial Hemp

**Abstract:**

Industrial hemp, a variety of Cannabis sativa and the same plant species as marijuana, is primarily grown for its high-quality fibers from bast and high-value Cannabidiol (CBD) oil from flowers and leaves. The inner core of the hemp plant stem, hemp hurd, is considered as cellulosic waste. The use of industrial hemp hurds for the production of 5-hyroxylmethyl furfural (HMF), a versatile platform chemical will bring benefits including: (1) revenues to farmers; (2) reduction of CO2 emission from the disposal of hemp hurds by landfilling and combustion; and (3) new technologies for conversion of lignocellulose. Herein, we have developed the one-pot conversion of hemp hurds into HMF using the imidazolium-based ionic liquid. Central Composite Design (CCD) and Response Surface Methodology (RSM) were used to design and optimize three processing parameters: reaction temperature (120-160°C), reaction time (1-4h), and solid loading (5-30 wt.%). We found that sugar release is sensitive to a change in temperature. Under the optimal saccharification condition, ~60 mol% of HMF was achieved from glucose dehydration. To understand the underlying physicochemical characteristics of hemp hurds that affect their hydrolysis, hemp hurds before and after dissolution in ionic liquid were characterized by X-ray Diffraction (XRD), 2D Heteronuclear Single-Quantum Correlation (HSQC) NMR, and Fourier transform infrared spectroscopy (FTIR).

# 499217 Catalytic Depolymerization of Lignin over Mesoporous Solid Lewis Acid Catalysts

**Abstract:**

**Catalytic Depolymerization of Lignin over Mesoporous Solid Lewis Acid Catalysts**

Kakasaheb Y. Nandiwale[1,2], Andrew M. Danby[1], Anand Ramanathan[1], Raghunath V. Chaudhari[1,2], and Bala Subramaniam[1,2]

[1]Center for Environmentally Beneficial Catalysis, The University of Kansas, Lawrence, Lawrence, Kansas 66047, United States

[2]Department of Chemical and Petroleum Engineering, The University of Kansas, Lawrence, Kansas 66045, United States.

**Keywords:** Biomass, Lignin depolymerization, Lewis acid

In this work, we report the performances of predominantly Lewis acidic, Zr-incorporated mesoporous silicates (Zr-KIT-5 and Zr-KIT-6)[1-2] for deconstruction of lignin into value added aromatic monomers. This work is motivated by several factors. First, homogeneous Lewis acidic catalysts metal triflates[3] have been shown to be effective in selective cleavage of ether linkages in lignin. Further, Brønsted acid sites are known to promote undesired condensation of the depolymerized products that may form during lignin deconstruction.[4] We therefore hypothesized that the use of a predominantly Lewis-acidic solid catalyst should increase the yield of the monomeric products compared to a predominantly Brønsted acidic catalyst such as H-ZSM-5. Indeed, high yields of monomeric and dimeric oxoaromatic compounds (up to 75 wt%) were observed during depolymenrization of dealkaline lignin on Zr-KIT-5 catalysts at 250°C. At similar operating conditions, the Zr-KIT-5 catalyst outperforms reported yields on H-ZSM-5 catalyst[5] with much lower formation of heavier residues. The effects of various solvents (1, 4-dioxane, acetic acid, methanol + water) and operating conditions on the depolymerization of various lignins [including corn stover lignin, PC 1369 hardwood lignin (MeadWestvaco), Indulin 80 and dealkaline lignin, (TCI chemicals)] along with the product selectivity and catalyst stability will be presented.

References:

(1) Ramanathan, A.; Zhu, H. D.; Maheswari, R.; Subramaniam, B., *Chem Eng J* 2015, *278*, 113-121.

(2) Pan, Q.; Ramanathan, A.; Snavely, W. K.; Chaudhari, R. V.; Subramaniam, B., *Ind Eng Chem Res* 2013, *52*, 15481-15487.

(3) Lancefield, C. S.; Panovic, I.; Deuss, P. J.; Barta, K.; Westwood, N. J., *Green Chem* 2017, *19*, 202-214.

(4) Li, C.; Zheng, M.; Wang, A.; Zhang, T., *Energ Environ Sci* 2012, *5*, 6383-6390.

(5) Deepa, A. K.; Dhepe, P. L., *Acs Catal* 2015, *5*, 365-379.

ADDIN EN.REFLIST

# 493651 Catalytic Conversion of Biomass Derived Lactones into Fuels and Chemicals

**Abstract:**

Catalytic conversion of biomass derived lactones into fuels and chemicals

Md. Imteyaz Alam*, Shelaka Gupta[a], Ashish Bohre[b], Ejaz Ahmad[a], Tuhin S. Khan[a], Basudeb Saha[b,c], M. Ali Haider[a]

[a.] REC Lab, Department of Chemical Engineering, Indian Institute of Technology Delhi, Hauz Khas, Delhi 110016, India

[b.] Laboratory of Catalysis, Department of Chemistry, University of Delhi, Delhi 110007, India

[c.] Catalysis Center for Energy Innovation, University of Delaware, Newark, DE 19713, USA

Tel: +91-11-26591016, Fax: +91-11-2658-2037
E-mail: imteyaz84@gmail.com

6-amyl-a-pyrone (6PP) is a fermentation derived unsaturated lactone having considerable potential to be developed as a platform chemical. Due to the presence of different functional moities and the position of double bond in the ring, myriad of reactions can be performed giving useful chemicals. For instance, ring-opening and decarboxylation of 6PP in water yielded non-2-en-4-one (1) (Figure 1a) with 95% conversion and nearly 60% selectivity at 498 K, without using any catalyst. The activation barrier was calculated to be 180 kJ/mol in presence of water using DFT, which compares well with the experimental apparent activation barrier of 166 kJ/mol. The selectivity of decarboxylated product was further enhanced using acidic silica-alumina and γ-alumina giving around 65% selectivity and 75% conversion at a reduced temperature of 463 K. The measured apparent activation energy for γ-alumina ( 99 kJ/mol) was comparable to the DFT calculated intrinsic endothermicity of the proton transfer step 98 kJ/mol (Figure 1b), which was essential for obtaining the ring-opened intermediate. The hydrogenation of non-2-en-4-one over Pd/C catalyst was further experimented to yield 4-nonanone (3) which can serve as a solvent or an intermediate for other reactions. In order to produce longer chain hydrocarbons of diesel and jet fuel ranged fuels, mixed oxide (CaO-MgO) was synthesized and utilized as a catalyst for the C-C coupling reaction via aldol condensation reaction. The condensation of nonanone with furfural and 5-hydroxymethyl furfural (HMF) over CaO-MgO yielded upto 40% (5) and 60% (4) of the branched aldol products at 443 and 393 K (Figure 1a), which may further be hydrogenated to give liquid transportation fuel. Furthermore, 6PP along with other 2-pyrones with its structural analogous was allowed for the direct hydrogenatiopn using alkali metal hydride. Up to 73% of corresponding hydric alcohol from 6PP was obtained when the reaction was performed in dry tetrhydrofuran under nitrogen atmosphere.

Figure 1. a) Reaction mechanism for the production of C15 fuel additives from 6PP, b) reaction energy diagram for the ring-opening of 6pp in water and c) acid catalysed system of γ-alumina.

# 504024 Advanced Membranes for PM2.5 Filtering, Ethanol Dehydration, and Air Dehumidification

**Abstract:**

Thermodynamically, the membrane process is the most efficient way to separate mixtures. It is evident that almost all the filtering and separation in living systems are conducted via a membrane. The membrane technologies are an attractive approach to addressing energy efficiency and environmental emission problems that are of significant concerns to the society. In this presentation, removal of PM2.5 particulates from air and separation of water molecules from a liquid of gas mixture are used as application examples in the environment and energy efficiency areas, respectively. The current technologies and recent development will be critically reviewed and compared for each of these technology subjects. Novel filtration membranes will be introduced for complete (>99%) removal of PM2.5 particulates in a very efficient way. Novel molecular sieve membranes will be shown for dehydration of ethanol/water with 98% more efficient than conventional distillation and adsorption, and for dehumidification of atmospheric air with more than 50% efficiency enhancement compared to conventional air conditioners.

# 503949 Large-Scale Removal of Atmospheric $CO_2$Â through Biomass Chemical Looping Combustion

**Abstract:**

Global climate change is understood to result largely from $CO_2$ emissions from combustion of fossil fuels. Several types of technologies are under development to minimize emissions of carbon dioxide, including post-combustion $CO_2$ scrubbing, oxy-fuel combustion, gasification with $CO_2$ capture and chemical looping combustion. Several studies indicate that chemical looping combustion has the lowest energy penalty and impact on cost of electricity. Consequently, interest in chemical looping combustion is increasing and today many groups around the world are researching the technology at lab and pilot scale. Most of this research focuses on coal or natural gas as fuels, since capturing fossil $CO_2$ for sequestration has the potential to make these fuels $CO_2$ neutral. However, if biomass is used as a fuel, carbon capture technologies make removal of $CO_2$ from the atmosphere possible, i.e. the process becomes carbon negative. This presentation discusses the development of chemical looping combustion technology and the potential for large-scale removal of $CO_2$ from the atmosphere by using biomass as a â€œ$CO_2$ filterâ€ and sequestering the concentrated $CO_2$ product resulting from chemical looping combustion.

# 497376 Greener Processing for the Conversion of Waste Plastics into Energy Storing Carbons

**Abstract:**

Environmentally *harmful plastic waste* is a major concern throughout the world. The developed novel autogenic process completely destroys unsorted plastic waste in a technologically beneficial, environmentally responsible manner before the plastic enters the waste stream. The one-step, low-energy, solvent-less process is called â€œupcyclingâ€ because it produces products having *greater value* than the original plastics, particularly structurally tunable carbon microspheres and carbon sheets that have important tribological and battery electrode applications

Rechargeable Li-ion batteries are the most promising energy storing sources in current generation of portable electronics, medical devices and electric vehicles. Developed upcycled sustainable carbon spheres and sheets are promising anode electrode materials for Li- and Na-ion batteries. This 2015 R and D 100 award winning technology produced hard carbon spheres comprised of disordered graphitic planes, that yields equivalent electrochemical capacity at higher potentials (vs. metallic lithium) thereby offering enhanced safety.

# 504135 Green Process Engineering: Fundamental Researches and Applications of Ionic Liquids

**Abstract:**

Sustainability is now showing up as one of the top challenges for the survival of human being that catches the attention of governments, industry, and academy. Green process engineering (GPE) is a useful tool and important research field that could make significant contributions in making wasteful and hazardous processes more green and sustainable for the benefit of the environment, economy, and society. Ionic liquids (ILs), as a kind of novel media for green chemistry and chemical engineering, have been studied extensively as green solvents, catalysts, and electrolytes due to their unique properties and high potentials. This talk will present recent case studies on the structure-property relationship, the formation and regulation rules of ionic clusters, mechanism of synergistic catalytic effect, as well as the methodology of IL-based process design and a couple of innovative green process technologies based on ILs. The talk concludes with open challenges in the fundamental researches and applications of ionic liquids.

# 504136 Unlocking Biomass Energy: Process Development and Scale-up of Biomass Conversion to Advanced Fuels and Chemicals

**Abstract:**

The sustainable production of both biofuels and bioproducts depends largely on the efficient and cost effective access to biomass conversion technologies. Due to the physical and chemical complexities of plant based biomass, the conversion process typically includes thermochemical pretreatment, enzymatic saccharification and microbial fermentation. This presentation will highlight the process development we have demonstrated in converting different types of biomass to fermentable sugars and fuel intermediates through various approaches. The scaling up challenges associated with some case studies will be addressed. An overview of the novel technologies, markets and substrate accessibility via efficient biomass utilization will be discussed. The research demonstrates how biomass conversion and upgrading via bioprocessing can influence key pathways to address global environmental sustainability barriers. Both technical and economic challenges of the conversion process will be discussed to provide some perspectives for future process development and optimization.

# 504137 Completing the New Biofuel and Bio-Derived Chemicals and Materials Research Challenge

**Abstract:**

The need to convert biomass into fuels and chemicals was one of mankind's earliest drivers for chemical and biochemical research. Indeed, at the turn of the century many industrial materials including inks, dyes, resins, paints, and synthetic fibers were made from bioresources. Yet, by the late 1960's most of these biobased routes had been displaced by petroleum processes. The energy crisis of the 1970's renewed interest and the need to develop new biorefining process technologies that were not depended upon petroleum reserves. Although these interests waned, the subsequent increased energy costs of the early 2000's and environmental concerns re-ignited the need for green sustainable engineering process to convert biomass to biofuels, biopower and bio-derived chemicals and materials. The recent petroleum glut has again challenged biorefining technologies but future population growth, environmental concerns and increased global GDP will only add focus on renewable resources as many conventional nonrenewable resources will eventually reach their limits of environmentally viable extraction. Along with the advances in genetics, biotechnology, modelling, green process engineering and chemical catalysts one of the most important developments of Green Chemical Process Engineering is a renewed interest by students, researchers, entrepreneurs, NGOs, and societal expectations that modern society minimize its impact on the environment while still providing value-added products. To address these needs, our group and others have focused on integrating many of these differing technical fields to develop the next generation of biorefining engineering processes. This presentation will review how the natural variation in plant resources and/or genetic manipulation of the plant cell wall can yield tailored bioresources that can be efficiently converted from cellulose, hemicellulose and lignin to biofuels, bio-based chemicals and materials.

# 504321 Methanol to Propylene Process: An Integration of Catalyst Development and Reactor Engineering

**Abstract:**

Methanol, which can be produced from coal or natural gas, is one of the most important C1 chemicals. The utilization of methanol to make light olefins including ethylene and propylene has been recently commercialized. However, in many applications, it is desired to maximize the yield of propylene, and thus methanol to propylene (MTP) process becomes increasingly interesting. It is normally hard to achieve high selectivity to propylene in methanol conversion. In this work, we have developed a multi-functional catalyst, over which three reactions, i.e. methanol conversion, ethylene alkylation, and C4+ cracking, can be catalyzed. By integrating these three reactions, we attempted to develop a MTP process with the selectivity to propylene higher than 75%. Evaluation of this catalyst has been carried out in laboratory-scale micro-fluidized bed reactor. It shows that the selectivity to propylene of 50% has been achieved in the methanol conversion reaction. We also studied the ethylene alkylation reaction and C4+ cracking reaction, and identified the optimal operation conditions. Based on the results obtained in the micro-fluidized bed reactor, we proposed an integrated process to maximize the selectivity of propylene. In achieving this goal, a turbulent fluidized bed reactor, in which the optimal operation conditions suitable for both methanol conversion and ethylene alkylation have been considered, has been designed and constructed in our pilot plant. The reactor can be operated with 300t/a methanol feed rate. Pilot plant experiments have been conducted for validating the process concept and scaling up the reactor. It shows the selectivity to propylene of 75%, in addition to a selectivity of ethylene of 10%, can be achieved in the pilot plant. This MTP process can be seen as a successful example of the integration of catalyst development and reactor engineering.

# 504320 Direct Production of Value-Added Chemicals Via Fischer-Tropsch Synthesis


**Abstract:**

Fischer-Tropsch synthesis (FTS) is a heterogeneous catalytic process for sustainable production of fuel and value-added chemicals via syngas derived from coal, biomass and natural gas. It is considered as an attractive alternative non-petroleum-based production route. FTS is a very complicated catalytic reaction process with co-production of paraffin, olefin, alcohol and other products. The direct synthesis of high value-added olefins and higher alcohols from syngas has drawn particular attention due to due to its process simplicity, low energy consumption and clean utilization of carbon resource. For the purpose of high selectivity to specific olefins or higher alcohols, it is very desirable to regulate the reaction network for certain reactions to take place preferentially. Obviously, it is necessary to control CO activation and then both carbon chain growth and termination for high selectivity to target product by development new catalyst systems and optimization of reaction conditions.

Reaction network during syngas conversion was firstly investigated by probe molecules techniques [1]. Alcohol, aldehyde, olefin as probe molecules were in-situ introduced into the reaction system. Several parallel reactions and second reactions were clarified. The existence of surface acyl species indicated that the insertion of CO rather than hydroxyl was the underlying mechanism for higher alcohols synthesis over modified FT catalysts along with the support from surface simulation. In addition, the dual active sites for higher alcohols formation was further investigated [2]. Several kinds of dual active sites ($Cu/FeC_x$, CuCo, and $Co/Co_2C$) were controllably prepared for higher alcohol synthesis. Both $Cu/FeC_x$ and $Co/Co_2C$ could afford high selectivity for long chain alcohols, while only short chain alcohols were produced for CuCo. The difference of the detailed surface structure and the corresponding catalytic performance was discussed in detail.

Fischer-Tropsch to olefins (FTO) by using new active phase was also investigated. $Co_2C$ nanoprisms with exposed facet of {101} and {020} were prepared and tested for syngas conversion [3]. It is widely accepted the formation of $Co_2C$ is considered to be one of the main reasons for deactivation of the Co-based FT reaction because it possesses very low activity for CO hydrogenation, with methane and $CO_2$ as the main products. However, due to the strong facet effect of $Co_2C$, $Co_2C$ nanoprisms afforded very high selectivity for the production of lower olefins (60.8 C%), while little methane (about 5.0 C%) was generated with the ratio of olefin/paraffin amongst the $C_{2-4}$ products being as high 30. The product distribution deviates markedly from the classical ASF distribution with the highest selectivity to propylene. The effect of several promoters and the model for modified ASF distribution were discussed [4]. In addition, the nanoeffects of $Co_2C$ were studied in details to reveal the structure-performance relationship.


References


[1] *Catalysis Science & Technology*ï¼Œ2015, 5, 4224

[2] *Catalysis Science & Technology*ï¼Œ2013, 3, 1591

[3] *Nature*, 2016, 538, 84..

[4] *ACS Catalysis*, 2017, 7, 3622.

# 504319 An Investigation into the Effect of Li and Mn Promotions on the Activity and Selecivities to Olefins and Alcohols of Co@$Co_2C$/Activated Carbon (AC) for Fischer-Tropsch Reaction


**Abstract:**

Abstract:

Utilization of non-precious transition metals for high alcohols and olefins synthesis is of a great importance in heterogeneous catalysis. We synthesized successfully Li and Mn promoted AC supported cobalt metal-carbide (Co@$Co_2C$/AC) catalysts, which present remarkable activity of more than 200g/kg-cat.h and selectivities of ca. 65 C% both for alpha-alcohols and alpha-olefins via the Fischer-Tropsch reaction at the mild conditions. The formation of the stable cobalt carbide and the Co@$Co_2C$ interface are found essential for the observed reactivity. The results of characterization from In-situ EXAFS, in-situ XRD, HRTEM, XSP, and chemo-adsorption of $H_2$ and CO experiments showed that Mn promotes the formation of $Co_2C$ species during Fischer-Tropsch reaction and Li increased the CO uptake, while both promoters have little impact on the amount of $H_2$ uptakes. Density functional theory calculations show that $Co_2C$ is highly efficient for CO non-dissociative adsorption, behaving as noble-metal-like, whereas the Co metal is highly active for CO dissociative adsorption and the subsequent carbon-chain growth. The interface between the cobalt metal and its carbide phase as well as the dual sites available at the interface for facile CO insertion to $C_nH_m$-groups formed on the Co sites. The remarkably high alpha-alcohol and alpha-olefins selectivities from the Li and Mn promotion effects, namely the formation of more interfacial active sites between Co and $Co_2C$ and enhancement of the ratio for chemo-adsorption CO to $H_2$ uptake on the surface of Co@$Co_2C$/AC catalysts, paves a new way for the design of the new catalysts for synthesizing various high value-products in syngas applications.

# 504154 Direct Production of Gasoline-Range Hydrocarbons from Carbon Dioxide over Iron-Based Multifunctional Catalysts

**Abstract:**

Converting $CO_2$ from a detrimental greenhouse gas into value-added liquid fuels not only contributes to mitigating $CO_2$ emissions, but also reduces dependence on petrochemicals. Unfortunately, the activation of $CO_2$ and its hydrogenation to hydrocarbons or other alcohols are challenging tasks because $CO_2$ is a fully oxidized, thermodynamically stable and chemically inert molecule. Most research to date, not surprisingly, is focusing on selective hydrogenation of $CO_2$ to short-chain products, while few studies on long-chain hydrocarbons, such as gasoline-range ($C_5$–$C_{11}$) hydrocarbons. The key to this process is to search for a high efficient catalyst.

Herein, we have succeeded in preparing high efficient, stable, and multifunctional Na–$Fe_3O_4$/Zeolite catalysts for the direct production of gasoline-range ($C_5$–$C_{11}$) hydrocarbons from $CO_2$ hydrogenation. This catalyst displayed record selectivity towards $C_5$–$C_{11}$ hydrocarbons (78%) as well as low $CH_4$ and CO selectivity under industrial relevant conditions, and gasoline fraction are mainly isoparaffins and aromatics thus favouring the octane number. Moreover, the composition of $C_5$–$C_{11}$ can be tuned by the choice of zeolite type and the integration manner of multifunctional catalyst. When Na–$Fe_3O_4$ nanocatalyst was matched with HZSM-5 zeolite and integrated by granule-mixing manner, up to 63% of aromatics in gasoline fraction were produced. While Na–$Fe_3O_4$ was matched with HMCM-22 zeolite, up to 60% of isoparaffins in gasoline fraction were obtained under the dual-bed configuration. Not only that, the multifunctional catalyst exhibited a remarkable stability for 1,000 h on stream, which definitely has the potential to be a promising industrial catalyst for $CO_2$ utilization to liquid fuels.

In-depth characterizations indicate that, during $CO_2$ hydrogenation reaction, this catalyst enables RWGS over $Fe_3O_4$ sites, olefin synthesis over $Fe_5C_2$ sites, and oligomerization/aromatization/isomerization over zeolite acid sites. The concerted action of the active sites calls for precise control of their structures and proximity. This study paves a new path for the synthesis of liquid fuels by utilizing $CO_2$ and $H_2$. Furthermore, it provides an important approach for dealing with the intermittency of renewable sources (sun, wind and so on) by storing energy in liquid fuels.

**References**

1. G. A. Olah, A. Goeppert and G. S. Prakash, *J. Org. Chem.*, 2008, **74**, 487-498.

2. R. W. Dorner, D. R. Hardy, et al., *Energy Environ. Sci.*, 2010, **3**, 884-890.

3. W. Wang, S. Wang, X. Ma and J. Gong, *Chem. Soc. Rev.*, 2011, **40**, 3703-3727.

4. F. Jiao, J. Li, X. Pan, et al., *Science*, 2016, **351**, 1065-1068.

5. P. Zhai, C. Xu, R. Gao, et al., *Angew. Chem. Int. Ed.*, 2016, **55**, 9902-9907.

6. J. Wei, Q. Ge, R.Yao, et al., *Nat. Commun.*, 2017, **8**, 15174.

# Innovative Syngas Production Catalyst for Utilization of $CO_2$ (CT-$CO_2$AR™)

*Takenori Kanda, Osamu Hirohata, Tomoyuki Mikuriya and Fuyuki Yagi*
*Chiyoda Corporation, Kanagawa, Japan*

## 1. Introduction

Production of synthesis gas from natural gas and conversion of synthesis gas to fuels and chemicals have become increasingly attractive from the viewpoint of energy resource diversification. The technology of synthesis gas production by the reforming of natural gas with $CO_2$ (hereafter referred to as "$CO_2$-reforming") is an effective solution for the issues of efficient energy resource utilization and environmental preservation.
Chiyoda has developed and commercialized a new noble metal based reforming catalyst named CT-$CO_2$AR™ which has a higher resistance to carbon formation and produces synthesis gas more effectively compared with commercial reforming catalyst.
In this paper, we will elaborate on the characteristics and the competitiveness of the CT-$CO_2$AR™ catalyst and the results of the operation of the commercial plant.

## 2. Characteristics of CT-$CO_2$AR™ catalyst

(1) The characteristics of $CO_2$/steam reforming

Synthesis gas is used for hydrogen production and as the feed stock of chemicals, and the $H_2/CO$ ratio of synthesis gas depends on the purpose.
For example, hydrogen production requires an $H_2/CO$ ratio of more than 3.0. For ethylene glycol production, the required $H_2/CO$ ratio is 2.0. For chemical production through oxo synthesis, the required $H_2/CO$ ratio is 1.0. For the feedstock of acetic acid or MMA requires only CO or as low as possible $H_2/CO$ ratio.
When hydrocarbons such as natural gas or naphtha and steam are used as reformer feedstocks, the steam reforming reaction shown in Eq. 1 is normal and the $H_2/CO$ ratio of produced synthesis gas becomes more than 3.0.

$$CH_4 + H_2O \Leftrightarrow 3\,H_2 + CO \qquad (1)$$

However, when $CO_2$ and steam are used as reforming agents, the $CO_2$ reforming reaction shown in Eq. 2 occurs.

$$CH_4 + CO_2 \Leftrightarrow 2\,H_2 + 2\,CO \qquad (2)$$

Therefore, by changing the ratio of feed $CO_2$ and steam, various $H_2/CO$ ratios of synthesis gas can be produced.

However, during the reforming reactions, the carbon formation reaction shown in Eq. 3 occurs on the catalyst surface. Deposited carbon causes the deactivation of the catalyst and the occlusion of the catalyst tube[1] leading to an unstable operation.

$$CH_4 \Leftrightarrow C\,(s) + 2\,H_2 \qquad (3)$$

To prevent carbon formation for a stable operation, commercial Ni based reforming catalysts require steam and $CO_2$ greater than the stoichiometric ratio.

Fig. 1 shows the relation between the operation condition and risk of carbon formation[2]. The bold line in the center represents carbon limit line at 600 ℃. In the black area to the left of this line, the process gas has enough high thermodynamic potential for carbon formation at 600 ℃. Operation around this line cannot be carried out with commercial Ni based reforming catalysts. They are usually operated within the excess steam carbon ratio (S/C) zone (blue area).

On the other hand, our new commercialized noble metal based reforming catalyst (CT-$CO_2$AR™) enables a stable operation under a lower S/C condition (red area) because it has a higher resistance to carbon formation.

In next section, the advantages of a lower S/C operation will be shown.

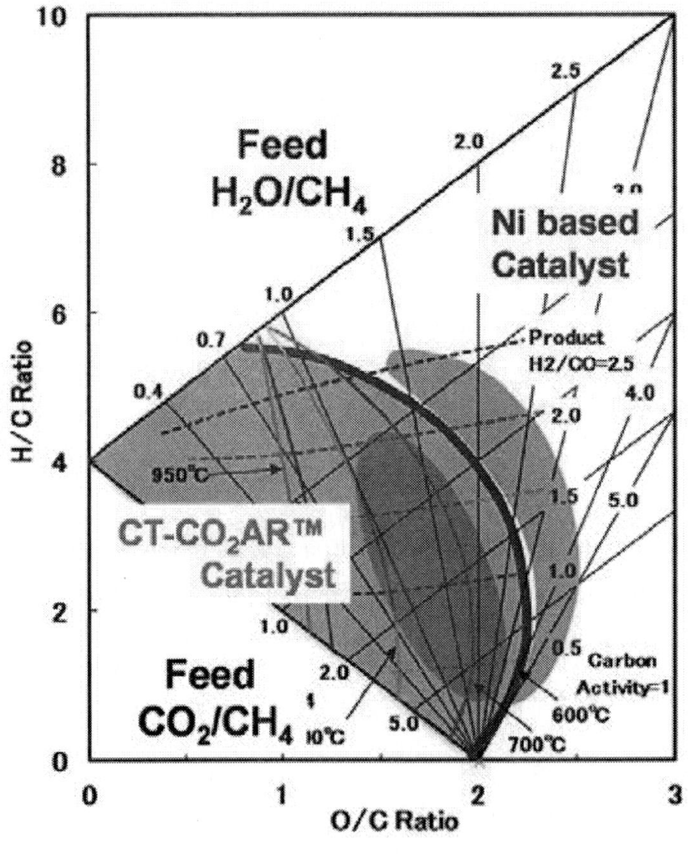


Fig. 1. Carbon Limit Diagram

(2) The advantages of a lower S/C ratio operation

In the case of the synthesis gas production process shown in Fig. 2, the advantages of a lower S/C operation are as follows:

1. Since the amount of steam and $CO_2$ is reduced, the heat duty to heat steam and $CO_2$ can be reduced. Therefore, the fuel consumption of the reformer can be reduced.
2. Fig. 3 shows the comparison of synthesis gas concentration at the reformer outlet between a high S/C (=2.2) operation and a low S/C (=1.0) operation in the case of $H_2/CO=1.0$
   In the case of a low S/C operation, the amount of $CO_2$ can be reduced and as a result, the system volume of $CO_2$ removal is downsized.
   In addition, because of lower $CO_2/C$, the capacity of the $CO_2$ compressor is downsized.
   Therefore, both the CAPEX and the OPEX of $CO_2$ removal and $CO_2$ compressor are reduced.
3. Fig. 3 shows that a lower S/C operation can reduce the amount of $H_2O$ at the reformer outlet and it is possible to reduce the amount of process condensate.
   In the case of a low S/C operation, the total amount of produced steam is less than that of a high S/C operation. However, the amount of steam uses as the reforming agent in a low S/C operation is significantly less than that in a high S/C operation.
   As a result, a low S/C operation can increase the amount of export steam.

From these reasons, the CT-$CO_2$AR™ catalyst is preferable to a commercial Ni based catalyst.

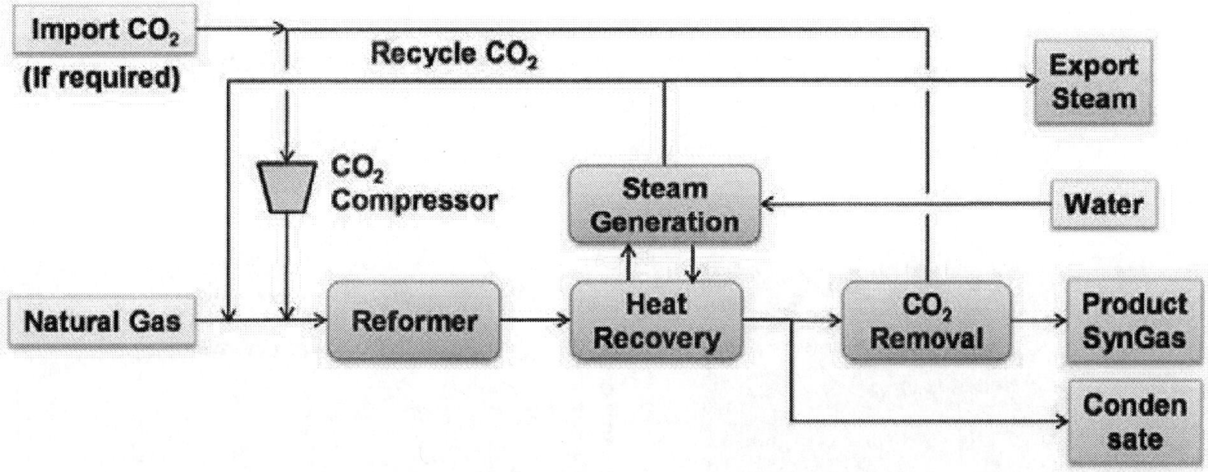


Fig. 2. Block Flow Diagram of Syngas production

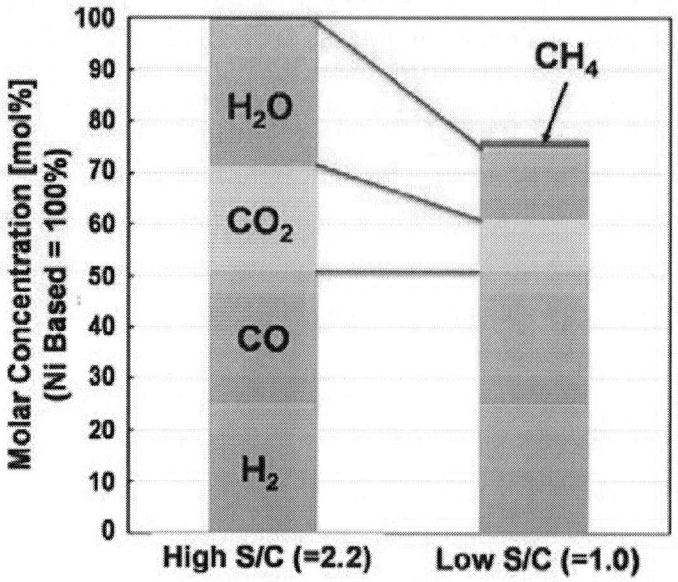


Fig. 3. Typical Example of synthesis gas composition in $H_2/CO=1.0$

(3) The characteristics of the CT-$CO_2$AR™ catalyst

Fig. 4 shows the result of EPMA (Electron Probe Micro Analyzer) of the CT-$CO_2$AR™ catalyst. Generally, the surface reaction on the reforming catalyst is the rate-determining step. Therefore, to reduce the amount of active metal, the active metal is supported on only the outer surface of the carrier like egg-shell.

Fig. 5 shows the accelerated experiment result. The CT-$CO_2$AR™ catalyst keeps its activity due to its high resistance to carbon formation.

We built the reaction kinetics model shown in Fig. 6 to estimate the behavior of the CT-$CO_2$AR™ catalyst.

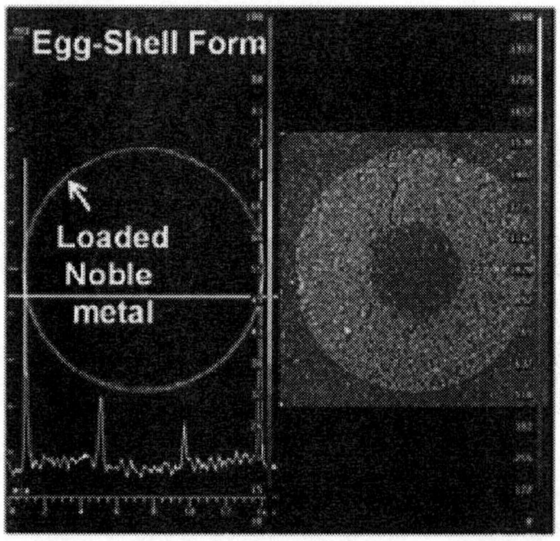


Fig. 4. EPMA result of CT-$CO_2$AR™ Catalyst

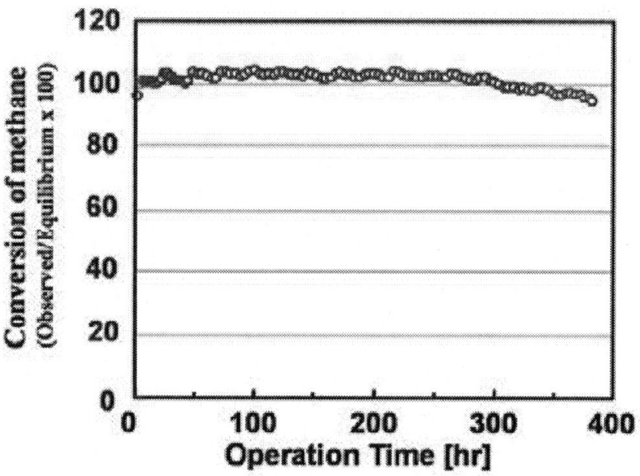


Fig. 5. Result of accelerated experiment

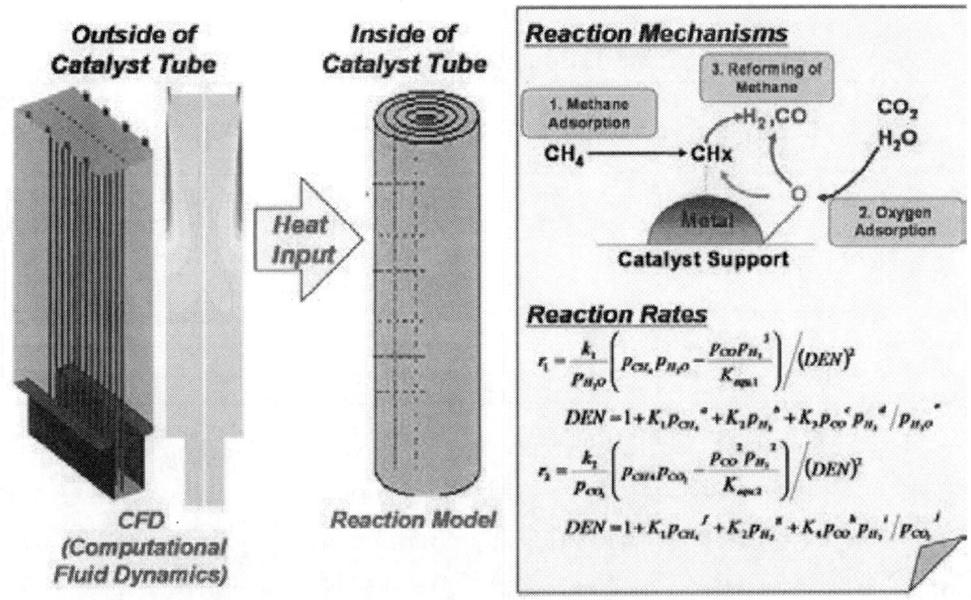


Fig. 6. Reaction model of CT-$CO_2$AR™ catalyst

## 3. Advantages of CT-$CO_2$AR™ catalyst

(1) Study Basis

Regarding the competitiveness of the CT-$CO_2$AR™ catalyst compared to a commercial Ni based reforming catalyst, we have compared the CT-$CO_2$AR™ catalyst and Ni based catalysts from the following three points of view based on the block flow shown in Fig. 2.

1. Total energy input (feed + fuel hydrocarbon + utility consumption)
2. Manufacturing costs of synthesis gas
3. $CO_2$ Emission per final products

[$CO_2$ Emission] = [$CO_2$ in Flue and Utility] - [Import $CO_2$] + [$CO_2$ in downstream process to produce the final products]

In this study we considered cases with $H_2$/CO ratios of 1.0 and 0.5. The conditions and estimated final product of each case are listed in Table 1.

(2) Study Result

From Fig.7-9 show the results of the process study. In the case of $H_2$/CO=1.0, compared to Ni based catalysts, the CT-$CO_2$AR™ catalyst can cut total energy input by 11% (Fig. 7) manufacturing cost by 12% (Fig. 8).Regarding $CO_2$ emission, the CT-$CO_2$AR™ catalyst can achieve a reduction of 22% compared with a Ni based catalyst (Fig. 9). In the case of $H_2$/CO=0.5, CT-$CO_2$AR™ catalyst is more competitive. Approximately 30% of both total energy input and manufacturing cost can be reduced and 76% of the amount of $CO_2$ emission can be reduced. From this study, the CT-$CO_2$AR™ catalyst is competitive compared with Ni based catalysts both in terms of CAPEX and OPEX.

Table 1. Process study conditions

| $H_2/CO$ Ratio | Catalyst | Feed S/C | Amount of ($H_2$+CO) | Final Products | Capacity |
|---|---|---|---|---|---|
| [-] | [-] | [-] | [Nm3/h] | [-] | [ton/hr] |
| CO (0.5) | Ni Based | 3.3 | 37,500 (25,000 as CO) | Acetic Acid | 500,000 |
| | CT-$CO_2$AR™ | 0.4 | | | |
| 1.0 | Ni Based | 2.2 | 25,000 | n-butanol (oxo synthesis) | 270,000 |
| | CT-$CO_2$AR™ | 1.0 | | | |

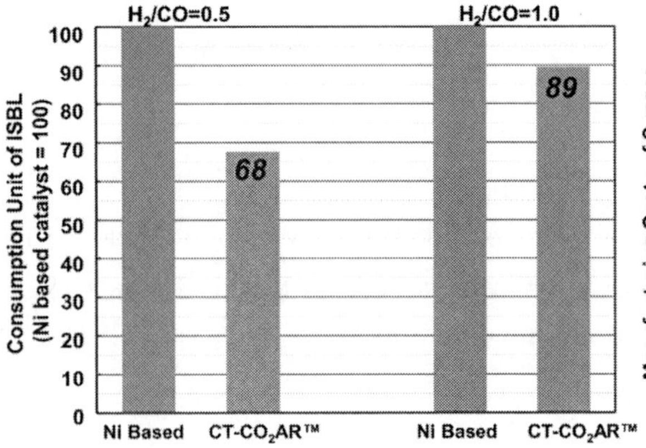


Fig. 7. Total Energy Input

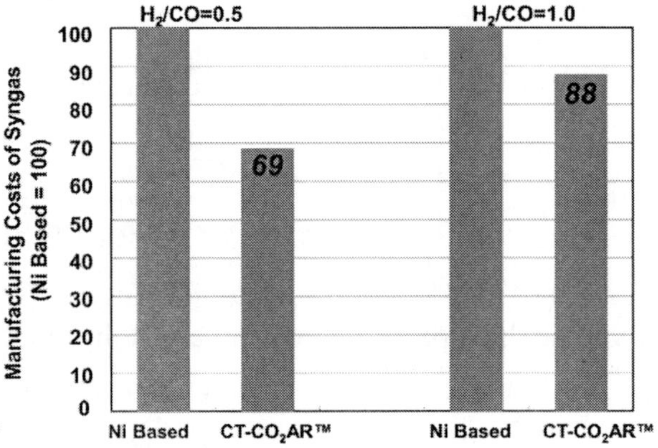


Fig. 8. Manufacturing Costs of Syngas

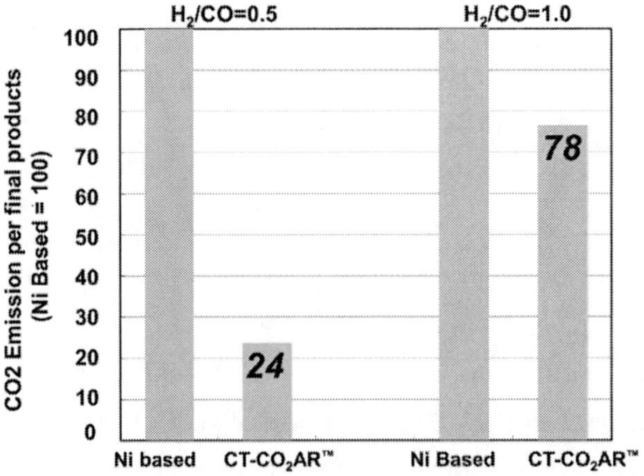


Fig. 9. $CO_2$ Emission per final products

## 4. Application to Commercial Plant

The CT-$CO_2$AR™ catalyst has been operated in a commercial plant in Japan since 2014. The information of the commercial plant is listed in Table 2. In this project, the reforming catalyst has been changed to the CT-$CO_2$AR™ catalyst and the S/C ratio at the inlet of the prereformer has reduced from 2.2 to 1.4 without changing the amount and specification of the produced synthesis gas.

Regarding the comparison between analysis result using the reaction model and the operation result, a pressure drop through the reformer tube is shown in Fig. 10 and the methane concentration of the synthesis gas is shown in Fig.11.

Since a significant increase in the pressure drop and methane concentration was not observed, a stable operation was achieved during the whole operation time.

In addition, our reaction model can accurately predict the performance of the CT-$CO_2$AR™ catalyst.

Table 2. Information about the commercial plant

| | |
|---|---|
| Client's Name : | KH Neochem Co., Ltd |
| Plant Location : | Yokkaichi, Japan |
| Product : | Synthesis gas $H_2$/CO ratio=1.0 |
| Product capacity : | 5,000 $Nm^3$/hr (as $H_2$+CO) |
| Project Scope : | Catalyst supply to the existing plant |
| Feed Hydrocarbon : | Natural gas or Naphtha |

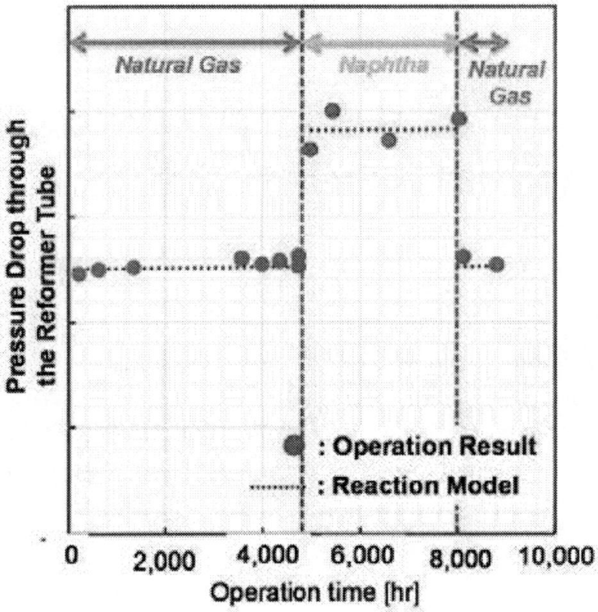


Fig. 10. Pressure difference of a reformer tube

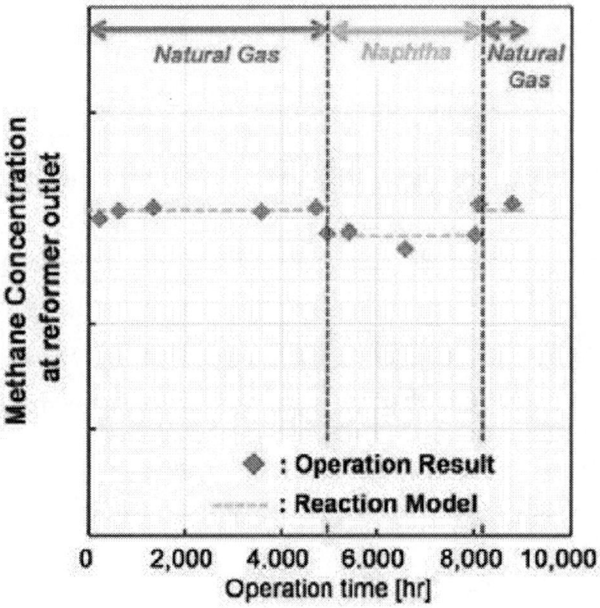


Fig. 11. Methane concentration at reformer outlet

Fig. 12 shows the comparison of the reformer duty and the amount of input natural gas between before and after replacing the CT-$CO_2$AR™ catalyst. By replacing the CT-$CO_2$AR™ catalyst, 10% of the natural gas input is reduced and this result matches the result of process study shown in section 3 (Advantages of the CT-$CO_2$AR™ catalyst).

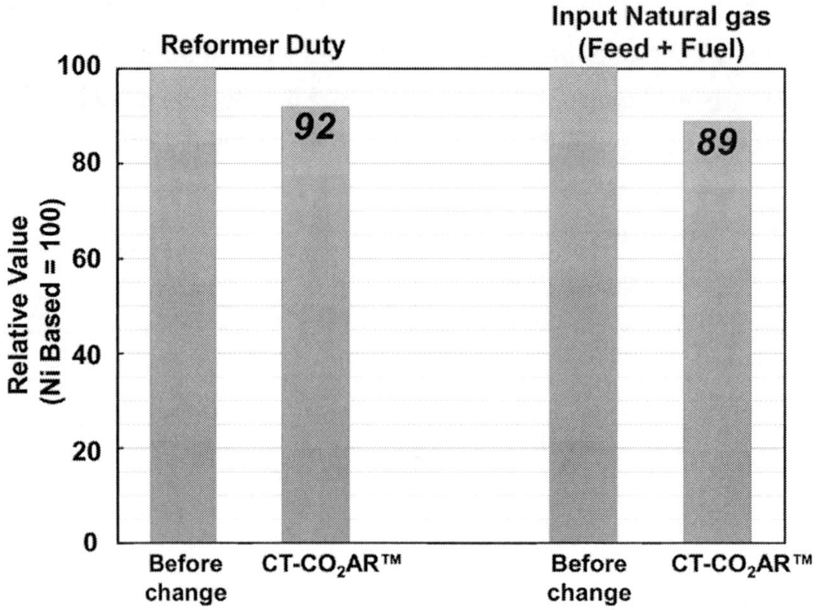


Fig. 12. Comparison result of commercial plant operation

## 5. Conclusion

The CT-$CO_2$AR™ catalyst enables a lower S/C condition operation because of its higher resistance to carbon formation.

Compared with a commercial catalyst, the CT-$CO_2$AR™ catalyst demonstrates high competitiveness in both manufacturing cost and $CO_2$ emission in the case of $H_2$/CO ratio = 0.5 and 1.0, especially, 0.5.

The CT-$CO_2$AR™ catalyst was installed in a commercial plant in 2014 and has been operating stably. 10% reduction of natural gas input was achieved in the commercial operation.

## 6. References


1) J.R. Rostrup-Nielsen, J. H. B. Hansen, *J. Catal.*, **144**, 38 (1993)

2) N.R.Udengaard, J.-H. Bak Hansen and D.C.Hanson, Oil & Gas J., Mar.9(1992)

# Solar Thermochemical Splitting of $H_2O$ and $CO_2$ using Nonvolatile Metal Oxides

**(Principal Author)**
**Rahul Bhosale**
Department of Chemical Engineering
College of Engineering
Qatar University, Doha, Qatar
rahul.bhosale@qu.edu.qa

**(Other Authors)**
**Gorakshnath Takalkar**
Department of Chemical Engineering
College of Engineering
Qatar University, Doha, Qatar

**Anand Kumar**
Department of Chemical Engineering
College of Engineering
Qatar University, Doha, Qatar

**Fares A. AlMomani**
Department of Chemical Engineering
College of Engineering
Qatar University, Doha, Qatar

Prepared for Presentation at
American Institute of Chemical Engineers
2017 Fall Meeting
Minneapolis, MN, USA
October 29 – November 3, 2017

# Solar Thermochemical Splitting of $H_2O$ and $CO_2$ using Nonvolatile Metal Oxides

**Rahul R. Bhosale, Anand Kumar, Gorakshnath Takalkar, Fares A. AlMomani**

Department of Chemical Engineering
College of Engineering
Qatar University, Doha, Qatar

**Keywords:** Thermochemical Reactions, Metal Oxides, Solar Fuels, $H_2O/CO_2$ Splitting

## Abstract

In this study, a Co ferrite/$ZrO_2$ ceramic nanomixture was synthesized by the propylene-oxide-assisted sol–gel method. Cobalt, iron, and zirconium nitrate salts were used as the metal precursors, and ethanol and propylene oxide were used as the solvent and gelation agent, respectively. The prepared gels were aged, dried, and annealed at 1000°C for different dwell times. The annealed Co ferrite/$ZrO_2$ powder was analyzed by powder X-ray diffraction (PXRD), Brunauer–Emmett–Teller surface area analysis (BET), and energy-dispersive spectroscopy (EDS),. The characterization results are reported in detail.

## 1. Introduction

Metal-oxide-based solar thermochemical water and carbon dioxide splitting is considered a promising technology to satisfy future energy demands [1]. Zinc oxide [2-5], iron oxide [5-6], tin oxide [7-8], doped iron oxides (ferrites) [9-14], ceria-based redox materials [15-19], and perovskites [20-24] are used for splitting. Among these oxides, ferrites have been used extensively for solar thermochemical water splitting; however, their use is extremely limited for $CO_2$ splitting.

However, the major issue limiting the use of ferrites in solar thermochemical cycles is the increased size of the ferrite grains obtained as a result of high-temperature processing. In this study, a Co ferrite/$ZrO_2$ ceramic mixture was synthesized by the propylene-oxide-assisted sol–gel method [25-27], exhibiting mixing of Co ferrite and $ZrO_2$ with a uniform distribution. The effect of annealing time on the phase and chemical composition, crystallite size, specific surface area (SSA), pore volume, and morphology of the Co ferrite/$ZrO_2$ ceramic nanomixture was investigated, and the results were reported in detail.

## 2. Experimental

### *2.1 Sol–Gel Synthesis*

Metal precursors of cobalt, iron, and zirconium were first mixed in a specific weight ratio and dissolved in ethanol using a sonication bath. Once a visibly clear solution was obtained, a predetermined amount of propylene oxide was slowly added, affording a gel. The gel was aged, dried, and annealed at 1000°C (at various annealing times) in air. Fig. 1 shows the flow chart for the synthesis of the Co ferrite/$ZrO_2$ ceramic nanomixture.

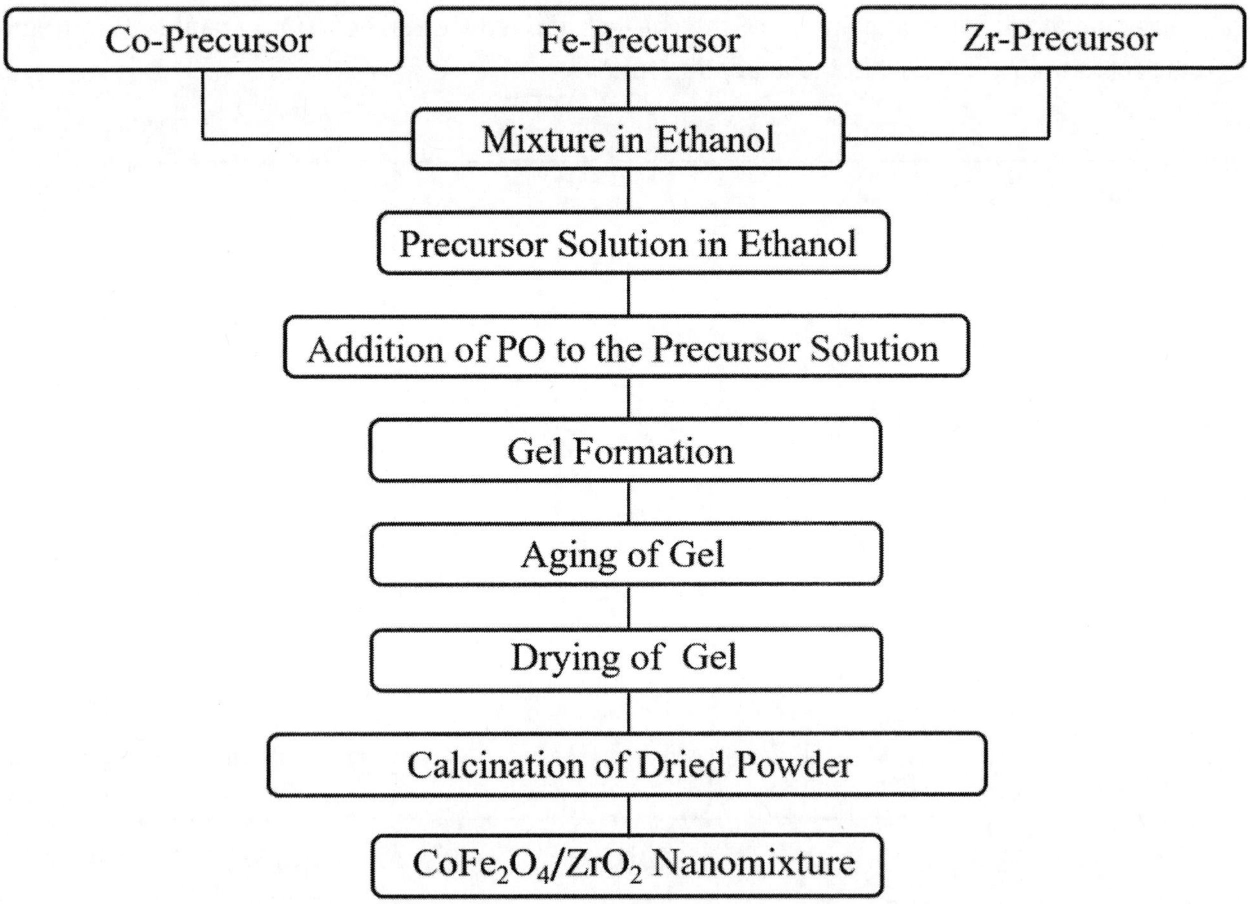


Fig. 1. Flow chart for the synthesis of the Co ferrite/$ZrO_2$ ceramic nanomixture by the sol–gel method.

### *2.2 Characterization*

To investigate the effects of annealing on the physiochemical properties of the Co ferrite/$ZrO_2$ ceramic nanomixture, the Co ferrite/$ZrO_2$ powder obtained after annealing was characterized by powder X-ray diffraction (PXRD), Brunauer–Emmett–Teller (BET) surface area analysis, scanning electron microscopy (SEM) with a field-emission gun, X-ray energy-dispersive spectroscopy (EDS), and high-resolution transmission electron microscopy (HR-TEM).

## 3. Results & Discussion

The effect of annealing time on the physicochemical properties of the sol–gel-derived Co ferrite/$ZrO_2$ ceramic nanomixture was investigated. For this purpose, the Co ferrite/$ZrO_2$ gels (prepared using 20 mL of propylene oxide, aged for 24 h, and dried at 100°C) were annealed at 1000°C for 1–5 h. The calcined powders thus obtained were characterized by PXRD and EDS to determine the phase and chemical composition. The results shown Fig. 2 clearly indicated that the phase and chemical composition of the sol–gel-derived Co ferrite/$ZrO_2$ ceramic nanomixture is unaffected by the increase in the annealing time.

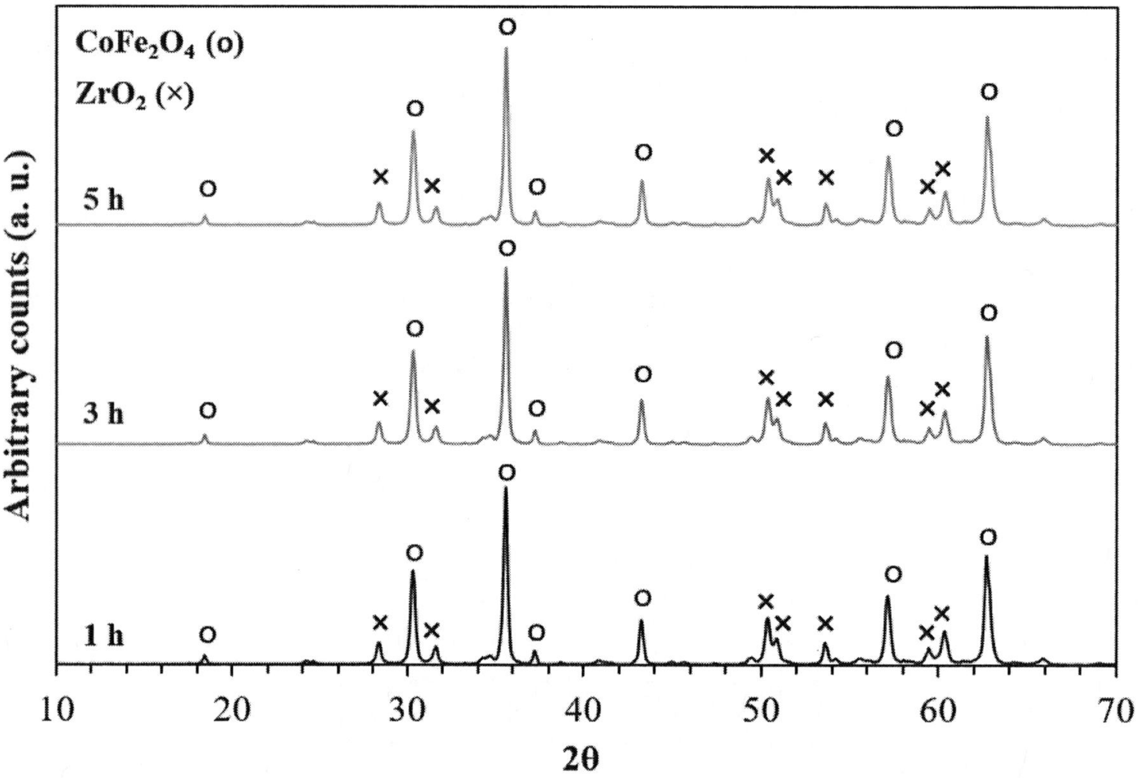


Fig. 2 – Effect of the annealing time on the phase composition of sol–gel-derived Co ferrite/$ZrO_2$ ceramic nanomixture.

The effect of the annealing time on the crystallite size of the sol–gel-derived Co ferrite/$ZrO_2$ ceramic nanomixture was also determined. Fig. 3 shows the results obtained. With the increase in the annealing time from 1 to 3 h, the crystallite size increases by 14.86 nm. With the further increase in the annealing time from 3 to 5 h, the crystallite size further increases by 22.83 nm. In summary, with the increase in the annealing time from 1 to 5 h, the size of the Co ferrite/$ZrO_2$ crystallites increases from 53.66 to 91.35 nm.

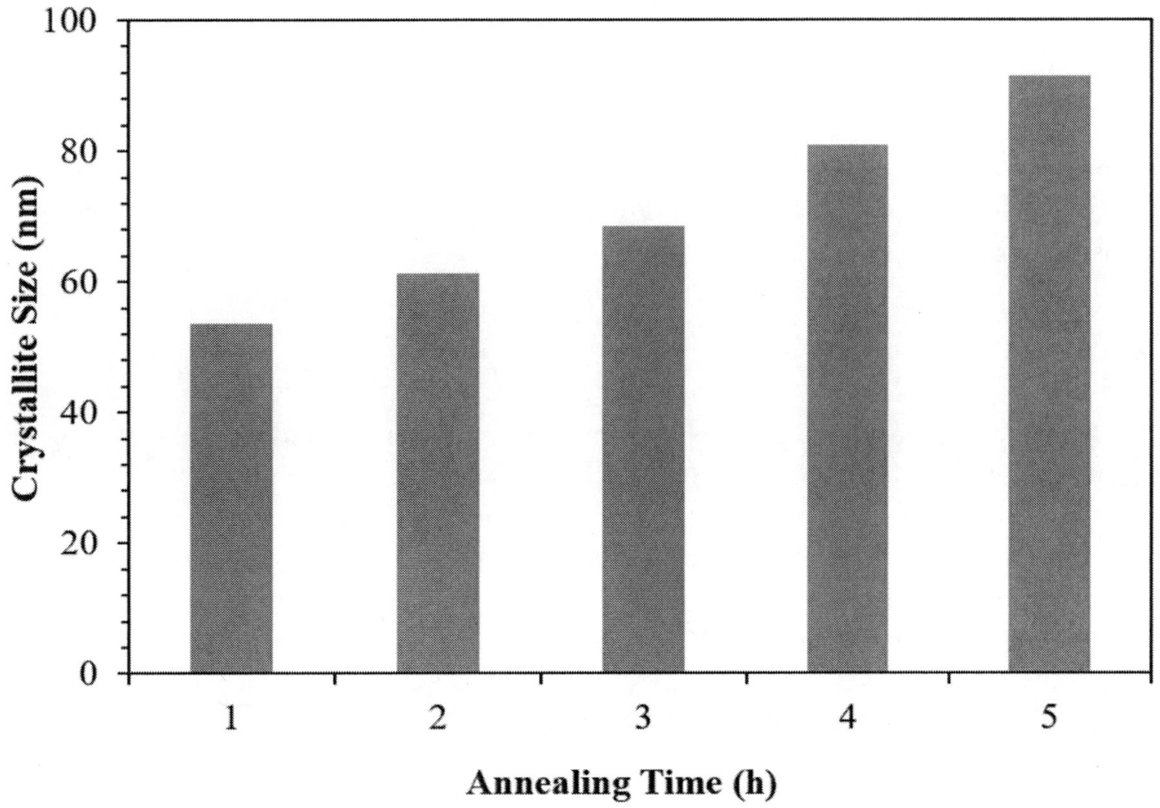


Fig. 3 – Effect of the annealing time on the crystallite size of the sol–gel-derived Co ferrite/$ZrO_2$ ceramic nanomixture.

In addition to the composition and crystallite size, the SSA and pore volume are also crucial parameters to be investigated. The results obtained by BET analysis indicated that with the increase in the annealing time from 1 to 3 h, the SSA decreases by 14.33%. Similarly, with the increase in the annealing time from 3 to 5 h, the SSA decreases by 13.57%. With the increase in the annealing time from 1 to 5 h, the pore volume decreases from $1.52 \times 10^{-3}$ $cm^3/g$ to $1.03 \times 10^{-3}$ $cm^3/g$.

## 4. Conclusion

In this paper, a Co ferrite/$ZrO_2$ ceramic nanomixture, with a pure phase composition, high SSA, pore volume, nanoparticulate morphology, and a uniform distribution of Co ferrite and $ZrO_2$ in the mixture, was synthesized by the propylene-oxide-assisted sol–gel method. The phase and chemical composition of the Co ferrite/$ZrO_2$ nanomixture was not affected under all of the annealing conditions. With the increase in the annealing time from 1 to 5 h at an annealing temperature of 1000°C, the crystallite size increased by 37.69 nm. with the increase in the annealing time from 1 to 5 h, the SSA and pore volume decreased by 2.59 $m^2/g$ and $0.484 \times 10^{-3}$ $cm^3/g$, respectively.

## 5. Acknowledgments

This publication was made possible by the NPRP grant (NPRP8-370-2–154) from the Qatar National Research Fund (a member of Qatar Foundation). The statements made herein are solely the responsibility of author(s).

## References

[1] J.R. Scheffe, A. Steinfeld, Oxygen exchange materials for solar thermochemical splitting of $H_2O$ and $CO_2$: a review, Materials Today 17 (2014) 341-348.

[2] P.G. Loutzenhiser, A. Steinfeld, Solar syngas production from $CO_2$ and $H_2O$ in a two-step thermochemical cycle via Zn/ZnO redox reactions: thermodynamic cycle analysis, International Journal of Hydrogen Energy 36 (2011) 12141-12147.

[3] A. Stamatiou, P. Loutzenhiser, A. Steinfeld, Solar syngas production from $H_2O$ and $CO_2$ via two-step thermochemical cycles based on Zn/ZnO and $FeO/Fe_3O_4$ redox reactions: Kinetic analysis, Energy & Fuels 24 (2010) 2716-2722.

[4] R. R. Bhosale, A. Kumar, F. AlMomani, R. B. Gupta, Solar thermochemical $ZnO/ZnSO_4$ water splitting cycle for hydrogen production, International Journal of Hydrogen Energy (2017) in-press.

[5] P. Loutzenhiser, M. Gálvez, I. Hischier, A. Stamatiou, A. Frei, A. Steinfeld, $CO_2$ splitting via two-step solar thermochemical cycles with Zn/ZnO and $FeO/Fe_3O_4$ redox reactions II: Kinetic analysis, Energy & Fuels 23 (2009) 2832-2839.

[6] R. R. Bhosale, A. Kumar, van den Broeke, Leo JP, S. Gharbia, D. Dardor, M. Jilani, J. Folady, M.S. Al-Fakih, M.A. Tarsad, Solar hydrogen production via thermochemical iron oxide–iron sulfate water splitting cycle, International Journal of Hydrogen Energy 40 (2015) 1639-1650.

[7] P. Charvin, S. Abanades, F. Lemont, G. Flamant, Experimental study of $SnO_2/SnO/Sn$ thermochemical systems for solar production of hydrogen, AIChE Journal 54 (2008) 2759-2767.

[8] R. R. Bhosale, A. Kumar, P. Sutar, Thermodynamic analysis of solar driven $SnO_2/SnO$ based thermochemical water splitting cycle, Energy Conversion & Management 135 (2017) 226-235.

[9] R. R. Bhosale, A. Kumar, F. AlMomani, U. Ghosh, P. Sutar, G. Takalkar, A. Ashok, I. Alxneit, Effectiveness of Ni incorporation in iron oxide crystal structure towards thermochemical $CO_2$ splitting reaction, Ceramics International 43 (2017) 5150-5155.

[10] C.C. Agrafiotis, C. Pagkoura, A. Zygogianni, G. Karagiannakis, M. Kostoglou, A.G. Konstandopoulos, Hydrogen production via solar-aided water splitting thermochemical cycles: Combustion synthesis and preliminary evaluation of spinel redox-pair materials, International Journal Hydrogen Energy 37 (2012) 8964-8980.

[11] R. R. Bhosale. A. Kumar, F. AlMomani, I. Alxneit, Propylene oxide assisted sol–gel synthesis of zinc ferrite nanoparticles for solar fuel production, Ceramics International 42 (2016) 2431-2438.

[12] M. Roeb, N. Gathmann, M. Neises, C. Sattler, R. Pitz-Paal, Thermodynamic analysis of two-step solar water splitting with mixed iron oxides, International Journal of Energy Research 33 (2009) 893-902.

[13] R.R. Bhosale, R.V. Shende, J.A. Puszynski, Thermochemical water-splitting for $H_2$ generation using sol-gel derived Mn-ferrite in a packed bed reactor, International Journal of Hydrogen Energy 37 (2012) 2924-2934.

[14] J.R. Scheffe, J. Li, A.W. Weimer, A spinel ferrite/hercynite water-splitting redox cycle, International Journal of Hydrogen Energy 35 (2010) 3333-3340.

[15] R.R. Bhosale, A. Kumar, F. AlMomani, I. Alxneit, Sol–gel derived $CeO_2$–$Fe_2O_3$ nanoparticles: Synthesis, characterization and solar thermochemical application, Ceramics International 42 (2016) 6728-6737.

[16] R.R. Bhosale, A. Kumar, F. AlMomani, U. Ghosh, S. Al-Muhtaseb, R. Gupta, I. Alxneit, Assessment of $Ce_xZr_yHf_zO_2$ based oxides as potential solar thermochemical $CO_2$ splitting materials, Ceramics International 42 (2016) 9354-9362.

[17] J.R. Scheffe, A. Steinfeld, Thermodynamic analysis of cerium-based oxides for solar thermochemical fuel production, Energy & Fuels 26 (2012) 1928-1936.

[18] W.C. Chueh, C. Falter, M. Abbott, D. Scipio, P. Furler, S.M. Haile, A. Steinfeld, High-flux solar-driven thermochemical dissociation of $CO_2$ and $H_2O$ using nonstoichiometric ceria, Science 330 (2010) 1797-1801.

[19] P. Furler, J.R. Scheffe, A. Steinfeld, Syngas production by simultaneous splitting of $H_2O$ and $CO_2$ via ceria redox reactions in a high-temperature solar reactor, Energy & Environmental Science 5 (2012) 6098-6103.

[20] J.R. Scheffe, R. Jacot, G.R. Patzke, A. Steinfeld, Synthesis, Characterization, and Thermochemical Redox Performance of $Hf^{+4}$ , $Zr^{+4}$ , and $Sc^{+3}$ Doped Ceria for Splitting $CO_2$, The Journal of Physical Chemistry C 117 (2013) 24104-24114.

[21] J.R. Scheffe, D. Weibel, A. Steinfeld, Lanthanum–strontium–manganese perovskites as redox materials for solar thermochemical splitting of $H_2O$ and $CO_2$, Energy & Fuels 27 (2013) 4250-4257.

[22] A.H. McDaniel, E.C. Miller, D. Arifin, A. Ambrosini, E.N. Coker, R. O'Hayre, W.C. Chueh, J. Tong, Sr-and Mn-doped $LaAlO_{3-\delta}$ for solar thermochemical $H_2$ and CO production, Energy & Environmental Science 6 (2013) 2424-2428.

[23] M. Gálvez, R. Jacot, J. Scheffe, T. Cooper, G. Patzke, A. Steinfeld, Physico-chemical changes in Ca, Sr and Al-doped La–Mn–O perovskites upon thermochemical splitting of $CO_2$ via redox cycling, Physical Chemistry Chemical Physics 17 (2015) 6629-6634.

[24] R. R. Bhosale, A. Kumar, A. Ashok, P. Sutar, G. Takalkar, M. Khraisheh, F. AlMomani, U. Ghosh, La-Based Perovskites as Oxygen-Exchange Redox Materials for Solar Syngas Production, MRS Advances (2017) in-press.

[25] R. R. Bhosale, R. Khadka, J. Puszynski, R. V. Shende, $H_2$ generation from two-step thermochemical water-splitting reaction using sol-gel derived $Sn_xFe_yO_z$, Journal of Renewable & Sustainable Energy 3 (2011) 063104-1 – 063104-12.

[26] R. Bhosale, R. Shende, J. Puszynski, $H_2$ generation from thermochemical water splitting using sol-gel derived Ni-ferrite, Journal of Energy and Power Engineering 4 (2010) 27-38.

[27] R.R. Bhosale, R.V. Shende, J.A. Puszynski, $H_2$ Generation From Thermochemical Water-Splitting Using Sol-Gel Synthesized Zn/Sn/Mn-doped Ni-Ferrite, International Review of Chemical Engineering 2 (2010) 852-862.

# 500493 Critical Issues in the Development of Commercial Natural Gas to Industrial Chemicals Bioprocesses

**Abstract:**

The low cost and abundant supply of natural gas make it America's greatest untapped energy resource, leading to the development of technologies for natural gas conversion to compounds of greater value. Intrexon has developed the first natural gas-to-liquids bioconversion platform powered by methantrophic fermentation, which applies synthetic biology to program bacteria and produce higher value materials. Our innovative genetic toolbox allows us to engineer the biology of the methanotroph to more efficiently convert natural gas into a wide range of valuable chemicals and fuels. These applications include isobutanol, farnesene, 1,4-butanediol, 2,3-butanediol, isoprene and isobutyraldehyde – the building blocks for industrial and consumer products like fuels, synthetic rubber, acrylics, resins and spandex.

Using a methanotrophic bioprocess to produce fuels and chemicals has several advantages. Natural gas bioconversion offers scaling opportunities beyond traditional gas-to-liquid (GTL) facilities, as well as biomass feedstock facilities. The theoretical yields of methanotroph/methane exceed that of yeast/sugar approaches by a wide margin. Biocatalytic processes that achieve high yields under milder processing conditions allow for reduced capital and operation expenditures, offering greater economic viability. To successfully bring this technology to commercial scale, consideration must be given not only to optimizing the bacterial metabolic pathway, but to key issues in the fermentor design, downstream processing, scale-up and overall plant economics.

At any scale, aerobic natural gas fermentation must overcome the challenge of gas-to-liquid transfer of low solubility gases to increase bioavailability of gas substrates to the bacteria. A low gas-to-liquid mass transfer rate reduces the bacterial productivity and process yield, which in turn increases fermentor size, capital cost, operating costs and may add additional equipment necessary to recycle gases. During commercial scale-up of gas fermentation processes, these factors can become cost prohibitive. Standard stirred tank reactors become unfeasible for use at larger scales, as the agitator power input and size become excessive. There are many fermentor designs currently available to increase the overall mass transfer coefficient of these gases. However, capital and operating expenditures of the reactor are critical metrics for evaluating commercialization feasibility. Unique fermentor configurations can greatly increase capital cost by raising the cost of fabrication and decreasing the ability to receive competitive bids. These unique fermentors also may not create a homogenous, well-mixed broth – which increases the cost of controls and instrumentation. The economic feasibility of gas transfer designs must be balanced with the biologic effects of reactor forces on the bacteria. Methods of increasing mass transfer often include increasing turbulence in the reactor, which can create detrimental effects on the bacteria through high shear forces, high pressure and vacuum areas, and dissolved gas substrate inhibition. Careful equipment selection and pilot trials are necessary to determine the various effects of gas transfer methods on mass transfer rates and cell viability.

Downstream process selection for commercialization is a balance of biologic limitations, technical feasibility, capital cost, operating cost and the risk of cutting-edge over well-established techniques. The developing bioprocess must be iterated upon the limitations of strain engineering and practicality of downstream processing techniques. Fermentation broth concentrations may be low to product inhibition on the cell, but this increases the size and energy requirement of downstream equipment. Cutting-edge technologies can increase the efficiency of low titer separation and reduce energy inputs, which can be capital intensive and unproven at commercial scales. Genetic engineering tools can increase the tolerance of the bacteria to products. However, the directed or natural evolution process requires time.

A minimum broth titer must be determined through techno-economic modeling of processes at each stage of process development and refinement. Bioprocesses also create metabolic byproducts which can add unit operations, and therefore cost, to the downstream purification process. Some byproducts may originate from the genetic pathway or from commercial feedstocks, as they are often lower in purity and have higher variability than those at the research scale. Additional downstream steps must be evaluated to determine if they are technically and economically feasible to meet final product specifications for sale, or if the byproduct genetic pathways can be manipulated to terminate production and not add cost to the overall process.

Furthermore, certain tools that aid in genetic engineering and strain development at the research and development scale must be eliminated or retooled to create commercially viable strains. During development, production pathways are often introduced to the bacteria through plasmids which carry antibiotic resistance. To avoid the material cost and downstream separation of the antibiotic at commercial scale, the genetic material must be integrated into the chromosome. This genomic integration may affect the strain's capability to produce product at previously observed yields, titers and rates. Another common development tool is that production pathways may be turned on or off using an externally applied chemical inducer. At scale, these chemical inducers can add material cost as well as affect downstream processes and product purity. Induction systems may be completely removed for commercial scale or based off an easily controllable environmental switch.

Favorable plant economics are necessary to move forward with commercialization of a technology. Scale is carefully selected based on a comparison of the production rate, fermentation productivity and yield targets to revenue, capital and operating costs. The plant location is an important selection factor, as costs are affected by utility rates in the area and/or co-location of the plant with shared infrastructure.

Traditional engineering workflows, specifically front-end loading strategy, were used to shape the commercial scale process design of Intrexon's natural gas to chemical plants. The biologic, technical and economic commercial feasibilities of the end-to-end natural gas to isobutanol process were critical to Intrexon's plant design. Laboratory scale learnings were applied to select technology that was commercially feasible at demonstration scale. This demonstration scale design was scaled down to pilot scale, where the operation will yield learnings into refinement of at-scale design, equipment selection, empirical data on heating and cooling, and controls strategies and feedback to further iterations on genetics.

# 495124 Direct, Single-Pass Thermocatalytic Upgrading of Biogas and Landfill Gas into Renewable Natural Gas over the Ultra-Low Loading Ru/Î³-$Al_2O_3$ Catalyst

**Abstract:**

**Direct, single-pass thermocatalytic upgrading of biogas and landfill gas into renewable natural gas over the ultra-low loading Ru/g-$Al_2O_3$ catalyst**

Yichen Zhuang and David Simakov

*Department of Chemical Engineering, University of Waterloo, Waterloo, ON N2L 3G1, Canada*

Converting $CO_2$ into synthetic renewable fuels is an attractive pathway to decrease $CO_2$ emissions and to reduce our dependence on fossil fuels. Thermocatalytic hydrogenation provides advantages of fast reaction rates and high conversion efficiencies (as compared to electro-catalytic and photo-catalytic routes), thus allowing for compact, high-throughput operation. One option is to convert $CO_2$ into synthetic methane, so-called renewable natural gas (RNG) via the Sabatier reaction ($CO_2 + 4H_2 = CH_4 + 2H_2O$). Water electrolysis using renewable or surplus electricity can be used to provide a non-fossil source of $H_2$ required for the reaction. There are multiple sources of concentrated $CO_2$-containing streams including biogas and landfill gas (up to 50 mol% $CO_2$). It is highly desirable to develop a technology that can convert such $CO_2$-rich streams directly, without costly upstream $CO_2$ separation.

One of the most fundamental challenges is the development of a catalyst, which is highly active, selective to $CH_4$ formation, stable against coking, and low-cost at the same time. The selectivity becomes a major challenge when a mixed feed (biogas or landfill gas contain up to 50 mol% $CH_4$ in addition to $CO_2$) is used instead of purified $CO_2$ feed. In this study, a highly selective Ru-based catalyst with ultra-low Ru loadings (less than 0.5 wt%) is reported. A series of catalysts with Ru loadings ranging from 0.05-0.5 wt% were synthesized and evaluated for their catalytic activity in selective conversion of $CO_2$ in the mixed $CH_4/CO_2$ feed. $CO_2$ conversions up to 80% were achieved with nearly complete selectivity to $CH_4$ generation (left plot in the figure below). Importantly, the catalysts showed good stability, without any notable decline in performance over 60 hours on stream at very high space velocity of GHSV = 90,000 1/h (right plot in the figure below). A variety of analytical techniques were employed in order to investigate the mechanism of the superior selectivity and activity, including electron microscopy, thermal gravimetric analysis, temperature programmed reaction, chemisorption, and X-ray photon spectroscopy. Our results point out at the strong structure-activity relationship and the importance of the active phase interaction with the support.

# 492217 Energy Quality Factor and Exergy Destruction Processes Analysis for a Proposed Polygeneration System

**Abstract:**

A new polygeneration system was established through combining coal pyrolysis with semicoke combustion, semicoke gasification and a steam Rankine cycle, and then the calculation methodology was developed to calculate the system. The energy and exergy efficiencies of the system are 65.9 % and 75.1 %, respectively; and the standard energy quality factor Î±Î, values of the involved 17 components and the energy quality factor Î± of each system stream were obtained. By comparing the differences among the standard energy quality factor Î±Î, values of the related 17 components, the features of the energy grade of the components at the standard conditions were elucidated. Then the effect of streamÂ¢s composition and temperature on the actual energy quality factor was discussed. Combining the concept of energy quality factor, the processes of exergy destruction in system blocks were visually expressed via a kind of diagram; and the reasons of exergy destruction were tried to be explained from the perspective of energy quality factor.

# 500591 Technical Economic Analysis of an Intensified Integrated Gasification Combined Cycle Plant Design Featuring Membrane and Adsorptive Reactors

**Abstract:**

Currently, coal-based power plants are the largest source of electricity both globally and in the United States, and despite there being an abundant supply of low-priced natural gas, the Energy Information Administration (EIA) projects that by 2040, coal-based power plants will still provide 254.1 GWe of the generation capacity of the United States, and will account for approximately 36% of the total electricity generation capacity globally1,2. For this reason, coal-based power plants have begun considering the adoption of carbon capture technology in order to reduce the quantity of CO2 emissions, since each year in the United States 6,870 million metric tons of greenhouse gases (GHG) are emitted into the atmosphere, and CO2 represents over 80% of these GHG emissions3.

Because of the environmental concerns these emissions evoke, several carbon capture technologies have been proposed and integrated into these coal power plants with the goal to reduce the CO2 emissions by at least 90%. The main process utilized for carbon separation and capture employs absorption with amine-type solvents, whereby CO2 is separated through the use of a Dual-Stage Selexol unit4.

In this work, the design of an Integrated Gasification Combined Cycle (IGCC) plant featuring membrane and adsorptive reactors is first developed and then analyzed from an economic viewpoint. The design features membrane and adsorptive reactors that have over 90% carbon capture capability without the use of a Dual-Stage Selexol unit. The resultant intensified design is able to produce power more efficiently than traditional IGCC plants with Carbon-Capture Storage (CCS) technology. The UNISIM software (Honeywell trademark) is used to create an intensified flowsheet for the proposed power production plant design, and heat integration studies are presented. A technical economic analysis (TEA) is carried out on the proposed flowsheet to determine economic feasibility of the novel design. The economics of the intensified designs will be compared to a traditional IGCC plant with CCS to identify the extent of process intensification over the baseline design.

1. Annual Energy Outlook 2013, Department of Energy, U.S. Energy Information Administration, 2013.
2. Annual Energy Outlook 2014, Department of Energy, U.S. Energy Information Administration, 2013.
3. Inventory of U.S. greenhouse gas emissions and sinks: 1990-2014. Washington, DC: U.S. Environmental Protection Agency, n.d. Print.
4. Department of Energy/National Energy Technology Laboratory. Cost and Performance Baseline for Fossil Energy Plants Volume 1b: Bituminous Coal (IGCC) to Electricity Revision 2b â€“ Year Dollar Update. Document Number: DOE/NETL- 2015/1727.

# 501648 An Electrochemical Method to Remove Aqueous Sulfide from Swine Manure


**Abstract:**

Sulfide is synthesized by sulfur reducing bacteria during the anaerobic degradation of swine manure. The year-long pit storage of swine manure allows the accumulation of sulfide, which eventually results in a high risk of hydrogen sulfide (H 2 S) releasing to the air during the manure pumping. H 2 S is an extremely hazardous and toxic gas; and a number of casualties have been reported due to H 2 S poisoning in the past several years. In order to minimize the sulfide level in swine manure and create a safe working environment, an effective way is presented in this study. By applying a certain voltage onto low carbon steel electrodes partially immersed in the swine manure storage, a 90% or above sulfide removal can be achieved after 1 to 2 weeks treatment. In our experiment, the optimal voltage was investigated to be 0.7 V, which allowed the majority removal of sulfide, and meanwhile a guarantee of less electrode material consumption. The effects of potential parameters, including manure solid content, electrode surface area, distance between electrodes, treatment temperature, biogas composition in the reactor headspace, and sulfide recovery after shutting down the power, were investigated. This electrochemical treatment has shown a significant aqueous sulfide removal from swine manure. Optimization based on mentioned parameters can further improve the treatment feasibility in reality.

# 499078 Pilot-Scale Testing of Electrochemical Removal of Hydrogen Sulfide in Deep-Pit Swine Manure Storage

**Abstract:**

The presence of hydrogen sulfide ($H_2S$) is a widespread issue in organic waste storage, and this toxic air contaminant causes serious concerns over environment, health and safety. One example is the deep-pit manure storage, a common practice for swine facilities in upper-Midwestern states. It occasionally experiences human or animal fatal overexposure to manure gas that contains $H_2S$. In this study, we developed an electrochemical-based oxidation method to remove or greatly reduce sulfide generation in the stored liquid manure. The treatment system needed some equipment installation and retrofit of the current storage pit with little maintenance requirement, low electricity consumption, and low chemicals usage. Lab-scale experiments were conducted in simulated swine manure storage, and systematically analyzed sulfide removal kinetics at varying total solids contents, initial sulfide concentrations, pH, applied voltages, electrode surface areas, electrode distances, and temperatures. An on-farm pilot-scale study was then performed to further confirm the treatment performance in close-to-real environment, and provide necessary data for techno-economic assessment. An electrochemical treatment device was designed and installed in two manure pump-out accesses in a testing barn that held about 200 growing-finishing pigs. Each pit access was divided to two compartments, one for control and the other for treatment installation. Perforated sheets of two electrode materials of low carbon steel (LCS) and stainless steel Type AISI 304 (SS) were installed in accesses and tested in 97 daysâ€™ operation. From the monitored $H_2S$ concentration in pit headspace during the treatment, it can be seen that the LCS treatment decreased a substantial amount of $H_2S$ concentration, achieving removal efficiency of 56%-78% depending on the applied voltage of between 0.6 and 0.7 V. SS treatment removed 13%-38% of $H_2S$. LCS treatment required a lower voltage application but achieved better removal performance, and is more economically feasible compared to SS installation according to techno-economic assessment. However, electrode material cost accounted for 76%-87% of the net present cost of the treatment assuming 20 yrsâ€™ life-time of the treatment device. Cheaper material, lower surface area, or more effective material will have to be optimized in order to decrease the treatment cost.

# 495035 Economic and Environmental Impact Analyses of Integrated Dehydrogenation and Hydroformylation in Gas-Expanded Liquid Media

**Abstract:**

**Economic and Environmental Impact Analyses of Integrated Dehydrogenation and Hydroformylation in Gas-Expanded Liquid Media**

Dupeng Liu,[1,2] Raghunath V Chaudhari[1,2] and Bala Subramaniam[1,2]

[1]Department of Chemical & Petroleum Engineering, [2]Center for Environmentally Beneficial Catalysis, University of Kansas

Lawrence, KS 66047

Increased shale gas production in the U. S. has also enhanced the availability of $C_2$-$C_5$ alkanes (the so-called natural gas liquids) as collateral products, and opportunities for their utilization in novel ways. For example, propylene production capacity via propane dehydrogenation (PDH) is slated to grow rapidly over the next several years.[1] However, for a typical PDH process, the per pass conversion is only about 50%. The separation of propane and propylene mixtures is highly energy and capital intensive. Previously, we had reported the beneficial effects of propane expanded liquids (with propylene/propane mixtures as feed) for propylene hydroformylation with Rhodium-based catalysts. At sufficiently high pressures (>10 bar), the pressurized propane/propylene mixture forms a gas-expanded liquid phase that promotes regio-selective hydroformylation of propylene with homogenous Rh complex. In addition, compared with conventional propylene hydroformylation process to produce butyraldehyde using polymer grade propylene as feedstock, the feed in the propane-expanded liquid (PXL) process is 50% cheaper.[3]

In this talk, economic and environmental analyses of the integrated dehydrogenation and hydroformylation in PXL process are presented and benchmarked against simulated conventional hydroformylation processes for which data were obtained from the Ruhrchemie/Rh™ne-Poulenc and Dow processes.[4-5] Preliminary simulation results indicate that capital investment and production cost associated with the PXL process is lower than the conventional processes (by up to 40% and 25%, respectively) due to the elimination of the $C_3$ separation section. Life cycle analysis (LCA) was also performed based on plant scale simulation of conventional and PXL processes. These analyses have provided guidance in catalyst design and in choosing materials and operating conditions that favor process economics while lessening environmental footprint.

**1.** Nawaz Z. Light alkane dehydrogenation to light olefin technologies: a comprehensive review[J]. Reviews in Chemical Engineering, 2015, 31(5): 413-436.

**2.**Nawaz Z. Dynamic modeling of catofin¨ fixed-bed iso-butane dehydrogenation reactor for operational optimization[J]. International Journal of Chemical Reactor Engineering, 2016, 14(1): 491-515.

**3.** Liu, D., Chaudhari, R.V., Subramaniam, B. Kinetic Modeling of Propene Hydroformylation in Propane-Expanded Liquid with Rh-Based Complexes, 2016 AIChE annual meeting San Francisco, CA.

**4.** Wattimena F, Heijmen H J. U.S. Patent 4,521,630[P]. 1985-6-4.

**5.** Devon T J, Phillips G W, Puckette T A, et al. U.S. Patent 5,332,846[P]. 1994-7-26.

# 492359 Bio-Energy with Carbon Capture and Storage (BECCS): Are Inefficient Power Plants a Better Option?

**Abstract:**

Negative emissions technologies (NETs) in general, and bio-energy with carbon capture and storage (BECCS) in particular, have been identified as key technologies in the context of limiting climate change to no more than 1.5 – 2°C (Fuss *et al.*, 2014). By actively removing $CO_2$ from the atmosphere, NETs provide a service, and thus investors would expect an appropriate risk-adjusted rate of return, varying as a function of the quantity of public money involved. In addition to providing negative emissions, BECCS facilities also generate firm low carbon power. However, as the electricity sector is integrating an increasing share of intermittent renewable energy (iRE), such as wind and solar, the dispatch load factor of such BECCS facilities may be small relative to their capacity. This has the potential to significantly underutilise these assets for their primary purpose of removing $CO_2$ from the atmosphere. In this study, we present a techno-economic-environmental evaluation of BECCS plants with a range of operating efficiencies, considering their full- and part-load operation relative to a national-scale annual $CO_2$ removal target, electricity price and negative emissions credit. We find that in all cases, a lower capital cost, lower efficiency BECCS plant is superior to a higher cost, higher efficiency facility from both environmental and economic perspectives. We show that it may be preferable to operate the BECCS facility in base-load fashion, constantly removing $CO_2$ from the atmosphere and dispatching electricity in a demand-following manner. We show that the use of this "spare capacity" to produce hydrogen for, *e.g.*, injection to a natural gas system for the provision of low carbon heating, can add to the overall environmental and economic benefit of such a system. The sensitivity of the model to biomass carbon footprint is also presented. The only point where our conclusions are altered is where the $CO_2$ emissions associated with the biomass supply chain are sufficiently large so as to eliminate BECCS $CO_2$ removal (Mac Dowell & Fajardy, 2017).

## References

Fuss, S., Canadell, J. G., Peters, G. P., Tavoni, M., Andrew, R. M., Ciais, P., Jackson, R. B., Jones, C. D., Kraxner, F., Nakicenovic, N., Le Quere, C., Raupach, M. R., Sharifi, A., Smith, P. & Yamagata, Y. (2014). Betting on negative emissions. Nature Climate Change, 4 (10), 850–853. https://doi.org/10.1038/nclimate2392

Mac Dowell, N., & Fajardy, M. (2017). Inefficient power generation as an optimal route to negative emissions via BECCS. Environmental Research Letters, at press: https://doi.org/10.1088/1748-9326/aa67a5

# APPLICATION OF SOME PREDICTIVE THERMODYNAMIC MODELS FOR INDUSTRIALLY IMPORTANT SYSTEMS

**Yizu Zhu**
**Mohammad Shafiei**

**AIChE Annual Meeting, Minneapolis, MN**

October 31, 2017

CHEMISTRY THAT MATTERS™

## THERMODYNAMICS CRITICAL TO INTEGRATED PROCESS ENGINEERING & ECONOMICS

Physical Properties play very important roles in sustainable process engineering and in improving companies' bottom line.

Critical phase equilibrium data are unavailable for compounds under many circumstances:

- Design new products
- Screen conceptual design processes
- Scaleup and commercialize new technologies
- Trouble-shoot & optimize manufacturing plants
- Support techno-economics

Reliable "Predictive Equation Of State" models can be very helpful in tackling the above challenges.

No. 1

## IMPACT OF PHYSICAL PROPERTIES ON SYSTEM ECONOMICS

A Close Boiling System

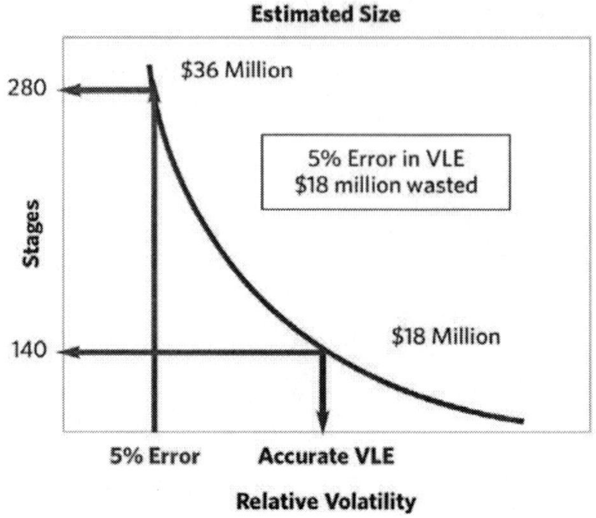


Dhole, et al, Aspen Tech White Paper, 2012

No. 2

## IMPACT OF PHYSICAL PROPERTIES ON SYSTEM ECONOMICS

Ammonia Reaction System using Aspen Plus V8.8

| | FEED | PRODUCT | | |
|---|---|---|---|---|
| | | "Best available" model with parameters regressed from Data | Using A+ PSRK Prediction | Using Ideal Gas Model |
| Temperature C | 450 | 450 | 450 | 450 |
| Pressure bar | 90 | 150 | 150 | 150 |
| Vapor Frac | 1 | 1 | 1 | 1 |
| Mole Flow kmol/hr | 4 | 3.272 | 3.263 | 3.3 |
| Mass Flow kg/hr | 34.06 | 34.06 | 34.06 | 34.06 |
| Volume Flow cum/hr | 2.67 | 1.36 | 1.36 | 1.32 |
| Enthalpy Gcal/hr | 0.012 | 0.003 | 0.003 | 0.003 |
| Mass Flow kg/hr | | | | |
| NH3 | | 12.39 | 12.55 | 11.92 |
| H2 | 6.05 | 3.85 | 3.82 | 3.93 |
| N2 | 28.01 | 17.82 | 17.69 | 18.21 |
| Mass Frac | | | | |
| NH3 | | 0.364 | 0.368 | 0.350 |
| H2 | 0.178 | 0.113 | 0.112 | 0.115 |
| N2 | 0.822 | 0.523 | 0.519 | 0.535 |
| Difference% | | 0.00% | 1.28% | -3.82% |

**The productivity impact is ~17.5$MM/year for a 1,000KTA Ammonia plant (Feb 2017 retail price).**

No. 3

سابك
sabic

## THERMODYNAMIC CHALLENGES AND PREDICTIVE MODELS

- Industry presents challenges for representing
  - Chemically non-ideal systems
  - Involving supercritical components
  - In multiphase regions at extreme temperature and pressures
  - Specialty chemicals and other systems of rare compounds with no VLE-LLE data

- Equation Of State approach is good for high pressure, but until recently, is not-applicable to highly non-ideal chemical systems.

- New Mixing Rules for EOS emerged in the 1990s to represent the interactions in liquid phases of complex mixtures:
  - Predictive SRK (Holderbaum and Gmehling, 1991)
  - Modified Huron Vidal (MHV2) (Dahl, Fredenslund, Rasmussen, 1991)
  - Wong-Sandler (1992)
  - Etc.

No. 4

سابك
sabic

## PENG - ROBINSON EOS

$$P = \frac{RT}{V_m - b} - \frac{a}{V_m(V_m + b) + b(V_m - b)}$$

**Mixing Rule**: $a = \sum_i \sum_j x_i x_j (a_i a_j)^{0.5}(1 - k_{ij}), \quad b = \sum_i x_i b_i$

$$a_i = \alpha_i 0.45724 \frac{R^2 T_{ci}^2}{P_{ci}}, \quad b_i = 0.07780 \frac{RT_{ci}}{P_{ci}}$$

$$\alpha_i(T) = [1 + m_i(1 - T_{ri}^{0.5})]^2, \quad m_i = 0.37464 + 1.54226\,\omega_i - 0.26992\,\omega_i^2$$

**Binary interaction parameter $k_{ij}$ needs to be regressed from phase equilibrium data**

No. 5

## APPLICATION OF PREDICTIVE EOS MODELS FOR HIGH PRESSURE AND HIGHLY NON-IDEAL SOLUTION SYSTEMS

**SRK:** $P = RT/(v-b) - a/v(v+b)$

**PSRK's Mixing Rule:**

$$b = \sum x_i b_i$$

$$a = b[g^E/A_1 + \sum x_i a_i/b_i + (RT/A_1)\sum x_i \ln(b/b_i)]$$

***$g^E$ : Excess Gibbs Energy from UNIFAC Group Contribution Activity Coefficient Model***

***$A_1$ : constant***

(Holderbaum and Gmehling, 1991; Dahl and Michelson, 1990)

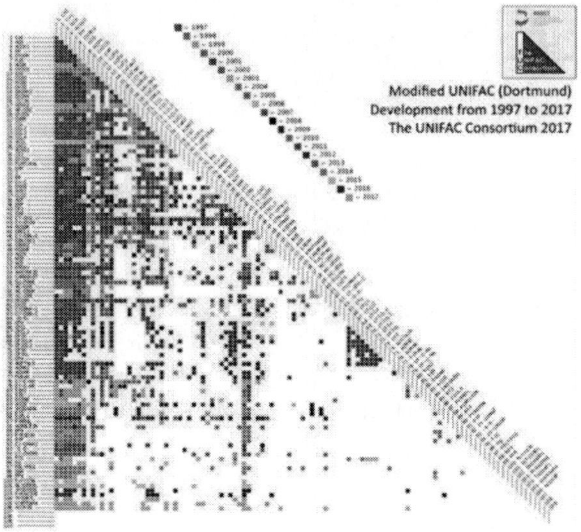


No. 6

## KEY FEATURES OF THE NEW MIXING RULES

- A composition-dependent term F(x)
- F(x) is related to Excess Gibbs Energy or Helmholtz Energy
- Allows a single Equation Of State model to represent the phase behavior of:
  - Low Density Fluid (usually by EOS)
  - High Density Fluid (usually by Activity Coefficient Models)
  - Polar Fluid including highly non-ideal LLE systems

No. 7

## PREDICTION FOR MEOH-H2S PHASE BEHAVIOR

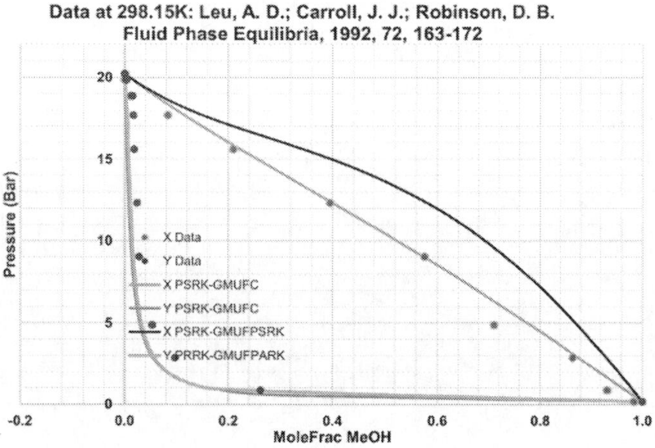


No. 8

## METHANE-N-BUTANE P-XY AT HIGH PRESSURE

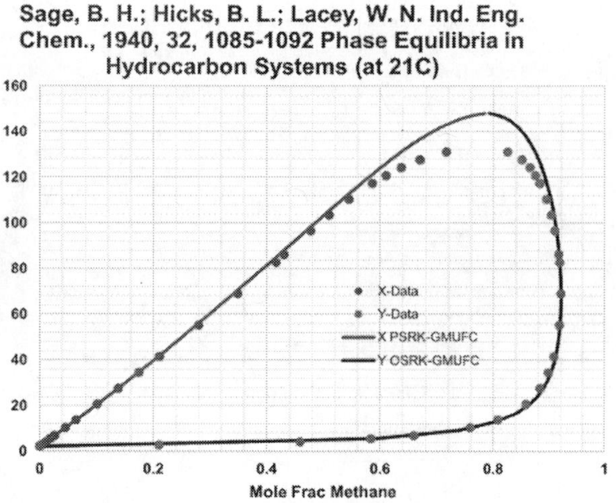


No. 9

## EtAc–HEXANE SYSTEM T-XY PHASE DIAGRAM

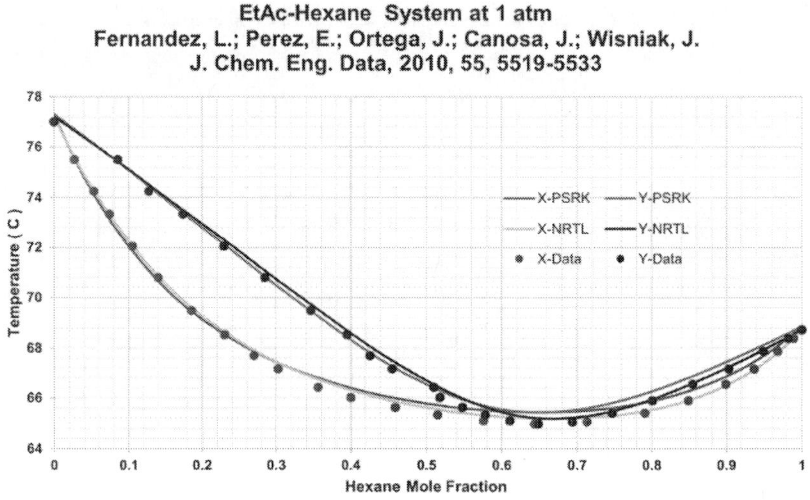


No. 10

## PSRK PREDICTION FOR HIGH P, STRONGLY NON-IDEAL SYSTEMS

MEOH-CO2 SYSTEM

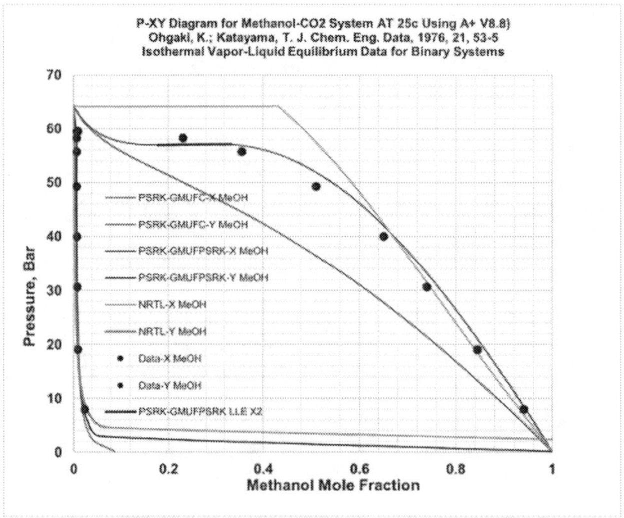


No. 11

# CO2-CYCLOHEXANE SYSTEM AT HIGH PRESSURE

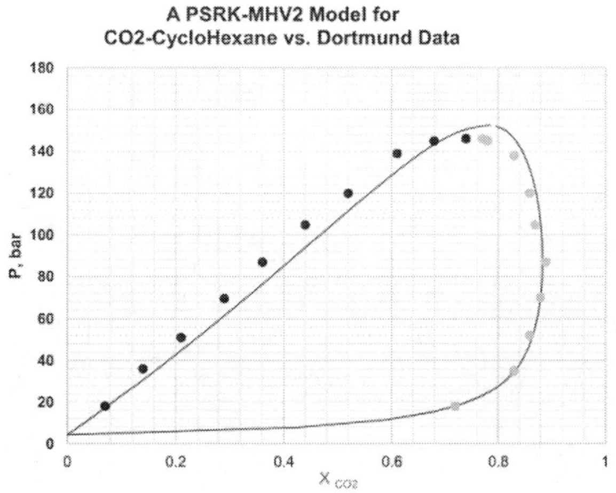


No. 12

# NH3-WATER SYSTEM AT TYPICAL PLANT OPERATION CONDITION

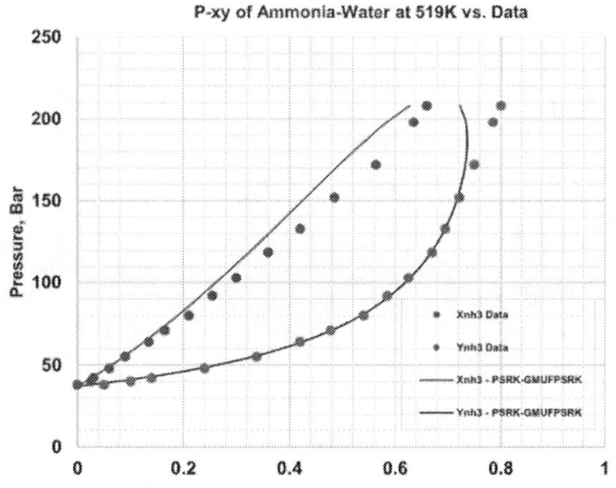


No. 13

# N-BUOH – WATER HETEROGENEOUS AZEOTROPE PREDICTION

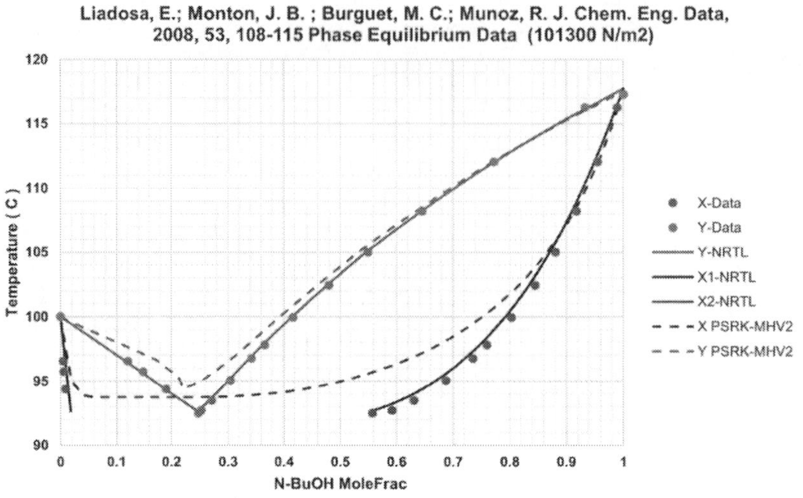


No. 14

# SEPARATION IN AN ETHYLENE PLANT WITH PSRK

| | Tower Reflux | Tower Bottom Liquid Rate | Tower Top Temperature |
|---|---|---|---|
| | t/h | t/h | C |
| Plant Data | 12.00 | 52.30 | -98.50 |
| RKS | 12.31 | 51.00 | -100.10 |
| PSRK | 12.99 | 52.40 | -99.50 |

No. 15

TEMPERATURE IMPACT ON HYDROCARBON CONCENTRATION IN RESIN USING POLYSRK FOR PRODUCT QUALITY IMPROVEMENT AND CAPACITY INCREASE

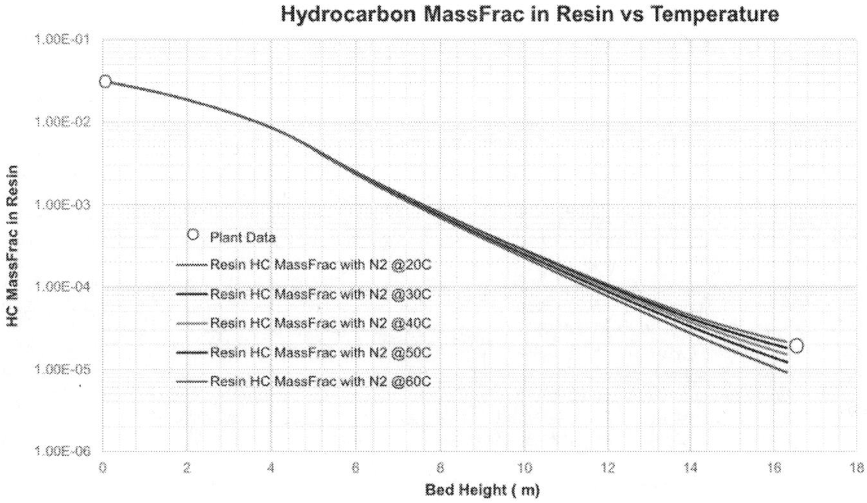




## CONCLUSIONS

- Physical properties have critical impact to the economics of process design and manufacturing operations.
- Predictive Equation of State models provide a powerful tool in representing highly non-ideal systems under high pressure where data are unavailable and other types of thermodynamic models are unable to handle.
- Predictive EOS models can be used with Excess Gibbs Free Energy of various liquid activity coefficient models available in simulation software such as Aspen Plus where binary interaction parameters are available (such as NRTL, WILSON, UNIQUAC, etc).
- This provides a powerful tool to handle a variety of complex systems where data are sparsely available:
  - Conceptual design of new products and processes
  - R&D
  - Scale-up and commercialization of innovative technologies
  - Manufacturing retrofitting and performance optimization

## DISCLAIMER

DISCLAIMER: THE MATERIALS, PRODUCTS AND SERVICES OF SAUDI BASIC INDUSTRIES CORPORATION (SABIC) OR ITS SUBSIDIARIES OR AFFILIATES ("SELLER") ARE SOLD SUBJECT TO SELLER'S STANDARD CONDITIONS OF SALE, WHICH ARE AVAILABLE UPON REQUEST. INFORMATION AND RECOMMENDATIONS CONTAINED IN THIS DOCUMENT ARE GIVEN IN GOOD FAITH. HOWEVER, SELLER MAKES NO EXPRESS OR IMPLIED REPRESENTATION, WARRANTY OR GUARANTEE (i) THAT ANY RESULTS DESCRIBED IN THIS DOCUMENT WILL BE OBTAINED UNDER END-USE CONDITIONS, OR (ii) AS TO THE EFFECTIVENESS OR SAFETY OF ANY DESIGN OR APPLICATION INCORPORATING SELLER'S MATERIALS, PRODUCTS, SERVICES OR RECOMMENDATIONS. UNLESS OTHERWISE PROVIDED IN SELLER'S STANDARD CONDITIONS OF SALE, SELLER SHALL NOT BE RESPONSIBLE FOR ANY LOSS RESULTING FROM ANY USE OF ITS MATERIALS, PRODUCTS, SERVICES OR RECOMMENDATIONS DESCRIBED IN THIS DOCUMENT. Each user is responsible for making its own determination as to the suitability of Seller's materials, products, services or recommendations for the user's particular use through appropriate end-use and other testing and analysis. Nothing in any document or oral statement shall be deemed to alter or waive any provision of Seller's Standard Conditions of Sale or this Disclaimer, unless it is specifically agreed to in a writing signed by Seller. Statements by Seller concerning a possible use of any material, product, service or design do not, are not intended to, and should not be construed to grant any license under any patent or other intellectual property right of Seller or as a recommendation for the use of any material, product, service or design in a manner that infringes any patent or other intellectual property right.

SABIC and brands marked with ™ are trademarks of SABIC or its subsidiaries or affiliates.



Any brands, products or services of other companies referenced in this document are the trademarks, service marks and/or trade names of their respective holders.

No. 18

## THANK YOU AND Q&A

No. 19

# 489973 Optimal Design of Renewable Energy Systems with Flexible Inputs and Outputs By the P-Graph Framework

**Abstract:**

The P-graph (process-graph) methodology with a new mathematical model for the operating units is used to optimize complex energy systems involving renewable energy sources. World energy consumption has been on the rise worldwide. If the current trends continue, we may face energy shortages in the future. Planning an efficient energy system for a small community or a city district is a challenging task. Often, the applied models are too simple, there are implicit assumptions and important constrains are disregarded. In other cases, the models are large and computationally difficult to handle, thus, only small examples can be solved. This paper proposes the application of the P-graph framework, introduced by Friedler et al, to design specific components of energy production systems. The P-graph framework is enhanced with a new mathematical model for the operating units to handle special constraints. A case study is presented to demonstrate the use of the framework.

# 499000 Discovering Heuristics for Sustainable Design By Multiobjective Evolutionary Optimization and Machine Learning

**Abstract:**

In sustainable process design (SPD), it is increasingly common to use multiobjective optimization for understanding the trade-off between monetary and life cycle environmental objectives. While these Pareto fronts are useful, what would be more valuable is the knowledge about what makes the solutions optimal. For example, it would be useful to know if there are specific decisions that must be made or avoided for ensuring optimality. Such knowledge can result in heurisitics that can be used to guide other SPD projects. The benefit of such knowledge for guiding conventional engineering design is well-known and widely used by practitioners. However, heuristics for SPD are not yet available. In order to obtain such insight, additional analyses are required to go beyond just finding the Pareto front.

Nonetheless, most current SPD generally neglect the ecological carrying capacity. The ignorance of these limits while designing systems has led to unintended harm, such as ecological degradation. Techno-Ecological Synergy (TES) framework [1] was developed to fill this gap to account for ecological carrying capacity by quantifying the demand (i.e. resource use and emissions) and supply (i.e. capacity of nature) of ecosystem services (ES). The sustainability metric has been defined based on the the difference between the supply and demand for ES.

This work integrates the recently developed approach of techno-ecological synergy (TES) design with multi-objective evolutionary optimization and classification and regression tree (CART) machine learning algorithm, with the main focus on obtaining insights to explain optimality. First, evolutionary optimization has been applied to obtain the Pareto fronts. Optimal points on the pareto fronts and suboptimal points in the solution space can be generated simultaneously with the optimization. The CART algorithm has been applied to classify the solutions in the decision space, which could potentially lead to the discovery of general heuristics about TES sustainability designs. The CART algorithm has the ability to detect interactions and identify groups that have similar outcomes along with the associated predictor variables[2]. The predictor variables that contribute to a more thorough classification can thus be determined and identified as a design heuristic.

The methodology has been applied to a case study describing the design of a residential system accounting for ecosystems like trees, lawn and a vegetable garden[3]. Technological, ecological and behavioural variables are considered in the design to simultaneously minimizing cost while maximizing environmental benefits. This multi-objective optimization problem is formulated as a simulation-based optimization by integrating EnergyPlus simulation software and evolutionary algorithm. The results show that the solutions obtained from TES-designed systems are economically superior compared to techno-centric solutions; they also have environmental benefits in terms of reducing overshoot, thus leading to a "win-win" scenario. In addition, based on initial work, the existence of shading trees has been identified as a design heuristic that will be "win-win". The discovery of such heuristics will be useful in guiding the future design of similar systems.

References

[1] Bakshi, Bhavik R., Guy Ziv, and Michael D. Lepech. "Techno-ecological synergy: A framework for sustainable engineering." Environmental science & technology 49.3 (2015): 1752-1760.

[2] Neville, Padraic G. "Decision trees for predictive modeling." SAS Institute Inc 4 (1999).

[3] Urban, Robert A., and Bhavik R. Bakshi. "Techno-ecological synergy as a path toward sustainability of a North American residential system." Environmental science & technology 47.4 (2013): 1985-1993.

# 501644 Life Cycle Modeling and Reactor Design for $CO_2$ Utilization

**Abstract:**

**Life Cycle Modeling and Reactor Design for $CO_2$ Utilization**

**Shaik Afzal[1], Mohammedsufiyan Challiwala[1], Debalina Sengupta[2], Nimir Elbashir[3] and Mahmoud El-Halwagi[1]**

Corresponding Author: debalinasengupta@tamu.edu

1 Artie McFerrin Department of Chemical Engineering, Texas A&M University, College Station

2 Gas and Fuels Research Center, Texas A&M Engineering Experiment Station

3 Chemical Engineering Department, Texas A&M University, Qatar

AIChE Annual Meeting, Minneapolis, MN, October 29 - November 3, 2017

The reaction of methane and carbon dioxide to produce syngas (mixture of CO and H2) is called Dry Reforming of Methane (DRM). DRM has received considerable interest in academia and industry due to CO2 being one of the reactant feed and its potential for fixing CO2. Most of the work done in DRM research involves novel catalysts to reduce the coking problem and thermodynamic studies to estimate product composition at different operating conditions. Syngas ratios are important for desired products, and the overall CO2 balance needs to be tracked for ensuring environmental sustainability.

This work focuses on modeling for maximum potential of the DRM reaction to reduce CO2 emissions of a process. In addition to reformer heat duty, this work considers upstream and downstream emissions from processes required for maintaining a desired syngas ratio. In the first part of this work, syngas generation at desired ratios is considered. Depending on the ratio of H2/CO, additional reforming methods like steam reforming or partial oxidation may be needed. A combination of two or more reforming methods constitutes combined reforming. With combined reforming methods, thermodynamic advantages can be achieved. In the second part of the study, a novel reactor scheme is proposed. This scheme makes use of the thermodynamic equilibrium estimates and creates more than one product of interest, at the same time achieving significant fixation of CO2.

The results from these studies are aimed at reducing CO2 through combination of reforming methods and novel reactor design. This study will help understand the impact of DRM unit on carbon emissions and the operating costs of a syngas generation unit. The results can then be compared with other CO2 reduction strategies in the industry to prioritize projects aimed towards a low-carbon emissions future. LINGO software has been used for equilibrium calculations and optimization. Appropriate reactor design, process simulation tools have also been used for this work.

# 491969 Methanol Synthesis Using Captured $CO_2$: Techno-Economic & Environmental Assessment

**Abstract:**

The purpose of this paper is to assess via techno-economic and environmental metrics the production of methanol (MeOH) using H2 and captured CO2 as raw materials. It evaluates the potential of this type of carbon capture and utilisation (CCU) plant on (i) the net reduction of CO2 emissions and (ii) the cost of production, in comparison with the conventional synthesis process of MeOH Europe. Process flow modelling is used to estimate the operational performance and the total purchased equipment cost; the flowsheet is implemented in CHEMCAD, and the obtained mass and energy flows are utilised as input to calculate the selected key performance indicators (KPIs). CO2-based metrics are used to assess the environmental impact. The evaluated MeOH plant produces 440 ktMeOH/yr, and its configuration is the result of a heat integration process. Its specific capital cost is lower than for conventional plants. However, raw materials prices, i.e. H2 and captured CO2, do not allow such a project to be financially viable. In order to make the CCU plant financially attractive, the price of MeOH should increase in a factor of almost 2, or H2 costs should decrease almost 2.5 times, or CO2 should have a value of around 222 â‚¬/t, under the assumptions of this work. The MeOH CCU-plant studied can utilise about 21.5% of the CO2 emissions of a pulverised coal (PC) power plant that produces 550 MWnet of electricity. The net CO2 emissions savings represent 8% of the emissions of the PC plant (mainly due to the avoidance of consuming fossil fuels as in the conventional MeOH synthesis process). The results demonstrate that there is a net but small potential for CO2 emissions reduction; assuming that such CCU plants are constructed inÂ Europe to meet the MeOH demand growth and the quantities that are currently imported, the net CO2 emissions reduction could be of 2.71 MtCO2/yr.

# 497565 A CFD-Based Dynamic Model of Absorption/Adsorption Process to Simultaneously Purify Landfill Gas and Treat Leachate

**Abstract:**

## Introduction

Greenhouse gases (GHGs) have been increasing as the human population has increased. As the human population continues to grow, GHGs will become an even more serious problem. The culprit that most people associate with GHGs is the petroleum and cement industries. However, an often-overlooked producer of GHGs is the waste sector. Landfills contribute both methane and carbon dioxide to the atmosphere in roughly equal proportions (collective gas is landfill gas (LFG)). Over a 100 year span, methane has a global warming potential of 25 (meaning it traps 25 times more heat than carbon dioxide) (Omar and Rohani 2015). To lessen the greenhouse effect of methane, landfills typically flare the LFG. However, there is a lot of usable energy that is wasted with this method. A new paradigm in landfill treatment has emerged: bioreactor landfills. Bioreactors accelerate the biodegradation process, producing methane and carbon dioxide at a faster rate. With the accelerated production of LFG there is potential for using it as a fuel.

Many methods exist to separate $CO_2$ and $CH_4$ (e.g. absorption, adsorption, membranes and cryogenic treatment), each with its own advantages and disadvantages. In our work, we use absorption to separate $CO_2$ and $CH_4$. We use water and landfill leachate, as absorbents, to dissolve the $CO_2$ while keeping the $CH_4$in the gas phase. For the absorption packing, zeolite particles are used. The benefit of zeolite is that while leachate is being used as the absorbent, the zeolite can remove heavy metals as an adsorbent. The separation, therefore, will proceed in both an absorption and an adsorption column. In the paper written by (Omar and Rohani 2017), experiments were performed to compare the usage of leachate versus water as an absorbent and zeolite versus glass beads as absorption packing. They found that leachate was comparable to water as an absorbent and zeolite was only slightly less effective (between 5 and 10% different) compared to spherical glass beads as the column packing. They also developed a stationary (steady-state) computational fluid dynamics model of the absorption process. In this work, the model is extended to the dynamic case while including the adsorption process.

## Results and discussion

The model consists of 3 parts: (1) the fluid flow of the 2 phases (liquid and gas), (2) the absorption of the carbon dioxide in leachate (taken as water in the CFD model) and (3) the adsorption of lead on the zeolite packing. The column geometry in the model was 6-cm diameter and 54.6 cm (21.5") height. Since the geometry is cylindrical, a 2-D axisymmetric model approximation was enforced. The model equations were solved using the finite element method available in the COMSOL Multiphysics® Software package.

The effects of the different parameters on the efficiency of absorption and adsorption expressed in terms of the dynamic time (time to reach steady-state) of the absorption, the steady-state $CH_4$ purity, as well as the concentration of lead in the liquid effluent stream, and the breakthrough time, were examined. The objective of the study was to separate $CH_4$from the gas stream, and remove heavy metals (Pb considered in this work) from the landfill leachate. Parameters examined were: column dimensions (height and diameter), the type of absorbent, column pressure and liquid/gas flow rates.

As the height of the column increased, the time to reach steady-state increased. The steady-state $CH_4$ purity also increased. At larger column heights, the residence time for mass transfer ($CO_2$ dissolution) is longer, therefore, more $CO_2$ dissolves, increasing the purity of $CH_4$ and the time required to reach steady-state. As for the adsorption, increased height of the column decreased effluent liquid concentration of Pb and increased the breakthrough time. A longer residence time allowed more adsorption. The breakthrough time increased since the volume of the column was larger and the volume of zeolite packing was also larger. With more zeolite, there was more surface area for adsorption, which corresponds to more adsorbate removed and a longer breakthrough time. When the diameter of the column increased, the purity of $CH_4$ increased. This is due to a decrease in the liquid velocity since the volumetric flow is constant. The time to reach steady-state also increased. The liquid effluent concentration of Pb decreased since the velocity of the liquid decreased. This resulted in an increase in residence time. The breakthrough time increased as the diameter of the column increased. An increase in column diameter corresponds to a greater volume, and therefore an increase in the breakthrough time.

As the mass transfer constant increased (i.e. for a better solvent), the time to reach steady-state increased. However, the steady-state $CH_4$ purity in the effluent gas stream also increased. Increasing the pressure also increased the purity of $CH_4$produced. Also, like increasing the mass transfer constant, the time to reach steady-state also increased. The effect of pressure on liquid effluent concentration of Pb and breakthrough time was negligible. Pressure does not affect liquid adsorption. For an increase in both mass transfer constant and pressure, the breakthrough time and the operation time were not affected.

Liquid and gas flow rates are possibly the two most important parameters controlling both the absorption and the adsorption processes. In the case of a landfill, the gas flow rate is constant since it is controlled by the slow biodegradation process. Therefore, in most cases, liquid is the parameter used to control the purity of the effluent gas. When liquid flow rate is increased, the steady-state purity of $CH_4$ increased. However, when the liquid flow rate increased, the liquid effluent concentration increased and the breakthrough time decreased. However, the operation time was decreased due to a shorter breakthrough time. The flow rate of liquid must then be manipulated to optimize the production of $CH_4$ while ensuring that the zeolite is not saturated too quickly. Manipulating the gas flow rate also has advantages and disadvantages. Increasing the gas flow rate decreases the effluent $CH_4$ purity and the time to reach steady-state. Therefore, to increase the purity, gas flow rate must be decreased. However, decreasing the gas flow rate decreases the overall $CH_4$ production rate. Therefore, both liquid and gas flow rates must be optimized to produce the maximum amount of high purity $CH_4$.

## Conclusions

An extension to the model developed by Omar and Rohani (2017) was presented to account for the transient behavior of a combined absorption/adsorption column. The column consists of zeolite particles used for column packing to increase mass transfer area. The gas being purified is LFG (50% $CH_4$/50% $CO_2$). The absorbent used is landfill leachate.

The model examines the effect of different parameters on the steady-state $CH_4$ purity produced from the LFG, the time to reach steady-state operation and the breakthrough time of Pb on the zeolite packing. Increasing the height of the column, increased both the purity and the breakthrough time of the column. Therefore, it is advantageous to increase column height since it increases purity of $CH_4$ and the time that absorption can occur due to an increase in breakthrough time. An increase in the column diameter increased the purity of the methane and increased the breakthrough time, meaning more gas can be produced before the operation is halted. However, the time to reach steady-state increased for both an increase in height and diameter of the column. Flooding must be avoided when calculating column diameter. By changing the solvent, the mass transfer coefficient can be increased to increase the $CH_4$ purity in the effluent gas stream. However, the goal of using a combined absorption/adsorption column is to remove heavy metals while purifying LFG. By increasing pressure, the purity of $CH_4$ increased. The mass transfer coefficient and pressure did not affect the breakthrough time. Therefore, higher purity of $CH_4$ can be achieved without sacrificing operation time. However, in both situations, the time to reach steady-state also increased. When liquid flow rate increased, the purity of the $CH_4$ increased. However, the breakthrough time decreased and thus the operation time decreased, meaning less $CH_4$ would be produced before operation is stopped. Therefore, there is a trade-off when increasing liquid flow rate. When the gas flow rate is decreased, the purity of $CH_4$ increased. However, less gas is purified. Therefore, there is also a trade-off when changing the gas flow rate. This model provides a method to optimize conditions for the $CH_4$production as well as treating leachate.

## References

Omar H, Rohani S (2015) Treatment of landfill waste, leachate and landfill gas: A review. Front Chem Sci Eng 9:15–32. doi: 10.1007/s11705-015-1501-y

Omar HM, Rohani S (2017) Removal of CO2 from landfill gas with landfill leachate using absorption process. Int J Greenh Gas Control 58:159–168. doi: 10.1016/j.ijggc.2017.01.011

Tontiwachwuthikul P, Meisen A, Lim CJ (1992) CO2 absorption by NaOH, monoethanolamine and 2-amino-2-methyl-1-propanol solutions in a packed column. Chem Eng Sci 47:381–390. doi: 10.1016/0009-2509(92)80028-B

# 500478 Modeling Methods for Concentrating a Formic Acid Product Generated from a Novel Electrochemical Reduction of $CO_2$ Cell Design

**Abstract:**

Formic acid is a commercial industrial chemical product that has various applications such as a chemical used in leather tanning, a preservative in silage (animal feed), non-salt deicer chemicals, oil field stimulation chemicals (replacing HCl), and as a chemical feedstock for producing various chemical derivatives. Annual worldwide production is currently about 900,000 tonnes/year.

One of the biggest potential markets is in the use of formic acid as a hydrogen storage chemical, It can be easily stored as a solution and can be readily converted to hydrogen using decomposition catalysts. A number of researchers are investigating formic acid for energy storage and in fuel cells. Alternatively, formic acid can also be converted to CO, which can then be used for producing renewable fuels if the $CO_2$ feed source is from biomass, power plants, or air.

Dioxide Materials has recently developed a novel cell design for producing a pure formic acid product from the electrochemical reduction of $CO_2$. The electrochemical cell utilizes a 3-compartment design, consisting of an anode compartment, center flow compartment, and a cathode compartment. The center flow compartment is bounded by a cation exchange membrane on the anode side and a PSTMIM based Sustainionâ„¢ X-37 anion membrane on the cathode side. A nanoparticle Sn-based GDE cathode in combination with a PSTMIM imidazolium-based ionomer are used in efficiently reducing the $CO_2$ feed to formate ions, which migrate into the center flow compartment. A pure formic acid product is recovered from the center flow compartment using a DI water carrier stream. A cation ion exchange media is employed as an efficient conductive electrolyte media in the center flow compartment. DI water is also used as an anolyte in the anode compartment. The electrochemical cell operates at a current density of 140 $mA/cm^2$ at a cell voltage of 3.50 V, with Faradaic efficiencies near 90%.

The pure formic acid product from the electrochemical system can range from 10 to 25 wt%. The next step in commercializing the technology is determining the best economic methods for concentrating the formic acid to a commercial product concentration of 85 to 99 wt%. Process modeling software, such as ChemCad and Aspen, is being used to model various distillation configurations to determine the most economically effective design. The results of the modeling will be discussed in the presentation.

# 497760 Carbon Dioxide Utilization in Geothermal Power Generation and Geologic Energy Storage

**Abstract:**

Flexible renewable power systems and large-scale energy storage are needed to ensure power grid reliability, reduce carbon dioxide (CO2) emissions to the atmosphere, and increase energy security. The increased installed capacity of wind and solar energy in many parts of the United States is underutilized due to the challenges of matching supply and demand with inflexible baseload power plants (e.g., nuclear, coal). These baseload plants have high capital costs, have limited flexibility in their power output, and require high capacity factors for economic viability. Diurnal misalignments between power supply and demand are addressed either by curtailing intermittent renewables or using highly flexible, but inefficient, natural gas peaking plants. Existing bulk energy storage facilities â€" such as batteries and pumped hydro â€" can partially mitigate these supply problems, but they have high capital costs, limited storage capacity, and cannot address seasonal supply/demand mismatches. Overall, existing energy storage technologies are not sufficiently developed and deployed to eliminate power curtailments.

CO2 Plume Geothermal (CPG) technology, together with multifluid Earth Battery geologic energy storage, a variation of CPG, address these challenges, providing baseload or dispatchable, renewable energy systems that can be deployed over far greater geographic extents than legacy geothermal and geologic energy storage approaches. Both technologies make use of CO2 to store and carry energy through geologic formations to power conversion systems at the surface, either harnessing geothermal heat to be used in power generation or storing energy from other sources in the subsurface. These technologies are the result of an ongoing research and commercialization effort amongst the University of Minnesota, Lawrence Livermore National Laboratory, Ohio State University, ETH in Zurich, and TerraCOH Inc.

By using existing well infrastructure and by taking advantage of the increased mobility of CO2 when it is used as a heat extraction fluid in saline aquifers in sedimentary basins, CPG decreases drilling risks and increases power production efficiency over traditional geothermal generation. The net result is the potential for a massive increase in economically viable geothermal power production, together with an expansion of intermittent renewable energy systems that require grid-scale energy storage to maximize implementation.

Here, we will describe a project underway to provide the first field demonstration of a multi-stage CPG and Earth Battery deployment. We have strategically divided deployment into multiple phases, as necessary to balance research, technological demonstration and financial investment.

# 497782 Efficiency Limits of an Integrated Solar-Driven $CO_2$ Capture and Reduction Systems

**Abstract:**

One of the grand challenges of the 21st century is the abatement of rising concentration of $CO_2$ in the atmosphere. The anthropogenic contribution of $CO_2$ from sources like power plants and chemical industries has led to the rise in $CO_2$ concentration at the rate of ~1.8 ppm per year in the atmosphere severely affecting the earthâ€™s climate. There is a dire need to develop efficient technologies for integrated $CO_2$ capture and reduction to sustainably produce carbon-neutral or carbon-negative fuels. Artificial photosynthesis is one of the promising techniques that utilizes solar energy and $H_2O$ to reduce $CO_2$ into fuels. However, the integration of a suitable $CO_2$ capture technique with the artificial photosynthetic system has not been evaluated or implemented. Here, we show the thermodynamic and realistic efficiencies of an integrated system that captures $CO_2$ from air/flue gas and simultaneously reduces it to produce carbon-neutral fuels using solar energy. Effectiveness and compatibility of various carbon capture technique such as an ideal process, adsorption, membrane separation, absorption, and electrodialysis in an integrated system have been evaluated using a real triple junction light absorber and ideal light absorbers of varying junctions (1 to 10) and band-gaps. The overall solar-to-fuel (STF) efficiencies for the $CO_2$ reduction products for the ideal multijunction light absorbers are in the range of 19-27% for single junction, 34-40% for double junction, and 24-34% for triple junction ideal light absorber. For the real triple junction light absorber, membrane separation coupled with $CO_2$ reduction was found to be the best configuration of an integrated system.

# 498190 A Process Integration Approach for a Sustainable Gtl Process Using Tri-Reforming of Methane

**Abstract:**

Reforming of methane to produce synthesis gas is one of the key components of a Gas to Liquid (GTL) process. This process can be carried out primarily using three technologies; namely the Steam reforming(SRM), Partial Oxidation (POX) and Dry reforming of methane(DRM). Each of the aforementioned processes have their advantages and limitations and differ in terms of oxidant used. Compared to SRM and POX reforming which utilize steam and oxygen, DRM uses carbon dioxide as oxidant. This capability of Dry reforming to utilize two greenhouse gases (methane and carbon dioxide) has therefore attracted due interest in many of the current research studies around the globe. Of the three aforementioned processes, SRM and POX are conventionally utilized processes, and have been implemented in industry since mid-20$^{th}$ century. While DRM is a relatively new process, and has not been implemented commercially due to critical process limitations like high endothermicity and coke formation. As an extension to our previous work in Challiwala et al. (2017) and Gabriel et al. (2014), the present work is targeted to utilize DRM process through a combined DRM/SRM/POX tri-reforming reaction with an objective to improve $CO_2$ footprints and water/energy nexus potential in a typical GTL superstructure. Some of the preliminary optimization results show that 47% fixation of $CO_2$ can be achieved per pass through tri-reforming reaction with energy requirements much lower than the individual SRM and DRM processes. The combined model also aims to incorporate transport effects in the reformer reactor in a separate CFD study in COMSOL® to make the process more realistic as opposed to thermodynamic equilibrium results. Therefore, a major effort of the current work is to combine the kinetics of the individual reforming reactions that could be compatible to accurately predict the behavior of a tri-reforming reaction. The developed combined tri-reforming based GTL superstructure is further targeted to compete with a base case scenario of Auto thermal reforming process (ATR, a combination of SRM and POX) in terms of water and energy requirements with overall optimization target to reduce $CO_2$ emissions. This work is a part of a broader $CO_2$ management project that aims for a multiscale experimental/simulation study from catalyst to reactor design in order to enhance $CO_2$ fixation in a reforming process.

References:

1. Challiwala, M. S., M. M. Ghouri, P. Linke, M. M. El-Halwagi, and N. O. Elbashir. 2017. 'A combined thermo-kinetic analysis of various methane reforming technologies: Comparison with dry reforming', Journal of $CO_2$ Utilization, 17: 99-111.
2. Gabriel, Kerron J, Patrick Linke, Arturo JimeÌnez-GutieÌrrez, Diana Yered MartÃnez, Mohamed Noureldin, and Mahmoud M El-Halwagi. 2014. 'Targeting of the water-energy nexus in gas-to-liquid processes: A comparison of syngas technologies', Industrial & Engineering Chemistry Research, 53: 7087-102.

# 498807 Ozonolysis in Liquid $CO_2$ As a Platform for Processing of Distributed Feedstocks

**Abstract:**

Given its high oxidation potential, there has been extensive interest in the use of ozone as an oxidant in the synthesis of various chemicals. Surging interest in making chemicals from biomass feedstocks has generated renewed interest in ozone as an oxidant. The ability to produce ozone on site is well aligned with the distributed biorefinery concept and eliminates problems associated with biomass transport and storage. However, wider application of ozone in industrial processes has been tempered by safety concerns often associated with ozonolysis.

We demonstrate a continuous reactor for performing safe ozonolysis of fatty acid methyl esters (FAMEs) using liquid $CO_2$ as solvent. The rapid nature of the reaction allows the use of small-volume reactors to completely convert the FAMEs, forming secondary ozonides as the primary products. The short residence times also help maximize the yields of the secondary ozonides by minimizing over-oxidation and the formation of oligomeric products. The liquid $CO_2$ medium promotes safe reactor operation by providing an essential fraction of overall reactor cooling and by diluting the vapor phase organics. We also demonstrate a continuous stirred reactor for the safe thermal decomposition of the secondary ozonides to their corresponding acids and aldehydes. Using a lumped kinetic model for the thermal decomposition of the ozonolysis products, we estimate activation energy values of $108.6\pm0.6$ kJ $mol^{-1}$ for the decomposition of secondary ozonides and $122\pm3$ kJ $mol^{-1}$ for the decomposition of the undesired oligomeric species.

The extension of this concept for valorizing lignin and stranded gas sources (such as natural gas liquids) will also be discussed.

# 499203 Understanding the Effects of Carbon Dioxide and Bicarbonate on *Chlamydomonas reinhardtii*

**Abstract:**

## Title:

Understanding the effects of carbon dioxide and bicarbonate on *Chlamydomonas reinhardtii*

## Authors:

Humeyra B. Ulusoy-Erol[1], Jamie A. Hestekin[1], Christa N. Hestekin[1], YuPo J. Lin[2], Catherine Atchley[1], and Benjamin Drewry[1]

[1] Ralph E. Martin Department of Chemical Engineering, University of Arkansas, Fayetteville,

AR, United States.

[2]Energy Systems, Argonne National Laboratory, 9700 South Cass Avenue, Argonne, IL, United States.

## Abstract

Inorganic carbon has an important role on algae growth and lipid production systems. Even though most studies focus on carbon dioxide ($CO_2$) as an inorganic carbon ($C_i$) source, in seawater inorganic dissolved carbon is in the form of bicarbonate ($HCO_3^-$). Furthermore, during intense photosynthesis of microalgae, pH in the media tends to increase which leads to low $CO_2$ concentration. For commercial production of algae in applications including biofuel production, carbonate salts can provide an adequate source of C for high rate algae growth. *Chlamydomonas reinhardtii*, a single-celled, highly adaptable green alga, was used to understand the effects of C source for algae growth system. Bold's Basal Medium (BBM), Tris-Acetate-Phosphate (TAP) medium, and BBM with the addition of sodium carbonate (BBMNa) on microalgae growth culture were analyzed. The results in different media indicated that the algae grown on BBMNa had higher biomass productivity (~20%) and produced more valuable fatty acids for the biodiesel industry than BBM and TAP mediums.

17AIChE Annual Meeting, Minneapolis, MN

# Power-to-gas : dynamic modeling of a catalytic methanation reactor

**Axel Fache, Laboratoire de thermique, énergétique, procédés (LaTEP), Pau, France**
a.fache@univ-pau.fr latep.univ-pau.fr

LaTEP/ EA 1932 LABORATOIRE de THERMIQUE ÉNERGÉTIQUE et PROCÉDÉS

RÉGION Nouvelle-Aquitaine

leaf

1

# Plan

**Context**
- General context
- Stakeholders
- Purpose of our project

**Mathematical model**
- Simulated reactor – main assumptions
- From the tube to the reactor
- Simulator capabilities

**Examples of dynamic simulation results**
- Feed gas re-injection after a 2h or a 20h idle-state

**Conclusion**

2

# Context

3

# General context

- Renewable electricity surplus ➔ electrolysis ➔ $H_2$
- Captured $CO_2$

**Catalytic methanation :**

$$CO_2 + 4.H_2 \rightleftharpoons CH_4 + 2.H_2O \ (\Delta_r H^0 = -165\ kJ.mol^{-1})$$

Context 4

## Stakeholders

→ **LEAF** : French startup company

→ **LaTEP** : technical study of the reactor (thermal issues, fluid mechanics, chemical efficiency...)

→ **French administrative region Aquitaine** : funding



## Purpose of our project

**Goal : Dynamic operation of the reactor** (Electricity surplus variations )

**Issues** :
- Thermal instability : positive feedback (reaction vs. safety; nonlinear transient behavior).
- Thermal inertia (catalytic medium, tubes walls, coolant) : which impact on dynamic operation?

# Purpose of our project

## Thermal instability and inertia - transient wrong-way behavior

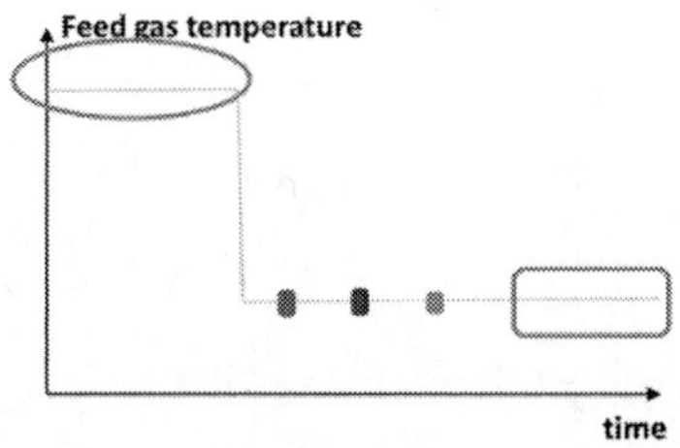


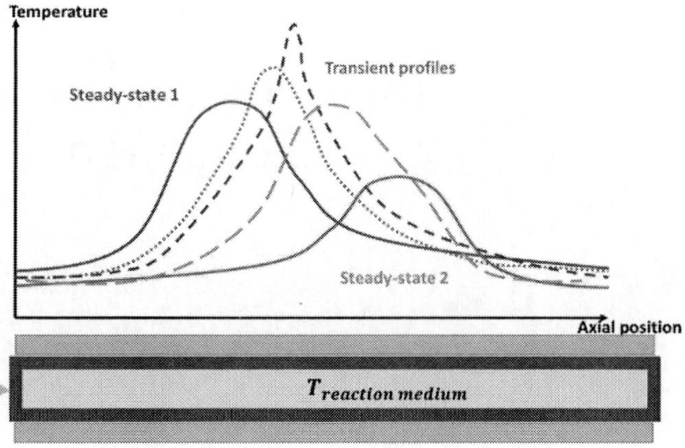


Context

7

# Purpose of our project

**Workplan :**

- **Simulating transient sequences to anticipate the <u>intrinsic</u> behavior of reactors, for the usual designs**
- **Dynamic optimization (sizing and control strategy)**
- **Technological solution proposal**

Context

8

# Mathematical model
# Simulation

9

## Simulated reactor – main assumptions

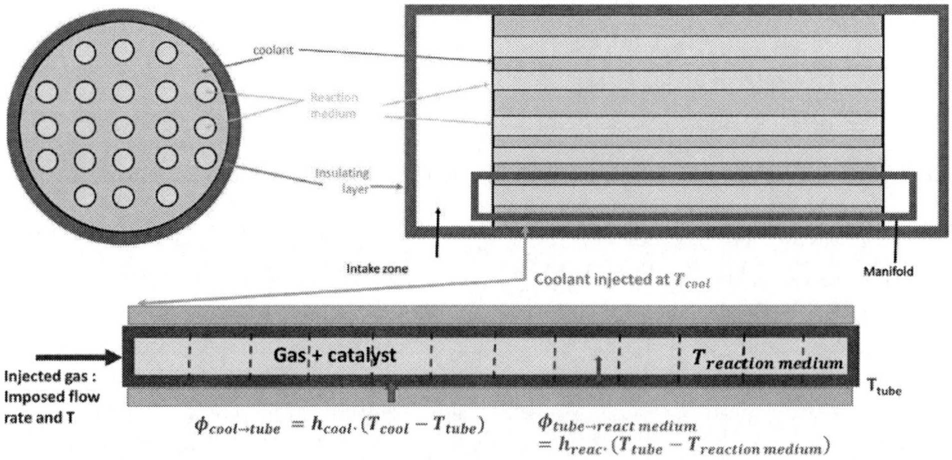


➢ **1-D discretization** in reaction medium, coolant, tube wall.

Mathematical model & simulation

10

## Simulated reactor – main assumptions

➢ **Pseudo-homogeneous model** : catalyst pellets and gas mixture are intimately mixed

❖ 3 reactions :

| | |
|---|---|
| 1- $CO_2$ methanation | $CO_2 + 4H_2 \leftrightarrow CH_4 + 2H_2O$ |
| 2- rWGS | $CO_2 + H_2 \leftrightarrow CO + H_2O$ |
| 3- CO methanation | $CO + 3H_2 \leftrightarrow CH_4 + H_2O$ |

**Kinetics by Xu & Froment**

Methane Steam Reforming, Methanation and Water-Gas Shift- 1. Intrinsic Kinetics, AIChE Journal vol.35 n°1, 1989

•5 species :

| | |
|---|---|
| 1 | $CO_2$ |
| 2 | $H_2$ |
| 3 | $CH_4$ |
| 4 | $H_2O$ |
| 5 | CO |

Mathematical model & simulation 11

## From the tube to the reactor

**Extensive quantity at reactor scale = quantity at tube scale x number of tubes (no side effect)**

➢ Mass, momentum and energy balance in the reaction medium
➢ Energy conservation in the coolant
➢ Energy conservation in the tubes walls

**When size effects exists : quantities are calculated based on the whole reactor size**

➢ Thermal loss
➢ Energy balance in miscellaneous lids/walls

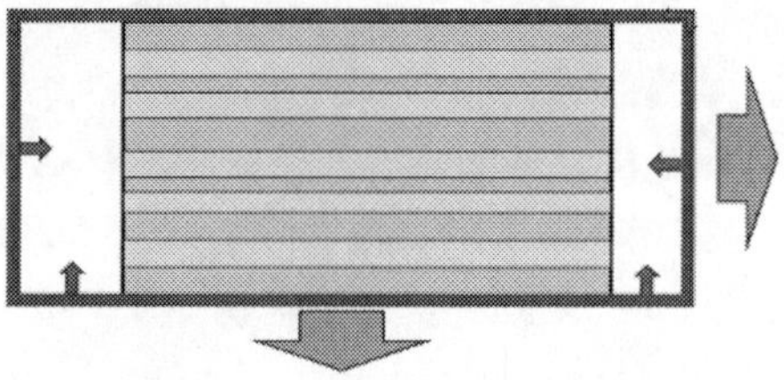

Mathematical model & simulation 12

# Simulator capabilities

Given:

- Design parameters ($R_{tube}$, $nb\ of\ tube,\ e_{tube},\ distance\ between\ tubes$ ... )
- A sequence to simulate ($Q_{v_{reactants}}$, $T_{cool}$ ... duration of flow ramps...)

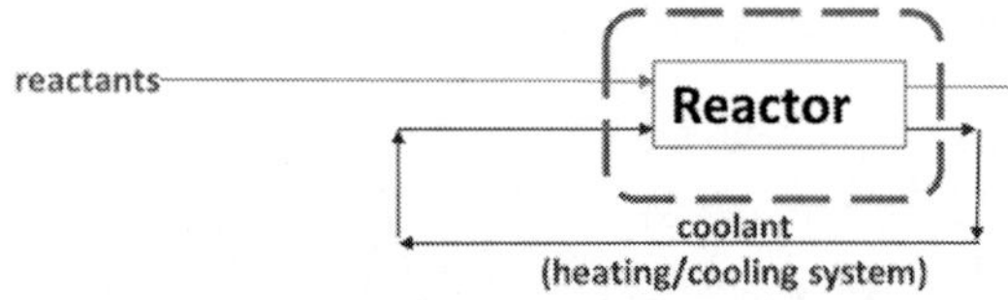


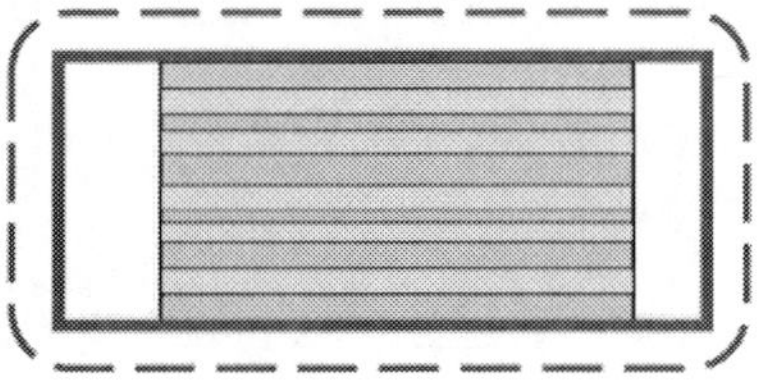

The simulation tool gives the **evolution, through time**, of :

- Local composition of the gas mixture (1D) in the reactor, outflow rates
- Local temperatures(1D) of reaction medium, tube wall, coolant
- Local and overall exchanged heat (including thermal loss from the reactor to the outside)
- Numerous quantities of interest (chemical conversion/yield, safety criteria...)



# Examples of dynamic simulation results

# 2 h idle-state followed by a steep re-injection of cold reactants

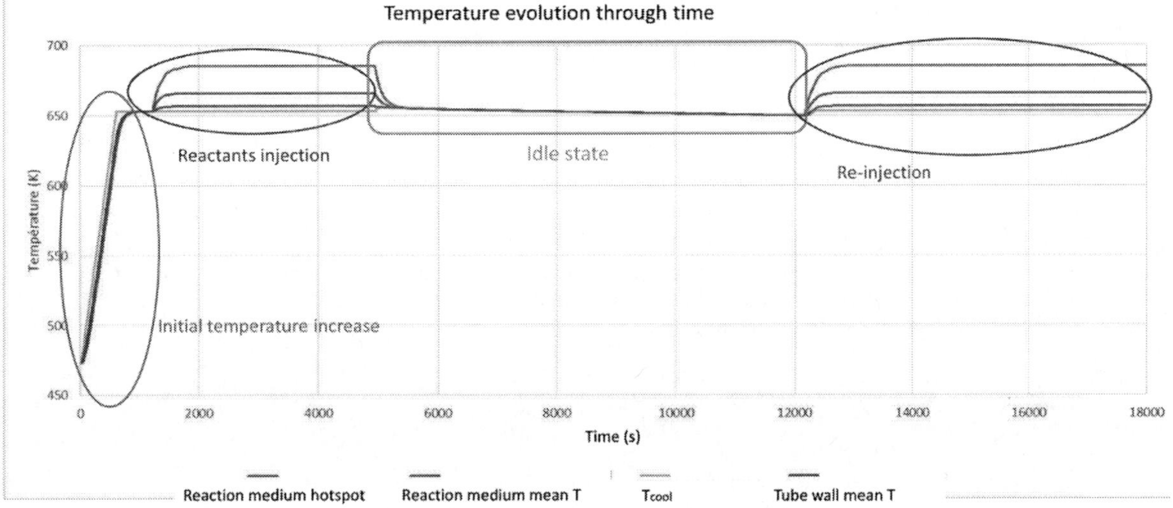


Examples of dynamic simulation results 15

# 20 h idle-state followed by a steep re-injection of cold reactants

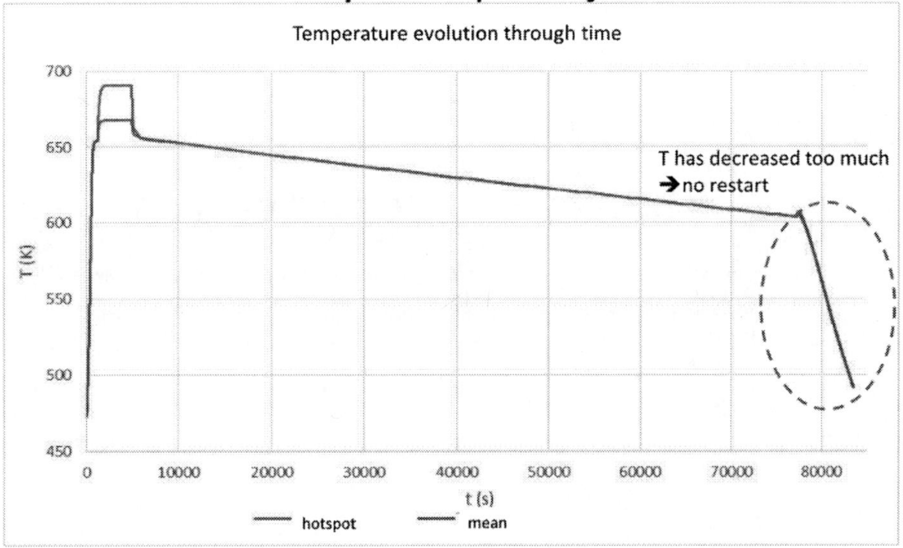


Examples of dynamic simulation results 16

# Conclusion

- The computation code simulates dynamic methanation, and evaluates chemical/thermal/safety performance through time

- Prospects :
  - Model improvement : using heterogeneous 2-D model, taking into account more transient specificity
  - Model calibration/validation, thanks to experimental data
  - Reactor optimization (sizing, control)

- A patentable technological solution has been found

17

# Thanks for your attention

**Axel Fache, Laboratoire de thermique, énergétique et procédés, Pau, France**
**Corresponding address :** a.fache@univ-pau.fr latep.univ-pau.fr

18

# 488287 The Effect of Water on $CO_2$ Capture By AHA Ionic Liquids


**Abstract:**

**THE EFFECT OF WATER ON $CO_2$ CAPTURE BY AHA IONIC LIQUIDS**



Gabriela Avelar, University of Notre Dame

Oscar Morales, University of Notre Dame

Joan Brennecke, University of Notre Dame



Key words: Ionic liquids, water, $CO_2$ capture.



In the context of postcombustion $CO_2$ capture, ionic liquids (ILs) are a promising alternative to traditional absorbents such as aqueous monoethanolamine solutions. One key advantage to ILs is their tunable enthalpy of reaction for $CO_2$ binding. ILs with aprotic heterocyclic anions (AHA) have been specifically designed for this application[1,2]. The AHA ILs are very promising, because they react stoichiometrically with $CO_2$, with modest enthalpy of reaction; and they do not present any increase in viscosity after $CO_2$ bonding, in contrast to the traditional ILs that chemically bind with $CO_2$.

Water is a very important impurity to consider in this process, since the flue gas contains a considerable amount of it, which is further complicated by the ability of water to change the IL viscosity, in addition to changing the chemistry of the $CO_2$ binding reaction. A full understanding of the reaction mechanism, kinetics and equilibrium values in the IL-$CO_2$-$H_2O$ system is therefore crucial to postcombustion $CO_2$ capture by AHA ionic liquids.

We propose a reaction mechanism where in addition to the reaction of the anion with $CO_2$ to form carbamate, the anion reacts with water, and is reprotonated, leaving $CO_2$ to react with hydroxide to form bicarbonate, as previously suggested for aminoacid[3], triazolide[4] and phenolic[5] ionic liquids. According to the proposed mechanism a decrease in the amount of water will correspond to the amount of reprotonated anion.

In order to quantify the amount of reprotonation, a proton NMR spectroscopy method was developed. The change in the amount of water was measured for different ionic liquids and water mixtures exposed to $CO_2$ by comparison of the integration of the water peak before and after $CO_2$ exposure. The rate of the IL-$CO_2$-$H_2O$ reaction was studied by proton NMR spectroscopy, as well. Finally, the $CO_2$ solubility in the AHA ILs in the presence of water and reversibility of the reaction were investigated using a volumetric method.

It was found that several AHA anions undergo reprotonation in the presence of water and $CO_2$, although the reaction appears to be reversible. The rate and equilibrium values, as well as the $CO_2$ uptake values in the presence of water, vary significantly with the nature of the anion.


References:


1. Seo S., Quiroz-Guzman M., De Silva A.M., Lee T.B., Huang Y., Goodrich B.F., Schneider W.F, Brennecke J.F. Chemically Tunable Ionic Liquids with Aprotic Heterocyclic Anion (AHA) for $CO_2$ Capture. *J. Phys. Chem. B.* 2014; 118:5740-5751.
2. Gurkan B., Goodrich B.F., Mindrup E.M., Ficke L.E., Massel M., Seo S., Senftle T.P., Wu H., Glaser M.F., Shah J.K., Maginn E.J., Brennecke J.F., Schneider W.F. Molecular Design of High Capacity, Low Viscosity,Chemically Tunable Ionic Liquids for $CO_2$ Capture. *J. Phys. Chem. B.* 2010; 1:3494-3499.
3. McDonald J.L, Sykora R.E., Hixon P., Mirjafari A., Davis J.H. Impact of water on $CO_2$ capture by amino acid ionic liquids. *Environ. Chem. Lett.* 2014; 12:201-208.

1. Thompson R.L., Shi W., Albenze E., Kusuma V.A., Hopkinson D., Damodaran K., Lee A.S., Kitchin J.R., Luebke D.R, Nulwala H. Probing the effect of electron donation on CO2 absorbing 1,2,3-Triazolide ionic liquids. *The Royal Society of Chemistry.* 2014; 4:12748-12755.
2. Wang C., Luo H., Li H., Zhu X., Yu B. and Dai S. Tuning the Physicochemical Properties of Diverse Phenolic Ionic Liquids for Equimolar CO2 Capture by the Substituent on the Anion. *Chem. Eur. J.* 2012; 18: 2153 â€“ 2160.

# 500252 Acrylonitrile Production from Biomass-Derived Intermediates

**Abstract:**

Carbon fibers are key materials in critical energy applications including light-weighting of vehicles for improved fuel efficiency and enabling larger wind turbine designs for increased power generation. Carbon fibers are produced through heat treatment of polyacrylonitrile fibers. Today, the acrylonitrile monomer is made primarily via ammoxidation of propylene with ammonia. However, price fluctuations in the feedstocks and the energy-intensive processes use to make carbon fibers have resulted in low-volume and niche adoption of carbon fibers into automotive applications, motivating the drive towards carbon fibers produced cheaply and reliably from renewable biomass sources. This talk will describe efforts in the Renewable Carbon Fiber Consortium to pursue acrylonitrile production from biomass-derived sugars via hybrid biological and chemo-catalytic processes. In particular, bench-scale efforts in two separate process schemes will be described that proceed via biological production of 3-hydroxypropionic acid or lactic acid, followed by subsequent separation and catalytic conversion to acrylonitrile.

# 501349 Multistream Integrated Biorefinery (MIBR) for Carbon-, Road-, and Bio-Materials

**Abstract:**

The success of a modern biorefinery heavily depends on the availability of diverse product streams and utilization of all fractions of input material. Traditional lignocellulosic biorefining focuses on ethanol, which has limited by-product streams that add value to the biorefinery. Essentially all current bioconversion platforms lead to a lignin-containing waste stream that needs further processing into valuable products. Although a certain amount of lignin (~30â€“40%) is needed for the thermal requirements of biofuel production, a modern cellulosic processing plant will have ~60% excess lignin. The utilization of the lignin-containing biorefinery waste as feedstock for renewable products offers a unique opportunity to achieve a multi-stream integrated biorefinery (MIBR), where the lignin-containing biorefinery waste will be utilized for various carbon-, road- and bio- materials to maximize economic return and sustainability. Specifically, we have advanced the fractionation, conversion, and processing technologies to enable different bioproduct streams. An efficient enzyme-mediator system was developed to fractionate lignin into low- and high-molecular weight fractions with more homogenous chemical characteristics. We have demonstrated that these lignin fractions are more amenable to different applications in bioconversion, asphalt binder modifier, and carbon fiber. The low molecular weight fraction is more amenable to bioconversion into PHA for bioplastics. In addition, the low molecular fraction can be used as unique asphalt binder modifiers to improve both high and low temperature performance of road pavement. The high molecular weight fraction can be used to fabricate carbon fiber with significantly improved mechanical performance. Together, MIBR will reduce ethanol production cost through the creation of high value bioproducts, produce asphalt binder modifier with unique features, develop an innovative, new pathway to quality carbon fiber; and create a means to utilize all carbon in the feedstocks.

# 492807 Improving Economics of Cellulosic Biofuels: An Integrated Strategy for Co-Producing 1,5-Pentanediol and Bioethanol

**Abstract:**

Lignocellulosic biomass is an abundant and sustainable feedstock that can be used to produce biofuels and biobased commodity chemicals. However, the low oil price and the challenges in applying biorefinery technologies at scale hamper the commercial production of biofuels. In this presentation we will discuss a strategy to integrate the production of a high-value coproduct, 1,5-pentanediol (1,5-PDO), along with a typical bioethanol production process.

1,5-PDO and two other conventional α,ω-diols (e.g. 1,6-hexandiol and 1,4-butanediol), are building blocks for polyurethane and polyester plastics. These α,ω-diols are currently derived from petroleum and, as a group, represent an annual market over $6 billion. An economically viable approach of producing 1,5-PDO from biomass-derived furfural has been developed recently (Brentzel et al., 2017; Huang et al., 2017), which offers the potential to coproduce 1,5-PDO as a high value co-product of bioethanol manufacturing process to improve cellulosic biorefinery economics.

In this work, we propose a novel strategy that integrates lignocellulosic biomass fractionation with simultaneous conversion of cellulose and hemicellulose into ethanol and 1,5-PDO, respectively. First, we fractionate biomass into its three primary constituents— hemicellulose, cellulose, and lignin—using γ-valerolactone (GVL) as a solvent (Alonso et al., 2017), which also enables excellent yields of furfural by dehydration of xylose. The purified furfural is then used to produce 1,5-PDO (Huang et al., 2017), and the cellulose portion is converted to ethanol via enzymatic hydrolysis and fermentation (Humbird et al., 2011), while the lignin is combusted to generate steam and electricity. We have developed an experimentally based process simulation model (mass and energy balances) to determine the economic potential of the integrated strategy. If we consider a plant processing to 1,000 dry metric tonne per day white birch feedstock and producing 38,520 ton per year 1,5-PDO and 19.3 MMgal per year ethanol, the integrated strategy leads to a minimum ethanol selling price of $1.52 per gallon when 1,5-PDO is sold at $3,100 per ton (Figure 1). It is shown that this strategy could lower the cost of ethanol produced from biomass by more than two dollars per gallon. Finally, we perform sensitivity analysis to identify the major economic drivers of the process and suggest future research directions.

**Figure 1. Minimum ethanol selling price as a function of 1,5-PDO selling price.** The "green" region (1,5-PDO selling price > $2,775 per ton) indicates that the integrated strategy has favorable economics for ethanol production compared with the biochemical conversion process used in NREL 2011 report (Humbird et al., 2011).

**References**

1. Alonso, D. M.; Hakim, S.; Zhou, S.; Won, W.; Hosseinaei, O.; Tao J.; Garcia-Negron, V.; Motagamwala, A. H.; Mellmer, M. A., Huang, K.; Houtman C. J.; Labbé, N.; Harper, D. P.; Maravelias, C. T.; Runge T.; Dumesic, J. A. Increasing the revenue from lignocellulosic biomass: Maximizing feedstock utilization. *Science Advances* 2017.

2. Brentzel, Z. J.; Barnett, K. J.; Huang, K.; Maravelias, C. T.; George, W.; Dumesic, J. A.; Huber, G. W. Chemicals from Biomass: Combining Ring-opening Tautomerization and Hydrogenation Reactions to Produce 1,5-Pentanediol from Furfural. *ChemSusChem* 2017, 133, 12689.

3. Huang, K.; Brentzel, Z. J.; Barnett, K. J.; George, W.; Dumesic, J. A.; Huber, G. W.; Maravelias, C. T. Conversion of Furfural to 1,5-Pentanediol: Process Synthesis and Analysis. *ACS Sustainable Chemistry & Engineering* 2017.

**4.** Humbird, D.; Davis R.; Tao L.; Kinchin C.; Hsu D.; Aden A. Process Design and Economics for Biochemical Conversion of Lignocellulosic Biomass to Ethanol: Dilute-Acid Pretreatment and Enzymatic Hydrolysis of Corn Stover. *NREL/TP-5100-47764*. Golden, CO: National Renewable Energy Laboratory, 2011.

# 498859 High Value Products from a Photobioreactor-Based Biorefinery


**Abstract:**

Though Algenol continues research and development activities aimed at biofuels (ethanol and biocrude), increasing emphasis is being placed on high value products and co-products that can both provide sustainable income and serve as a market entry/risk reduction strategy for eventual development of biofuels. Algenol has successfully built and operated a large pilot scale (2 acre) integrated biorefinery (IBR) based on its proprietary photobioreactors (PBRs) and genetically modified cyanobacteria to produce fuel grade ethanol via an EPA approved pathway (70% carbon footprint reduction compared to gasoline). Ethanol productivities in the range of 4000-5000 gal/acre-yr (annualized) have been achieved. However, the system is not economic under current market conditions. The IBR and R&D support infrastructures are now being applied to higher value products, including a biocrude co-product. Though several products are under active consideration, algae based soil amendment is one application that is well advanced in our product development pipline. The soil amendment contains nitrogen-fixing cyanobacteria as the active component (based on *Anabaena* and/or *Nostoc* strains). The formulated product provides enhanced yields as a biofertilizer, can be used as a soil amendment for reclamation of degraded soil, and can result in high carbon storage levels in the soil.[1,2] This talk will use the soil amendment application as an example and provide a description of the selection process for product development (e.g., techno-economics, life cycle impacts, compatibility with PBR growth systems) and the development process from the lab, through outdoor testing, and then to pilot scale deployment. In particular, the alignment of indoor and outdoor performance will be demonstrated for several strains relevant to the above application, along with a modeling system that has been developed for productivity prediction for any rational position on the earth's surface, and for various configurations for Algenol's vertically-oriented PBR arrays. Harvesting and downstream processing will be described and experimentally demonstrated for several cyanobacterial strains, including strains designed for biocrude production. A comparison of the relative productivities and economics of PBR systems versus open pond systems will be discussed.


[1]Accelergy Corporation, *"Montana Integrated Carbon to Liquids Study,"* US DOE Award Number: DE-FE0003595, December 2013.

[2]US Patent2014/0345341, November 27, 2014; *"Integrated Process for Producing Fuels and BioFertilizers from Biomass and Products Produced."*

# 499440 A Highly Selective Dehydration of D-Xylose and Wheat Straw $C_5$-Sugars into Furfural By Using Supercritical $CO_2$ as Catalyst: A Green and Efficient Approach

**Abstract:**

Crude oil is the principal feedstock for production of liquid fuels and also affords a platform for synthesis of a wide range of added-value chemicals and materials. Due to price fluctuation of crude oil-derived products coupled to increasing concerns about the impact of fossil emissions on environment, society is becoming more environmentally conscious. Thus, traditional chemical technologies for production of fuels and chemicals from biomass are being replaced with innovations. This includes the use of novel and green solvents. In this work, the selective dehydration of both D-xylose and lignocellulose-derived $C_5$-sugars into furfural, using supercritical $CO_2$ as catalyst and tetrahydrofuran (THF) as extracting solvent, is for the first time reported. The scrutinised approach involves the benefits from the use of auto-generated carbonic acid from dissolution of $CO_2$ in aqueous phase, assisted by THF, guiding to the improvement of furfural yield and reaction selectivity. This novel approach allowed to achieve a xylose conversion into furfural above 83 %mol, furfural yield of 70 %mol and selectivity up to 84 %mol at 180 ºC for 60 min with 50 bar of initial $CO_2$ pressure.[1] At aforementioned operational condition, furfural yields and selectivities up to 40 %mol and 44 %mol with respect to wheat straw hemicellulose hydrolysate were reached.[2] First, the influence of various operating conditions on pentose conversion and furfural yields were investigated. Then, various volumetric ratios of water and THF (v/v) were investigated to help define reaction conditions that favour the highest yield of furfural. In an effort to understand how addition of $CO_2$ improves furfural yield and its impact by the presence of THF, different pressures of $CO_2$were also investigated.

**Acknowledgments**

IF/00471/2015 grants and by BBRI - Biomass and Bioenergy Research Infrastructure (ROTEIRO/0189/2013) and the Associated Laboratory for Sustainable Chemistry–Clean Processes and Technologies–LAQV which is financed by national funds from FCT/MEC (UID/QUI/50006/2013) and co-financed by the ERDF under the PT2020 Partnership Agreement (POCI-01-0145-FEDER-007265).

**References**

1. A. R. C. Morais and R. Bogel-Lukasik, *Green Chem.*, 2016, **18**, 2331-2334.

2. A. R. C. Morais, M. D. J. Matuchaki, J. Andreaus and R. Bogel-Lukasik, *Green Chem.*, 2016, **18**, 2985-2994.

# 493990 Process Design for the Synthesis of Biodiesel, Glycerol, 3-Hydroxypropionic Acid and 1,3-Propanediol


**Abstract:**

The current tendency towards the use of renewable biomass feedstocks instead of the use of non-renewable resources (for example, petroleum), has created the need for development of new, innovative production processes within the chemical and biochemical industries. Therefore, alternative solutions able to reduce the consumption of non-renewable resources and consequently minimize effects on climate change, waste, while, improving product yields, should be promoted. This replacement of petroleum with biomass as the primary material for fuel and chemical production is basically the idea behind the biorefinery concept.

This project is focusing on the ways that glycerol (an intermediate or by-product in a biorefinery) could be converted into value added platform chemicals so that its surplus production is not ending up as waste. The effective utilization of glycerol will ultimately contribute to the viability of biodiesel production since it will significantly lower its already high production cost. Thus, the design and development of a sustainable process to utilize this organic raw material becomes imperative.

The case of the simultaneous production of two platform chemicals with glycerol as a starting material has been investigated, namely of 3-hydroxypropionic acid (3HP) and 1,3-propanediol (1,3PDO). This process is part of a bigger one, where, by using vegetable oil as feedstock, the aim is to produce biodiesel. However, as glycerol is obtained as a by-product, it is then subjected to biotransformation with recombinant resting cells of *Lactobacillus reuteri* in an aqueous solution. The resting cells initially grow in the fermenter at 37°C and then the biotransformation of glycerol is performed in a stirred-tank reactor. This biotransformation has 3 main products, 3-hydroxypropionic acid (3HP) and 1,3-propanediol (1,3PDO), that are of interest as well as significant amounts of 3-hydroxypropionaldehyde (3HPA) as by-product (that is not of interest and needs to be reduced).

*L.reuteri* is selected for biotransformation for several reasons, such as, it is a safe bacteria, has a great potential as a biocatalyst, and it cannot be used as a carbon source by glycerol, unlike other bacteria. Last but not the least, apart from 3HPA, no other by-products are formed during the bioconversion of glycerol by *L.reuteri*, thereby making the downstream process much simplified.

The goal of the project is to investigate the optimal production process towards 3HP and 1,3PDO simultaneously, in comparison to the existing methods that are oriented to the production of either only 3HP [Ramakrishnan Gopi et al.,2015] or only 1,3PDO [Nakamura and Whited, 2014]. The research is orientated towards a setup of the production process that produces equal amounts of the desired products, is economically and environmentally feasible and sustainable. In addition, the existence of significant amounts of the aldehyde by-product (3HPA) needs to be avoided. Two scenarios are investigated; one where 3HPA is isolated, purified and sold in the market as a by-product, and the other, where it is recycled back to the biotransformation reactor such that the production of 3HP and 1,3PDO are favoured.

According to the results regarding the production of 3HP and 1,3PDO, it is possible to determine the quantity of glycerol that needs to be allocated for the production 3HP and 1,3PDO and the remaining that will be allocated to the market. That is, the operation will depend on the function of the demand for glycerol, its current price, as well as the demands and prices of 3HP and 1,3PDO.

The presentation will provide the modelling and design details of the integrated process with a special focus on the increase of the atom efficiency of the processes involved, on the sustainability of the production process, the environmental impact versus the production and capital costs for different scenarios of product demand and price.

# 499629 Thermally Robust Molten Salts & Ionic Liquids: Thermodynamics Properties and Phase Behavior

**Abstract:**

Ionic liquids have enjoyed much interest in the past 20 years because of their unique and tunable properties. The vanishingly low vapor pressures have lead them to be considered as environmentally benign replacement solvents for reactions and separations as they have very low risk for fugitive emissions. Additionally, as they are often comprised of organic ions, their physical and chemical properties can be tuned at the molecular level using the full palette of organic chemistry.

In addition to being solvents, there has been a great deal of interest in using ionic compounds as heat transfer fluid due to the large range of temperatures where they remain liquids. Unlike traditional molten salt heat transfer fluids, ionic liquids (which formally have melting points below 100°C) can be liquids at room temperature, and even far below 0°C. Because of this versatility, our group has begun studying ionic liquids and molten salts (those with melting points between 100 and 200°C), that are also stable at temperatures up to and above 350°C in air for weeks. Such thermal stability is achieved by limiting the functional groups to arene and substituted arenes. In doing so, the structural diversity is highly constrained and molecular design choices are limited. Previously, we have described these challenges in terms of how they limit structure/property analysis, in that it is difficult to develop homologous series which are consistently thermally stable.

In this work, we examine a new and expanded class of phosphonium- and sulfonium-based ionic compounds, present their synthesis, thermophysical properties (melting points, enthalpy and entropies of fusion, heat capacity) and the liquid-liquid equilibrium of several of these compounds with molecular aromatics. Additionally, kinetic thermal gravitational analysis of the compounds will be presented to understand the thermal stability as a function of temperature and time. These compounds are of potential interest in a number of areas related to industrial use, including use as heat transfer fluids, lubricants and separations media.

2017 AIChE Annual Meeting

Ionic Liquid: Thermodynamic and Properties

Nov. 1, 2017, Minneapolis

# Structure and Application of Hydrogen Bond in Ionic Liquid


Kun Dong

Center of Ionic Liquids and Green Engineering

Institute of Process Engineering, Chinese Academy of Sciences


## Motivation: Quantitative Relationship between Property and Structure

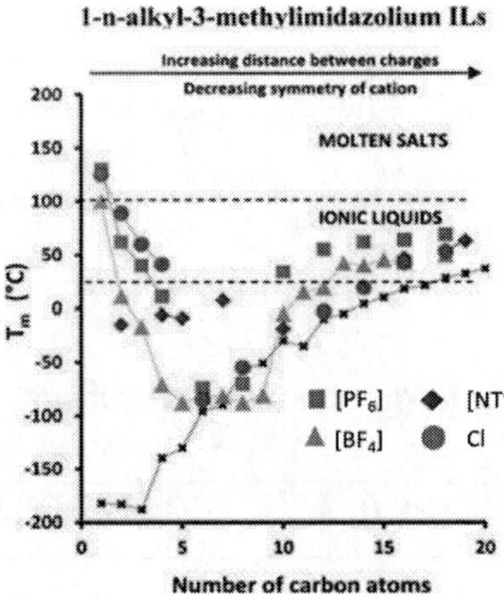

**Not only asymmetry but also ion–ion noncovalent interactions**

- ✓ *Low melting point*
- ✓ *Low vaporization pressure*
- ✓ *High thermal and chemical stability*
- ✓ *High conductivity and viscosity*
- ✓ *Tuned by combining different ions*

The enhanced asymmetry of cation reduce Coulomb forces and disrupt lattice packing the $m_p$ decrease, but when the cation alkyl chain is long enough (n > 10) cohesive energy increases, this will increase $m_p$ of the salts.

# Hydrogen bond is Possible in Ionic Liquid

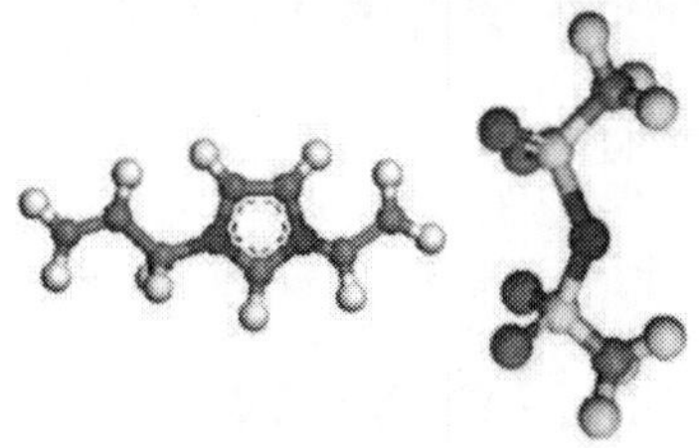

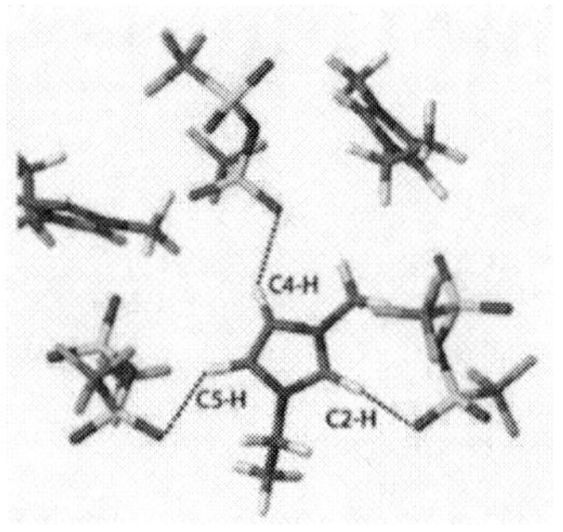


Cations are the H-bond donors, anions are the H-bond acceptors.

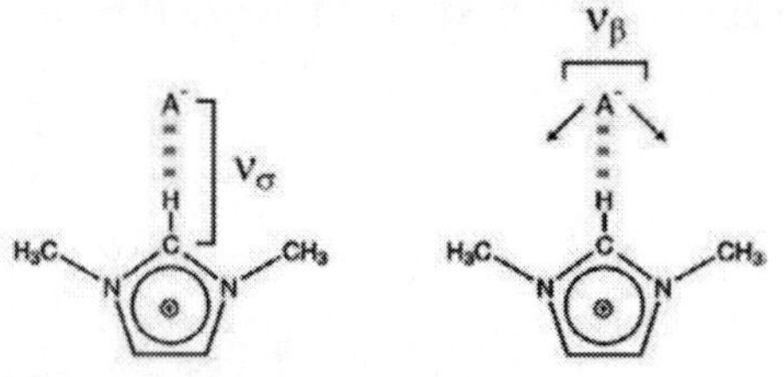


The stretching and bending vibrational modes for the $C_2$–H···A H-bond

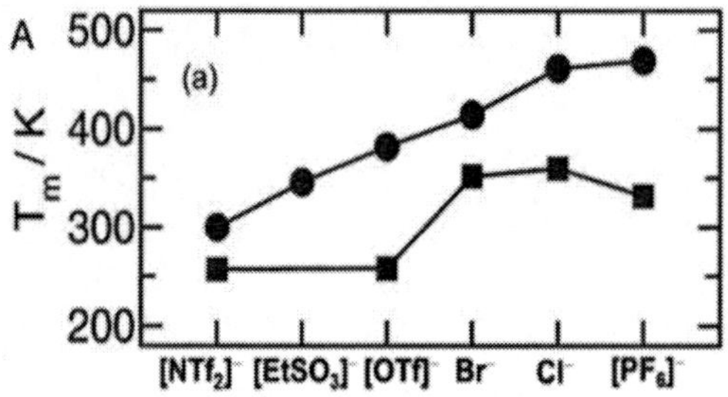


The properties of imidazolium-based ILs can be tuned by adjusting the ratio between H-bond energies and Coulomb interactions

Fumino, K. et al. *Phys. Chem. Chem. Phys.* 2011, 13, 14064

# H-bonding Geometries in EmimCl Ion-pair

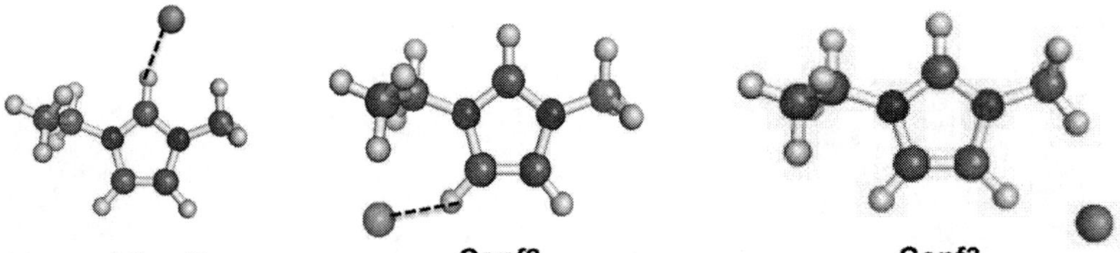

**Conf1** **Conf2** **Conf3**

**Hydrogen-bonded complexes** MP2/6-311+G(d,p)

| | R(C···Cl), Å | ∠(C-H···Cl), ° | ΔR(C-H), Å |
|---|---|---|---|
| Conf1 | 3.025 | 159 | 0.046 |
| Conf2 | 3.095 | 148 | 0.024 |
| Conf3 | 3.129 | 146 | 0.025 |
| $CH_3$-H···$Cl^-$ | 3.755 | 180 | 0.015 |

**Shorter distance, smaller angle, larger bonding change indicate the stronger H-bond**

Dong, K. et al. *J. Phys. Chem.* A 2006, 110, 9775-9782; Hunt, P. A. et at. *Chem. Eur. J.* **2006**, 12, 6762-6775

# Energetic Components

SAPT (Symmetry-Adapted Perturbation Theory) Energy are in kJ mol[-1]

| | $E_{Coul}$ | $E_{Exch}$ | $E_{Ind}$ | $E_{Disp}$ | $E_{Elst}$(tot) | $E_{Int}$ | Elst% | Ind% | Disp% |
|---|---|---|---|---|---|---|---|---|---|
| Conf1 | -434.8 | 179.7 | -63.7 | -37.8 | -359.3 | -397.1 | 90 | 16 | 9.5 |
| Conf2 | -400.3 | 163.6 | -55.7 | -36.0 | -322.0 | -358.0 | 90 | 15.5 | 10.1 |
| Conf3 | -405.2 | 166.7 | -56.7 | -40.6 | -322.2 | -362.8 | 89 | 15.6 | 11.2 |
| $H_3C$-H ··· $Cl^-$ | -12.0 | 22.7 | -11.0 | -3.0 | -8.5 | -11.5 | 26 | 95 | 26 |

$E_{Int} = E_{Elst}(tot) + E_{Disp}$ $\quad E_{Elst}(tot) = E_{Coul} + E_{Exch} + E_{Ind} + E_{Exch-Ind} + \delta E_{int}^{HF}$

ILs are not as ionic as one imagine, besides the coulomb force, **other interactions** are not negligible, the different forces compensate each other, consequently, the system is highly flexible and liquid-like.

# Structures of H-bond in Typical Ion-pairs

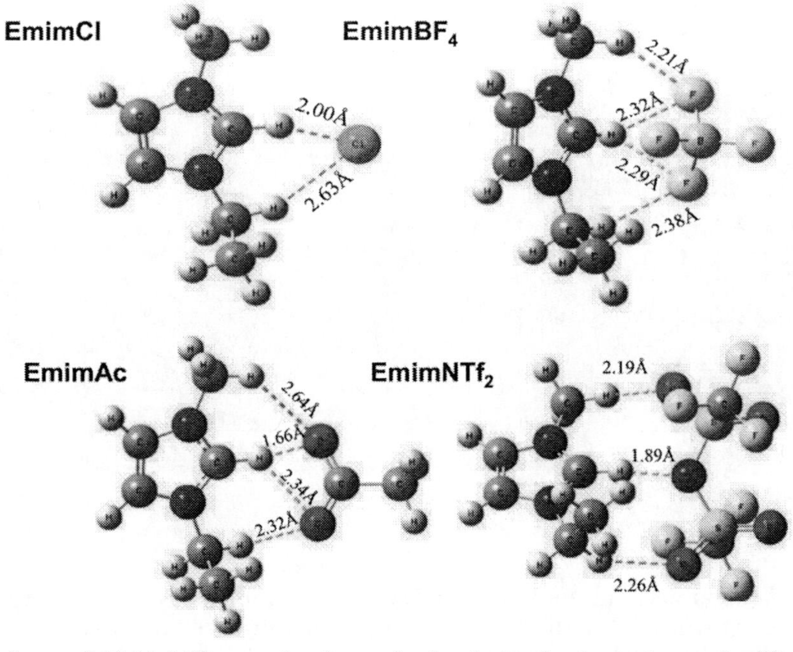


**H-bonds exhibit bifurcated and chelated structure in the ion-pairs**

Dong, K., et al. *J. Phys. Chem.* B 2012, 1007-1017; Chem. Commun. 2016, 52, 7644-7664 6

## Efficient Charge Transfer by H-bond

| | $Q_{BI}$ | **NCT** |
|---|---|---|
| EmimCl | 0.753 | 0.247 |
| $EmimBF_4$ | 0.870 | 0.130 |
| $BmimBF_6$ | 0.880 | 0.120 |
| $EmimNTf_2$ | 0.850 | 0.150 |
| $BmimNTf_2$ | 0.820 | 0.180 |

*The charges based on the BlÖchl partial charges;

Dong, K., et al. *J. Phys. Chem.* B 2012, 1007-1017

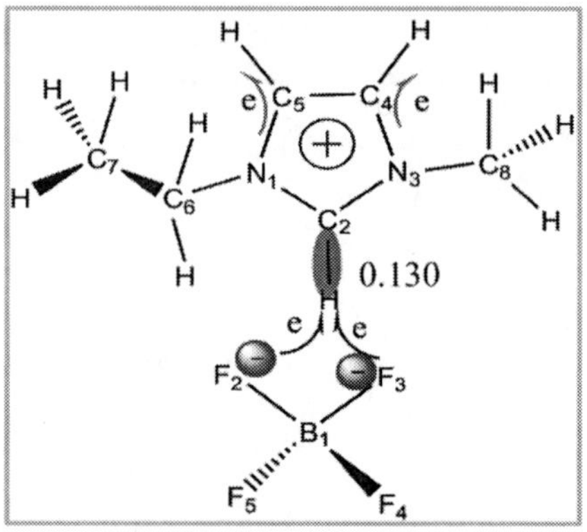


**[Emim][BF4]**

Efficient charges transfer by the H-bond paths cause these reduced charges
Halogen anions > $NTf_2$ > $PF_6$ and $BF_4$ anions

7

## Qualitative Molecular Orbital Description

**MmimCl**

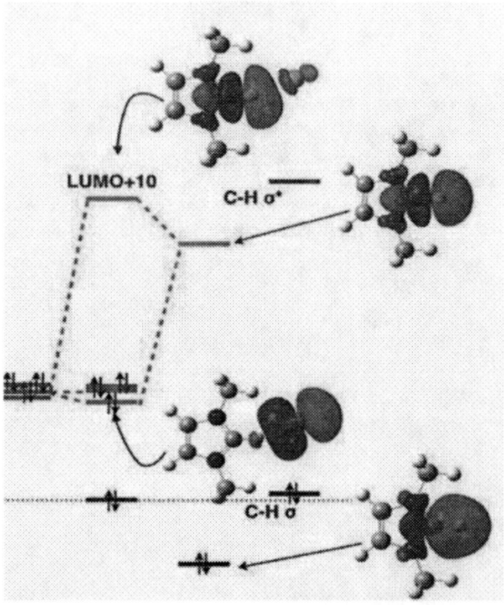


**$EmimBF_4$**

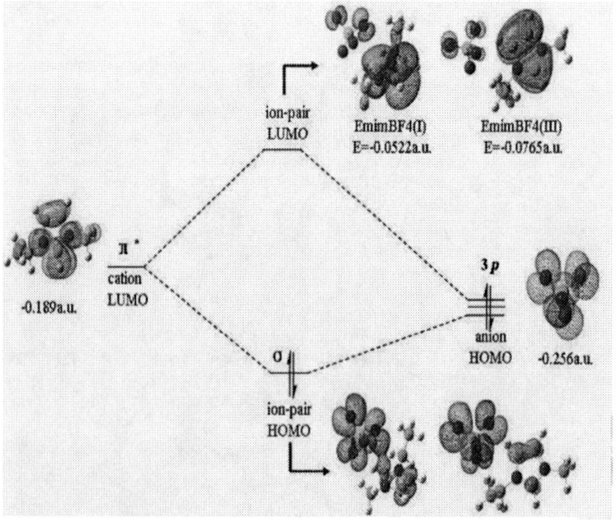


Counter ions will experience a stronger electrostatic attraction to reduce energy gap. The key effect of H-bond is to lower of the X–H σ* orbital energy and increased population of this orbitals, and amplify the covalent orbital overlap

**This coupling of electrostatic and H-bond is nature for stabilizing the ion-pair**

8

## NBO Donor-acceptor Stabilization Energy

$$E^{(2)}_{n\to\sigma^*} \cong -2\frac{\langle n|\hat{F}|\sigma^*\rangle^2}{E_{\sigma^*}-E_n}$$

Be correlated with the strength of H-bond

| ILs | H-bonds | ΔE(kJ/mol) | $E^{(2)}$( kJ/mol) | % |
|---|---|---|---|---|
| MmimCl | C-H···Cl | 385.71 | 173.00 | 44.8 |
| EmimCl | C-H···Cl | 379.83 | 179.00 | 47.1 |
| $MmimNTf_2$ | C-H···N, C-H···O | 363.30 | 119.00 | 37.8 |
| $EmimNTf_2$ | C-H···N, C-H···O | 357.30 | 96.15 | 27.0 |
| $BmimNTf_2$ | C-H···N, C-H···O | 356.20 | 110.95 | 31.2 |
| $BmimNO_3$ | C-H···O | 358.52 | 111.00 | 31.0 |
| $MmimBF_4$ | C-H···F | 350.03 | 50.00 | 14.3 |
| $EmimBF_4$ | C-H···F | 348.14 | 56.00 | 16.1 |
| $BmimPF_6$ | C-H···F | 319.00 | 55.00 | 17.2 |
| $EtNH_3NO_3$ | N-H···O | 556.47 | 320.00 | 57.5 |
| $PropNH_3NO_3$ | N-H···O | 470.44 | 250.00 | 53.1 |

15%-30% stronger for the aprotic ILs
50% stronger for the protic ILs
H-bonds are anion-dependent, the order is $Cl^-$> $[NTf_2]^-$ > $[NO_3]^-$ > $[BF_4]^-$/$[PF_6]^-$

9

## The Linear Relationship between NCT and $E^{(2)}$

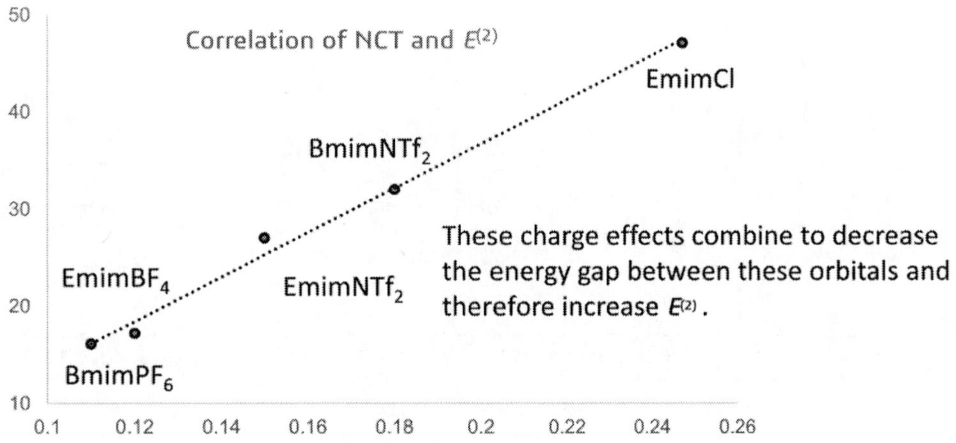


The substantial increase in NCT and $E^{(2)}$ is the origin of the for the special H-bond with strong charged attraction

Hunt, P. "doubly ionic H-bond"
Zhang, S. "Z-bond"

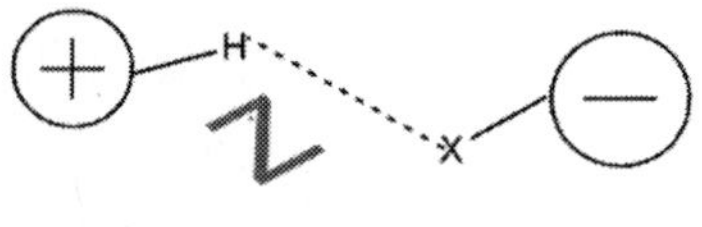


Dong, K. et al. *Chem. Rev.* 2017, 6636-6695.

Z-bond: Electrostatic + H-bond 10

# H-bonding Network

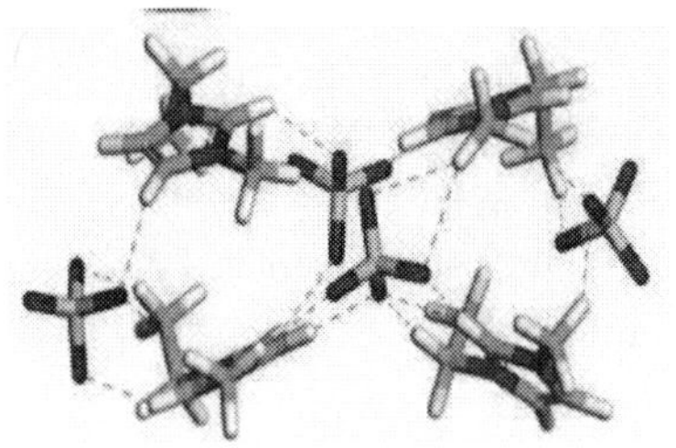

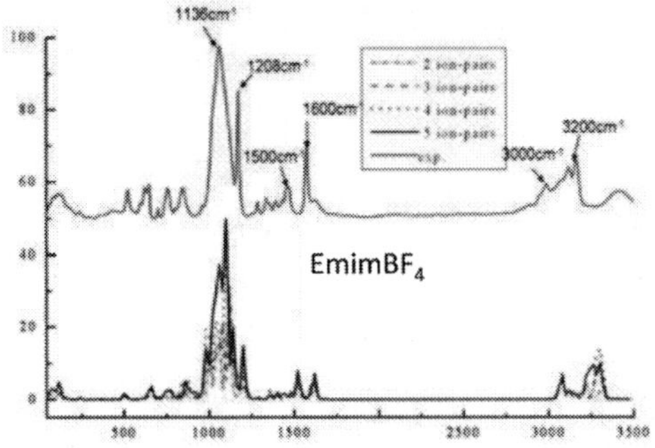


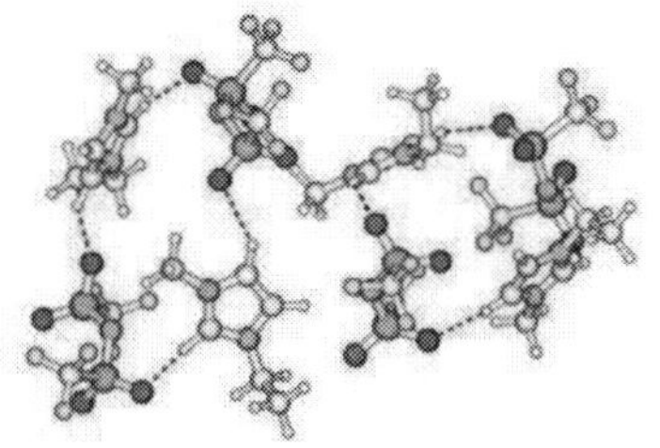

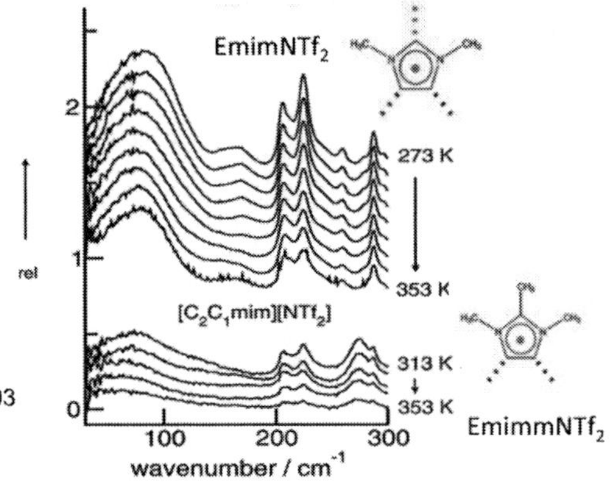


Dong, K. et al. *J. Phys. Chem.* B, 2012, 116, 1007-1017.

Fumino, K. et al. *Phys. Chem. Chem. Phys.* 2014, 16, 21903

**H-bonding network that can be characterized by Far-infrared the region 0~500 cm$^{-1}$**

11

# H-bonding Networks Induce the Larger clusters in ILs Aqueous Solution

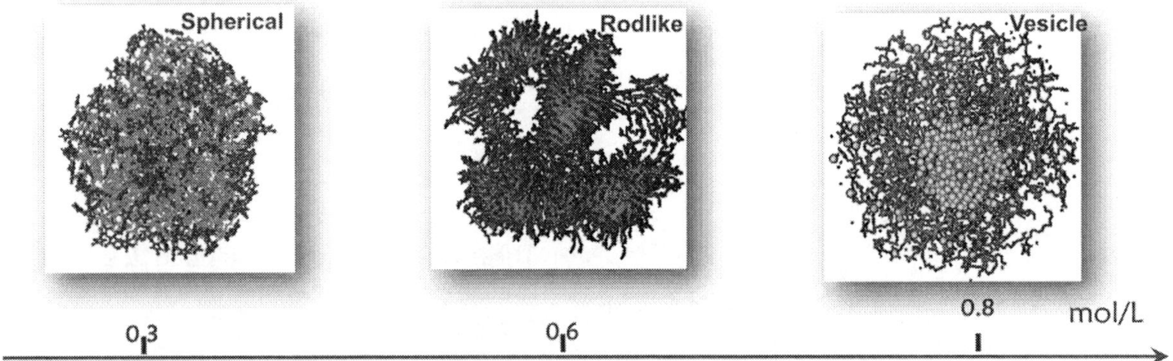


**Molecular dynamics simulation based on the developed force field**

| | | H5-Br | H4$^m$-Br | H4$^a$-Br | Total |
|---|---|---|---|---|---|
| Solution | HB No. | 17.9 | 10.2 | 7.7 | 35.8 |
| | CO No. | 0.36 | 0.34 | 0.25 | 0.95 |
| neat | HB No. | 123.3 | 77.9 | 94.2 | 295.4 |
| | CO No. | 1.68 | 1.36 | 1.27 | 4.31 |

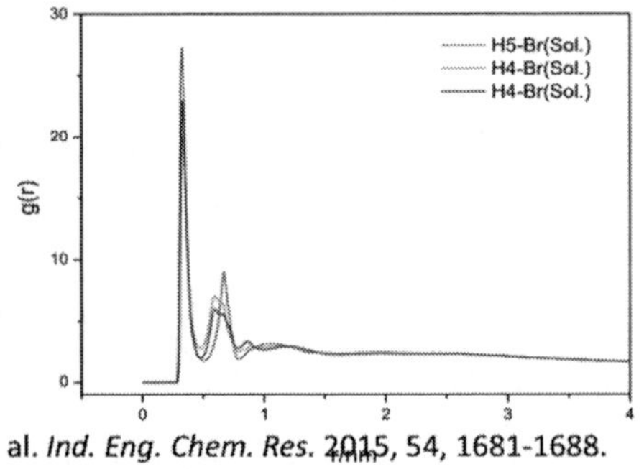


Dong, K. et al. *Chem. Commun.* 2016, 52, 6744-6764.   Liu, X. et al. *Ind. Eng. Chem. Res.* 2015, 54, 1681-1688.

12

# Prediction of Enthalpy of Vaporization

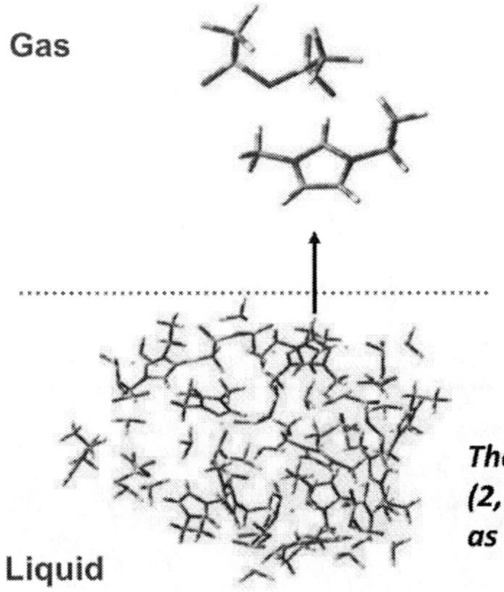


$$\Delta_{vap}H(T,P) = H_{gas}(T,P) - H_{liquid}(T,p)$$

$$\Delta_{vap}H(T,P) = H_1(T,P) - (\frac{H_n(T,P)}{n})$$

***The condensed phase is aggregates with different sizes (2, 4, 6, 8, 10 and 12 ion-pairs). The gas phase is assumed as the single ion, ion-pair and neutral molecule.***

The ILs are [Emim][$NTf_2$] with strong H-bonds and [Emmim][$NTf_2$] with weak H-bonds. Calculation was performed at M06/3-21G* level.

K. Dong, et al. *Phys. Chem. Chem. Phys.*, 2013, 15, 6034

13

# $\Delta_{vap}H$ *(298K, 1bar)* of [Emim][$NTf_2$] and [Emmim][$NTf_2$]

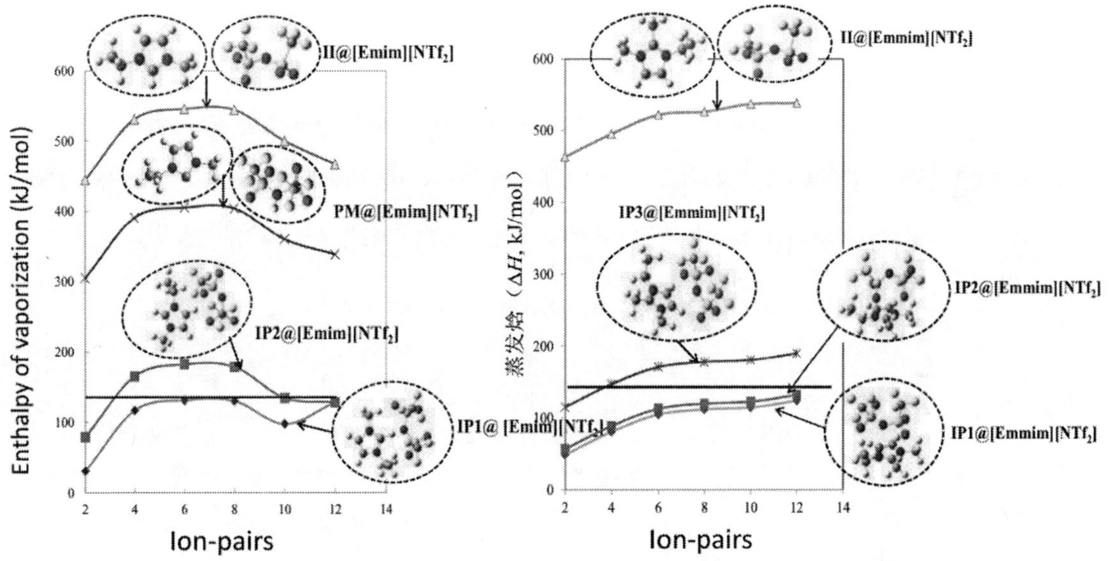


**It is found that when the gas species are ion-pair the enthalpies of vaporization are agreeable with the experimental measurement,** suggesting that the gas phase should consist purely of ion-pairs

14

## Effect of H-bonds on Enthalpy of Vaporization

| Size | 2 | 4 | 6 | 8 | 10 | 12 | Average |
|---|---|---|---|---|---|---|---|
| IP1@EmimNTf$_2$ | 30.82 | 116.67 | 131.20 | 130.09 | 97.62 | 128.39 | 105.80 |
| IP1@EmimmNTf$_2$ | 81.33 | 116.91 | 132.07 | 148.90 | 149.19 | 206.98 | 139.23 |
| IP2@EmimNTf$_2$ | 79.22 | 165.07 | 182.23 | 178.45 | 134.34 | 128.26 | 144.59 |
| IP2@EmmimNTf$_2$ | 114.91 | 146.67 | 170.54 | 177.64 | 180.35 | 189.64 | 163.29 |

Because of the ideal gas approximation the enthalpies of vaporization show a cluster size behavior comparable to the binding energies

The enthalpies of EmimNTf$_2$ are decreased compared with the EmmimNTf$_2$, which is opposite to expectation, the H-bond decrease the enthalpy

$$\Delta_{vap} H(T,P) = \Delta E + RT$$

$$\Delta E = nE_{IP} - E_{cluster}$$

EmmimNTf$_2$ interaction in gas phase is increased by H-bonding, resulting in the increasing $\Delta E$, and increasing the $\Delta H$.

15

## MD simulation for Dissolution of Cellulose in ILs

**Our system:**

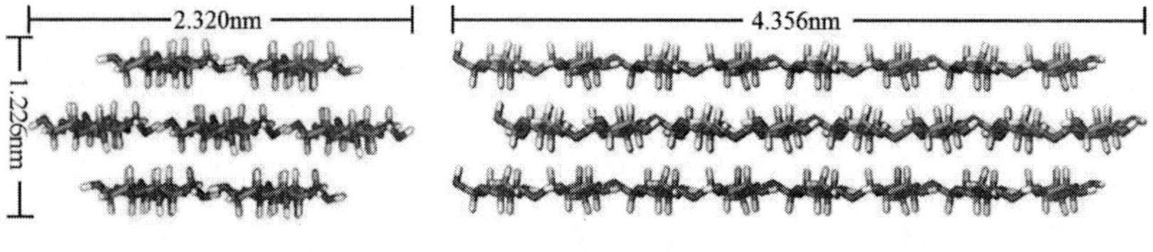


**A bunch of cellulose chains（7*8）**

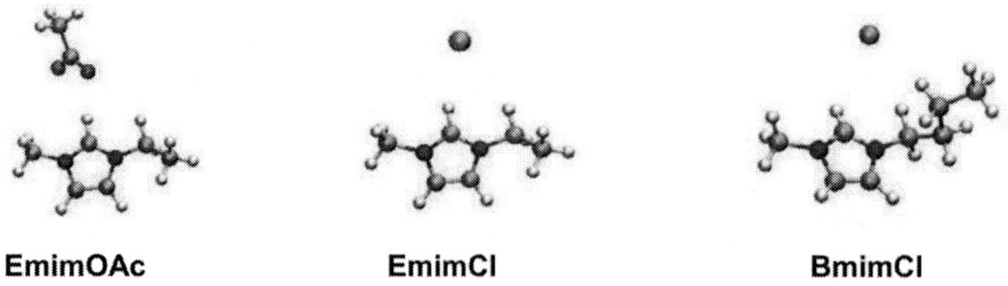


Dong, K. et al. *Chem. Rev.* 2017, 6636

Li, Y. et al. *Phys. Chem. Chem. Phys.* 2015, 17, 17894

16

# The Dissolution Process and Interaction

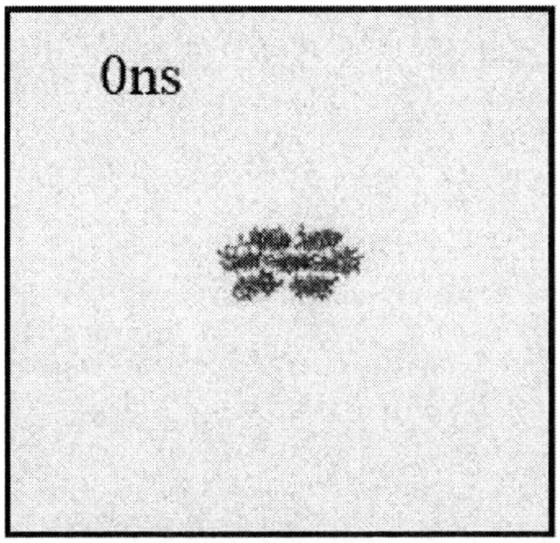


[Emim][OAc]

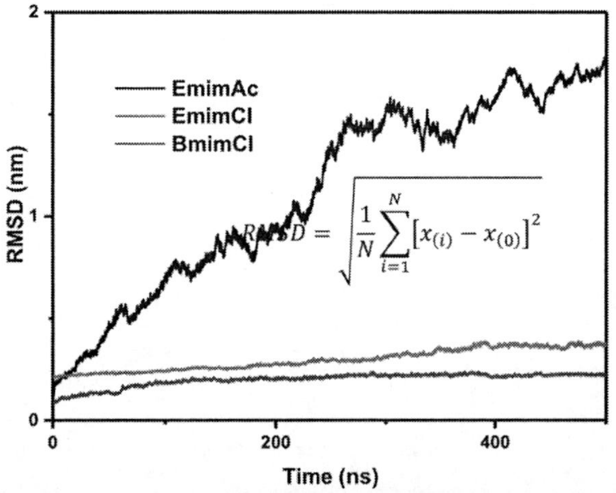


**The dissolution rate EmimAc > EmimCl > BmimCl**

| E(kJ/mol) | $Emim^+$ | $OAc^-$ | $Emi^+$ | $Cl^-$ | $Bmi^+$ | $Cl^-$ |
|---|---|---|---|---|---|---|
| $E_{coul}$ | -299.57 | -1562.64 | -227.93 | -1553.82 | -74.88 | -1396.72 |
| $E_{LJ}$ | -555.69 | -19.86 | -639.80 | 180.15 | -633.73 | 168.38 |
| $E_{tot}$ | -855.26 | -1582.50 | -867.73 | -1373.67 | -708.62 | -1228.34 |

17

# H-bonds with Anions and Cations

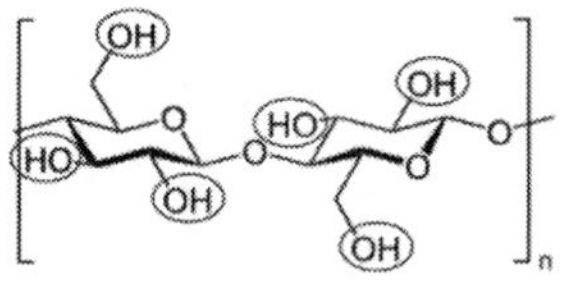

donor

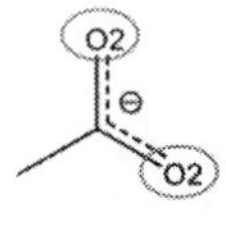

acceptor

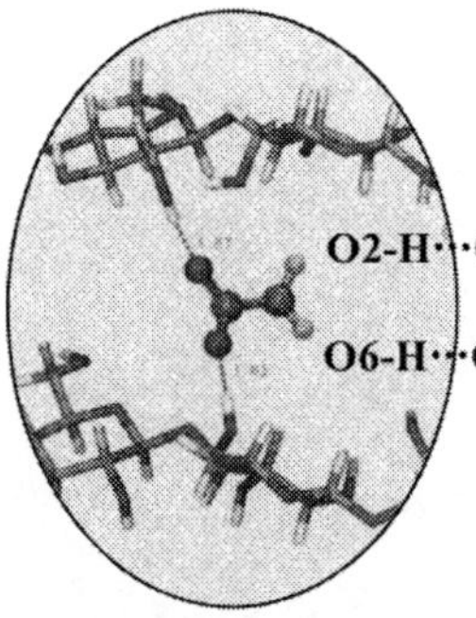


$Cl^-$

acceptor

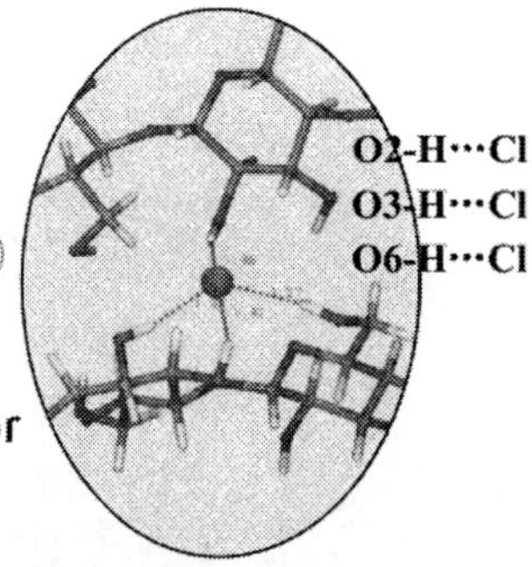


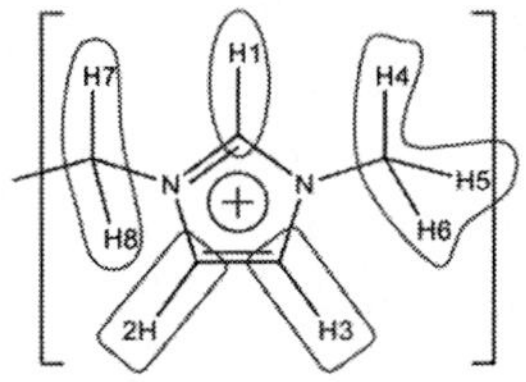

donor

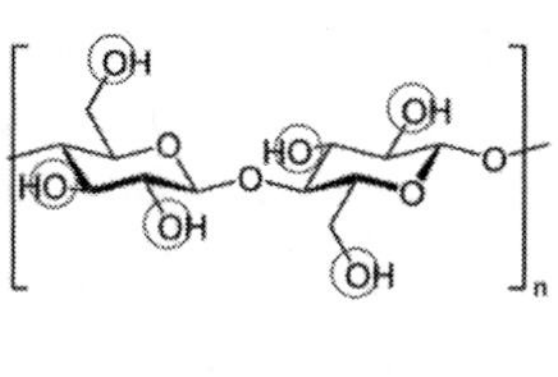

acceptor

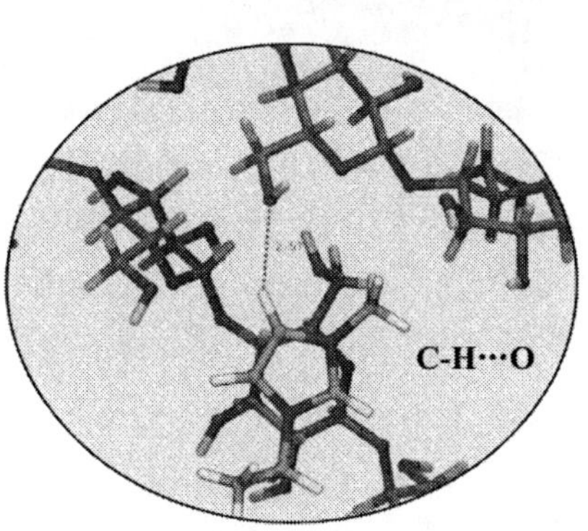


18

## Structure of H-bonds with Anions and Cations

| Anion | $O(OAc^-)$ | $Cl^-(Emi^+)$ | $Cl^-(Bmi^+)$ |
|---|---|---|---|
| cellulose | 142.13 | 102.39 | 94.17 |

| Cation | $Emim^+(Ac^-)$ | $Emim^+(Cl^-)$ | $Bmim^+(Cl^-)$ |
|---|---|---|---|
| cellulose | 22.67 | 10.80 | 16.42 |

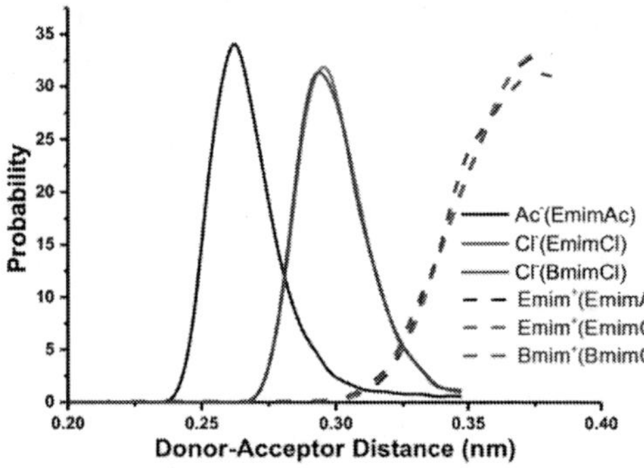


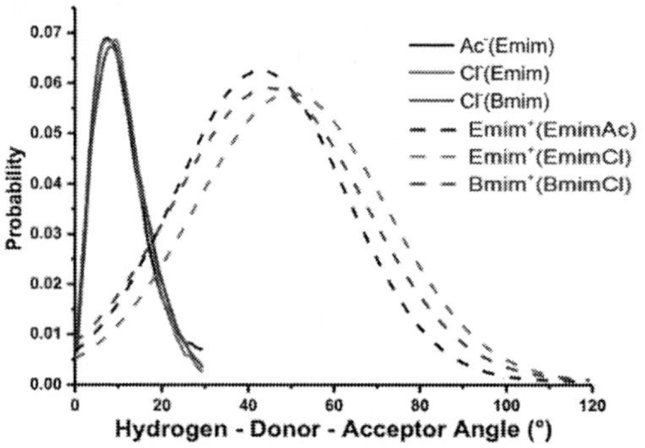


**Distance:**
**O-H···OAc: 0.26~0.28 nm**
**O-H···Cl: 0.3 nm**
**Cations have longer distances**

**Angle:**
**O-H···OAc and O-H···Cl: 10°**
**Cations have larger angles (40~50° )**

From structure and number, the H-bond between cellulose and anion is stronger than those between cellulose and cation.

19

## Proposing Dissolving Mechanism

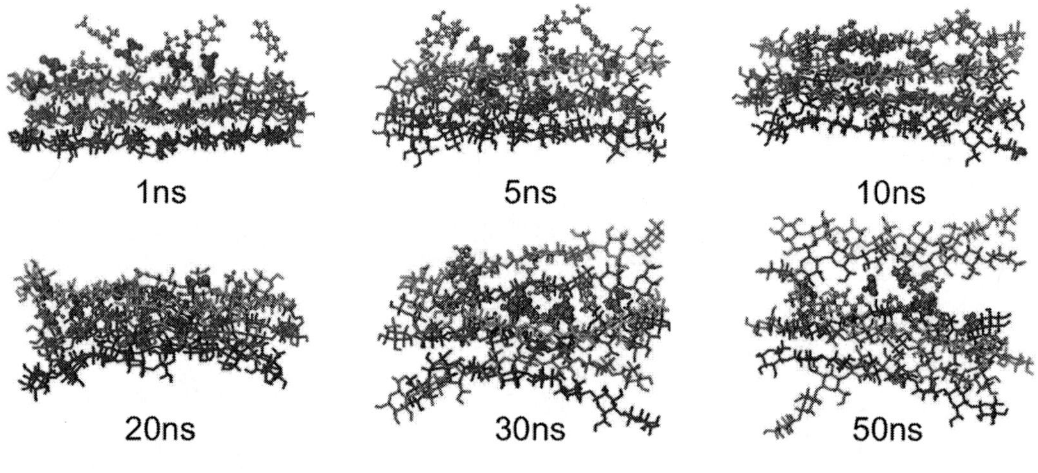


Cations bind to the cellulose bunch surface by VDW interactions
Anions insert into the cellulose bunch and form H-bonds with cellulose.
More anions insert into the cellulose chains, cations start to intercalate into cellulose bunch due to their strong electrostatic interaction with anions and cellulose dissolution begins.

20

## Conclusions

- ✓ H-bonds in the ILs show some different features with the conventional H-bond in geometry, energy, charge transfer and orbital overlap. As a structure-directing force, H-bonds in ILs are responsible for stacking and self-assembling
- ✓ H-bonds have effect on enthalpy of vaporization and lead to a decrease of the enthalpies unexpectedly
- ✓ H-bonds are predominant interaction in dissolution of cellulose in ILs

Potential Applications:

- ✓ Tuning Properties
- ✓ Designing New ILs in catalysis, electrochemistry, biomass pretreatment and
- ✓ separation

## Acknowledgement

Prof. Suojiang Zhang, Xiaomin Liu, Xiaochun Zhang, Xiaoqian Yao, IPE

Prof. Peter T. Cummings, Vanderbilt University

NSFC

VANDERBILT

21

# 501046 A Molecular Understanding of Cellulose Dissolution in Aqueous Ionic Liquids

**Abstract:**

A major challenge in the pretreatment of biomass using ionic liquids deals with the removal of water from the solvent. Residual water that is present in biomass releases upon dissolution, commonly inhibiting the ability of the ionic liquid to further dissolve cellulose. Recently however, a class of phosphonium hydroxide-based ionic liquids have been found that can rapidly dissolve cellulose at concentrations between 30% to 60 wt% water. Interestingly, optimal dissolution occurs at intermediate concentrations and any excess water in these systems naturally evaporates, returning the solvent to the optimal dissolution conditions. The unique ability of this ionic liquid to dissolve cellulose in the presence of large amounts of water is not well understood. In this work, molecular dynamics simulations are used to provide a molecular-level understanding of this process. Mixtures of alkylphosphonium hydroxides and water were studied over the entire water concentration range to determine the character and behavior of the pure solvent. Additional simulations of the IL/water mixtures and cellulose were performed and analyzed to provide an in-depth view of ionic liquid structure, hydrogen bond dynamics and thermodynamic changes of state, providing insight into the molecular driving forces of this unique class of aqueous ionic liquids.

# 503081 Freezing Point Determination of Water-Ionic Liquid Mixtures

**Abstract:**


**Freezing point determination of water-ionic liquid mixtures**



Yanrong Liu,[a,b,c] Anne S. Meyer,[a] Yi Nie,[c] Suojiang Zhang,[c] Yongsheng Zhao,[c] Philip L. Fosbøl,[b] Kaj Thomsen*[b]

[a]*Center for Bioprocess Engineering, Department of Chemical and Biochemical Engineering, Technical University of Denmark, Søltofts Plads 227, 2800, Lyngby, Denmark*

[b]*Center for Energy Resources Engineering (CERE), Department of Chemical and Biochemical Engineering, Technical University of Denmark, Søltofts Plads 229, 2800, Lyngby, Denmark*

[c]*Beijing Key Laboratory of Ionic Liquids Clean Process, Key Laboratory of Green Process and Engineering, State Key Laboratory of Multiphase Complex Systems, Institute of Process Engineering, Chinese Academy of Sciences, P. O. Box 353, Beijing, 100190, China*



In many instances ionic liquids (ILs) end up being IL-water mixtures. No data are available on the freezing points of such mixtures, and assessment of the freezing point curves of IL–$H_2O$ mixtures can also be used to understand the eutectic behavior of the solutions. In this work, freezing points of aqueous solutions of HOEtpyBr, HOEtmimBr, AmimCl, EtOMmimCl, EmimDep and EmimAc were measured using a modified Beckmann apparatus with automatic data logging. These ILs were chosen because they are potential ILs for dissolving cellulose. From the freezing point curves, it can be observed that the behavior of dilute solutions of ILs in water is very similar to that of dilute solutions of $MgSO_4$ and $MgCl_2$. Water activities in IL aqueous solutions were predicted by COSMO-RS and COSMO-SAC and compared to water activities derived from the experimentally determined freezing points. The COSMO-RS predictions were closer to the experimental water activities than the COSMO-SAC predictions. The experimental results indicate that the freezing points of IL–$H_2O$ systems are affected by the nature of both cations and anions. However, according to the COSMO-RS excess enthalpy prediction results, the anions have a relatively higher influence than cations on the IL–$H_2O$ interaction.


**Figure 1.** Freezing points of the studied IL solutions and of salt solutions compared at higher concentrations.


[1] Liu, Y. R.; Thomsen, K.; Nie, Y.; Zhang, S. J.; Meyer, A. S. Predictive Screening of Ionic Liquids for Dissolving Cellulose and Experimental Verification. *Green Chem.* **2016**, *18*, 6246–6254.

[2] Fosbøl, P. L.; Pedersen, M. G.; Thomsen, K. Freezing Point Depressions of Aqueous MEA, MDEA, and MEA–MDEA Measured with a New Apparatus. *J. Chem. Eng. Data* **2011**, *56*, 995–1000.

[3] Diedenhofen, M.; Klamt, A. COSMO-RS as a Tool for Property Prediction of IL Mixtures-A Review. *Fluid Phase Equilibr.* **2010**, *294*, 31–38.

**Effects of the dielectric contrast and hydrogen bond between polymer and ionic liquid on the shift in the critical point of the spinodal from a polymer-poor to a polymer-rich region**

*Issei Nakamura*
*Department of Physics*
*Michigan Technological University*

Michigan Technological University

## Key Experiments for the mixture of polymer and ionic liquid

- Unusual LCST phase behavior of PEO in ionic liquids ([EMIM][BF4]).

H. N. Lee, N. Newell, Z. Bai, and T. P. Lodge. Macromolecules, 45, 3627 (2012)

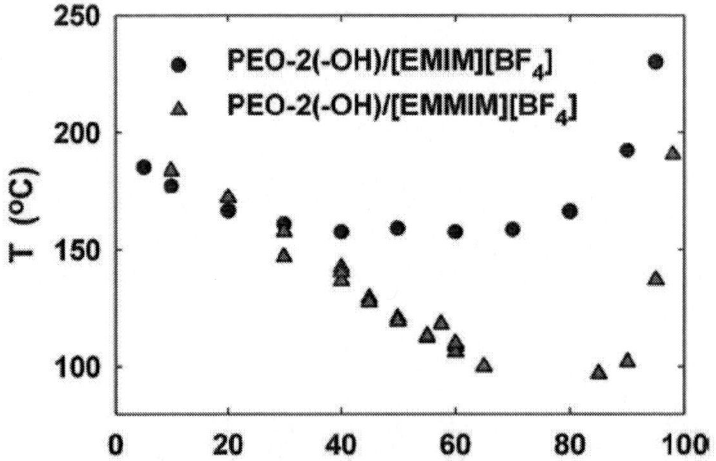


Critical point is unlikely to scale with $\phi_c \sim 1/\sqrt{N}$.

Michigan Technological University

## Key analysis and prediction

- Coarse-grained mean-field theory that accounts for free volume and non-random mixing

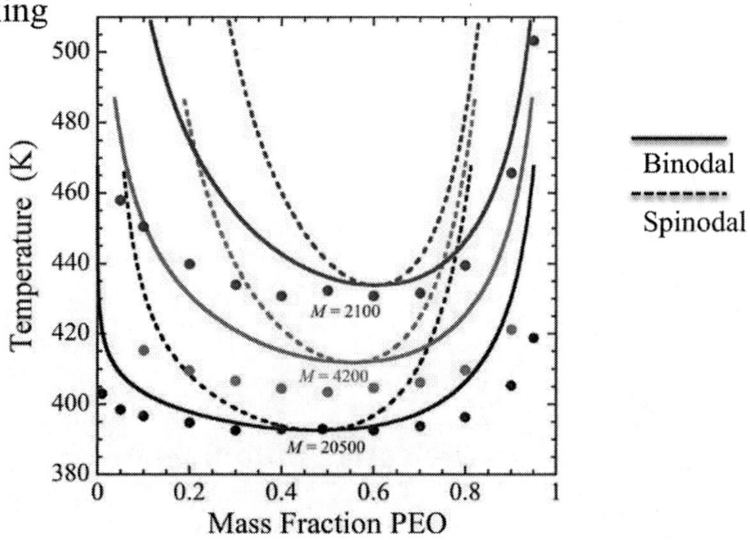


R. P. White and J. E. G. Lipson, Macromolecules, 46, 5714 (2013)

- Unfavorable entropic effects in the mixture
- *Effects of hydrogen bonding between the PEO and the IL?*

Michigan Technological University

## Origin of LCST

LCST → Changes in entropy due to

(1) changes in free volume (or **equation of state**)
[*R. P. White and J. E. G. Lipson, Macromolecules, 46, 5714 (2013)*]

(2) **the effects of hydrogen bonding**
[*R. P. White and J. E. G. Lipson, Macromolecules, 46, 5714 (2013)*;
*E. E. Dormidontova, Macromolecules, 35, 987 (2002)*]

→ **Release of solvents from (or binding of solvents on) polymers**

# Motivation

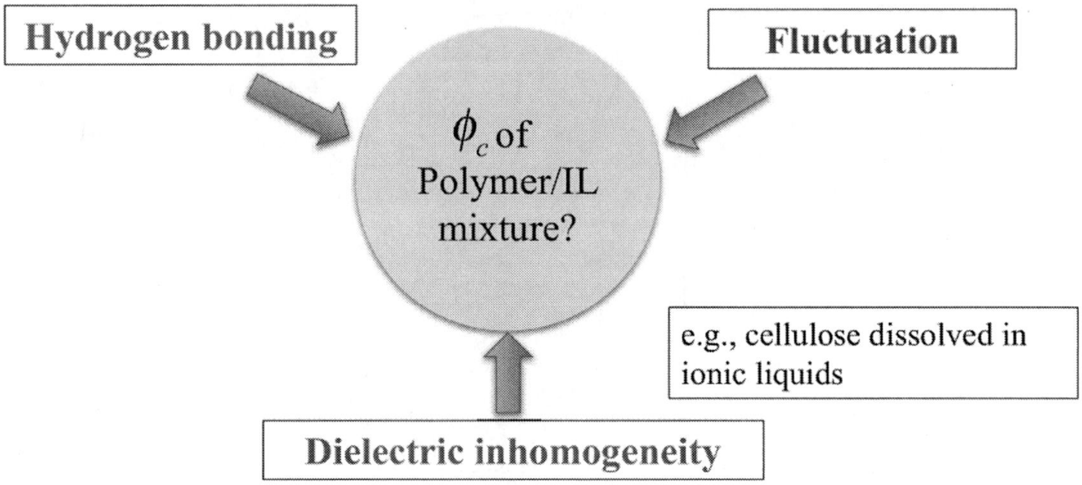


**Do effects of hydrogen bonding, dielectric inhomogeneity, and fluctuation cause significant shifts in $\phi_c$?**

Michigan Technological University

# Coarse-grained mean-field Theory

- Translational entropy
- Flory-Huggins interaction
- Hydrogen bond free energy
- Born solvation energy

$$\varepsilon_r = \varepsilon_{\mathrm{PEO}}\phi_{\mathrm{PEO}} + \varepsilon_{\mathrm{IL}}\phi_{\mathrm{IL}}$$

$$\frac{v_p f_{tot}}{k_B T} = \frac{\phi_p}{N}\ln\phi_p - \frac{\phi_p}{N} + \frac{v_p}{v_0}\phi_0\ln\phi_0 - \frac{v_p}{v_0}\phi_0 + \frac{v_p}{v_0}\phi_0\ln(\phi_0-\phi_b) - \frac{v_p}{v_0}(\phi_0-\phi_b) + \phi_p\ln\left(1-\frac{v_p\phi_b}{v_0\phi_p}\right) + \chi\phi_p(1-\phi_p) + \frac{l_0 v_p(1-\phi_p)}{2a\varepsilon_r v_0}$$

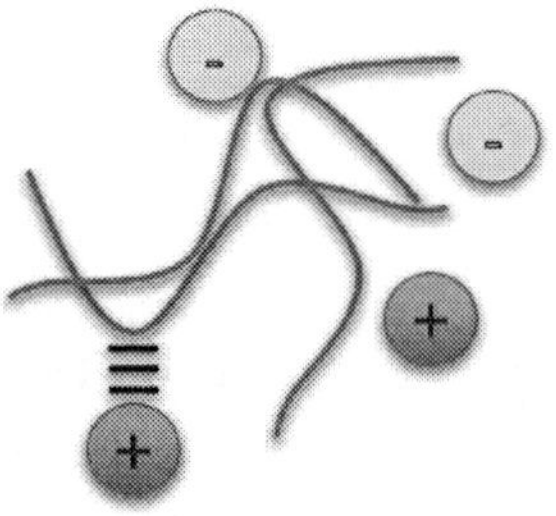

$$\frac{n_b}{(Nn_p-n_b)(n_0-n_b)} = \tilde{K}_{eq}$$

(Langmuir adsorption equation)

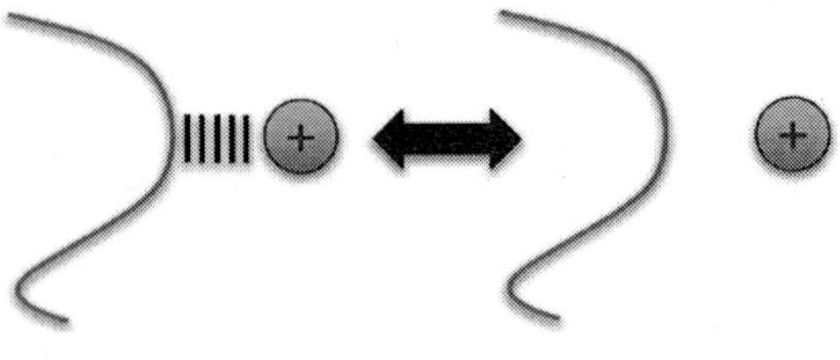


- E. E. Dormidontova, Macromolecules, 35, 987 (2002)
- R. Chun-Lai, I. Nakamura, and Z.-G. Wang, Macromolecules, 49, 425 (2016)
- I. Nakamura, Macromolecules, 49, 690 (2016)


Michigan Technological University

## Results: Born solvation energy

- Translational entropy
- Flory-Huggins interaction
- ~~Hydrogen bond free energy~~
- Born solvation energy

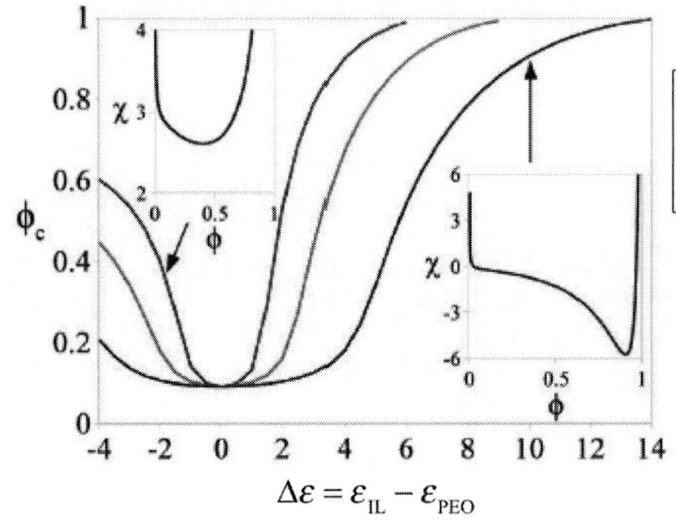


Critical point shifts towards polymer-rich regions.

- I. Nakamura, Macromolecules, 49, 690 (2016)

Michigan Technological University

## Results: Born solvation energy + Hydrogen bond free energy

- Translational entropy
- Flory-Huggins interaction
- Hydrogen bond free energy
- Born solvation energy

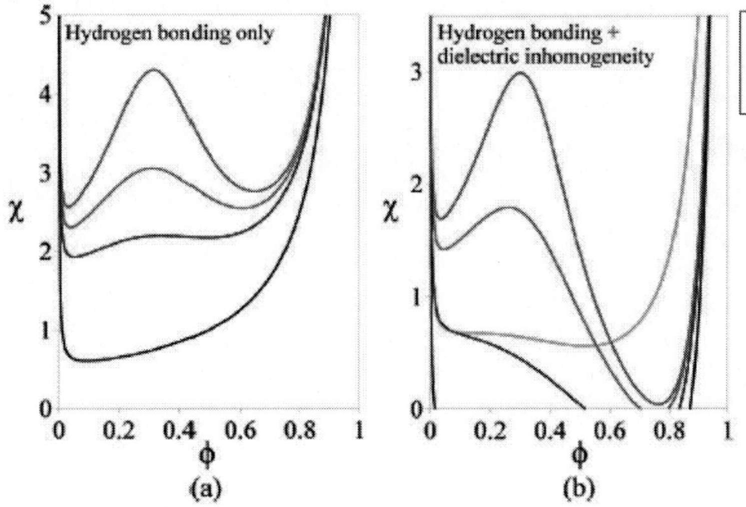


**"W" shape or flat structure**

***Cf. sulfobetaine polymers in an aqueous solution***

*D. N. Schulz et al., Polymer, 27, 1734 (1986)*

- I. Nakamura, Macromolecules, 49, 690 (2016)

Michigan Technological University

## Coarse-grained mean-field Theory + Fluctuation

**Fluctuations** in the local density and electrostatic potential around the homogeneous phase.

- M. Olvera de la Cruz, S. F. Edwards, and I. C. Sanchez, J. Chem. Phys., 89, 1704 (1988)
- G. H. Fredrickson, A. J. Liu, F. S. Bates, Macromolecules, 27, 2503 (1994)
- Z. -G. Wang, J. Chem. Phys., 117, 481 (2002)

$$H_p = \sum_{i=1}^{n_p} \int_0^N dt \left(\frac{3}{2b_p^2}\right)\left[\frac{d\vec{R}_i(t)}{dt}\right]^2$$

$$Z = \sum_{n_p, n_+, n_-} \frac{\lambda_p^{n_p}}{n_p!}\frac{\lambda_+^{n_+}}{n_+!}\frac{\lambda_-^{n_-}}{n_-!} \int \prod_{i=1}^{n_p}\prod_{j=1}^{n_0}\prod_{k=1}^{n_0} \mathcal{D}\vec{R}_i(t)$$
$$\times\, d\vec{r}_j^{(+)}\, d\vec{r}_k^{(-)}\, \delta[v_p\hat{c}_p(\vec{r}) + v_0\hat{c}_+(\vec{r})$$
$$+ v_0\hat{c}_-(\vec{r}) - 1]\exp(-H_p - H_c - V_{FH})$$

$$H_c = \frac{1}{2}\int d\vec{r}\, d\vec{r}'\, \hat{\rho}(\vec{r})v(\vec{r}-\vec{r}')\hat{\rho}(\vec{r}')$$

**Field-theoretical transformation**

Effective free energy to the one-loop order

$$Z \approx \exp\left[-\frac{F_{FH}}{k_B T}\right]\int \mathcal{D}\Omega\delta[v_p\delta c_p(\vec{r}) + v_0\delta c_+(\vec{r}) + v_0\delta c_-(\vec{r})]$$
$$\exp\left[-\frac{\Delta F^{(2)}}{k_B T}\right]$$

Michigan Technological University

## Results: Fluctuation effects

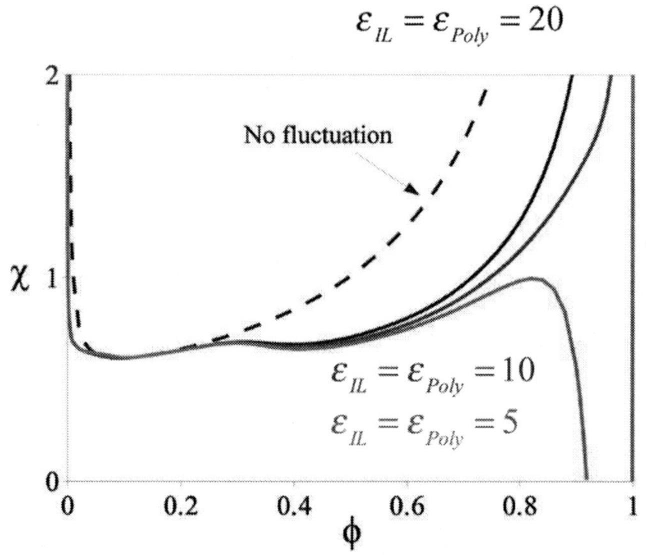


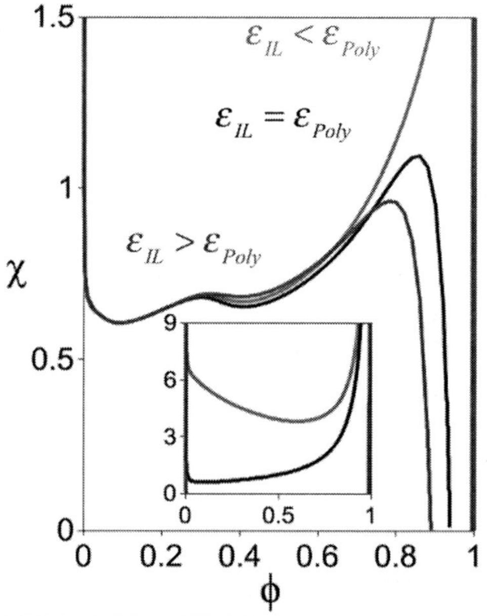


**Both fluctuation and dielectric inhomogeneity are of equal significance for the phase behavior.**

- I. Nakamura, Macromolecules, 49, 690 (2016)

Michigan Technological University

## Summary and Conclusion

- IL-complexed PEO ~ Polyelectrolyte chain

- Hydrogen bonding causes the "W" shape of the spinodal curve.

**Dielectric contrast + Hydrogen bonding** ➡ **Flat structure of the spinodal curve**

- Fluctuation effects alter the geometrical trend of the spinodal curve.

$\Delta\varepsilon$ ⇨ **Shift $\phi_c$ towards polymer rich regions.**

**Acknowledgment:**
Start-up funding of Michigan Tech
NSFC (21474112)

Michigan Technological University
1885

# 486983 Assessing the Reliability of Computing Ion Pair Lifetimes and Diffusivity to Predict Experimental Viscosity Trends of Ionic Liquids

**Abstract:**

One of the most important properties for ionic liquids is viscosity. A common method being used to predict the viscosity of ionic liquids computationally is quantitative structure-property relationship (QSPR) analysis, which is extremely quick to run computationally, but is difficult to extend beyond the training set used. Another method used to predict viscosities is classical molecular dynamics simulations, which can be applied to any ionic liquid. Unfortunately, viscosity is a computationally expensive property to calculate using molecular dynamics, and thus is not feasible for a high-throughput screening process. Instead, two other properties, ion pair lifetime and self-diffusivities, can be calculated using molecular dynamics with much less computational expense. These properties were calculated for 24 ionic liquids, covering a range of cation and anion classes. All calculations were performed at 400 K to reduce the computational expense. The values of self-diffusivity and ion pair lifetime were then compared with the experimental viscosities of the ionic liquid at 298 K. Experimental viscosities were found to have a strong correlation with both the ion pair lifetime and the inverse diffusivity. For both properties, the ionic liquids were divided into two groups based on the nature of the anion, one group being linear/bent anions and the other containing planar/spherical anions. It was found that both the ion pair lifetime and the diffusivity can be used to predict the ionic liquids with the lowest viscosity with an 80% success rate, with 1/15th the computational cost of calculating viscosity directly. This predictive capability with low computational expense would allow for rapid screening of ionic liquids to determine those with low viscosities.

# 484484 Ammonia Absorption in Ionic Liquids

**Abstract:**

The history of ionic liquids (ILs) effectively started in 1914, when the physical properties of ethylammonium nitrate ($[CH_3CH_2NH_3^+][NO_3^-]$, m.p. 13-14 °C) were first reported. ILs are generally defined as salts composed of discrete cations and anions with melting points below 100 °C, and many are liquid at ambient temperature. IL research has grown rapidly over the past decade due to the realization that these materials have many unique properties such as negligible vapor pressure and outstanding solvation potential. Ionic liquids have been further emphasized by the fact that their physical and chemical properties can be finely tuned by varying both the cation and anion. Our research has focused on accurately measuring vapor-liquid equilibria (VLE) and vapor-liquid-liquid equilibria (VLLE) and using thermodynamic models to understand the phase behavior of binary and ternary gas mixtures in ILs. We published for the first time in 2006-2007 the first ammonia ($NH_3$) gas solubilities in 8 room temperature ionic liquids (RTILs): 1-butyl-3-methylimidazolium hexafluorophosphate ([bmim][$PF_6$]), 1-hexyl-3-methylimidazolium chloride ([hmim][Cl]), 1-ethyl-3-methylimidazolium bis(trifluoromethylsulfonyl)imide ([emim][$Tf_2N$]), 1-butyl-3-methylimidazolium tetrafluoroborate ([bmim][$BF_4$]), 1-ethyl-3-methylimidazolium acetate ([emim][Ac]), 1-ethyl-3-methylimidazolium thiocyanate ([emim][SCN]), 1-ethyl-3-methylimidazolium ethylsulfate ([emim][$EtOSO_3$]), and N,N-dimethylethanolammonium acetate ([DMEA][Ac]). The VLE experiments were performed using static equilibrium cells over a temperature range from 283 to 373 K. Now, for the first time a gravimetric microbalance (XEMIS) is available from Hiden Isochema (United Kingdom) which is capable of measuring VLE using corrosive gases such as $NH_3$ over broad temperature (77 to 773 K) and pressure (0 to 170 bar) ranges. This presentation will focus on our new VLE measurements with $NH_3$ in RTILs using the highly accurate XEMIS gravimetric microbalance. Pressure-temperature-composition (*PTx*) data for $NH_3$ in RTILs at isothermal temperatures will be reported and correlated using the Peng-Robinson equation of state. Knowledge of $NH_3$ and RTIL phase behavior has led to several practical applications which will be highlighted.

# 491675 Solubilities and Diffusivities of Oxygen in Ionic Liquids for Electrochemical Applications

**Abstract:**

Ionic liquids (ILs) have been recognized as potential electrolytes for various electrochemical applications due to their favorable properties, such as good ionic conductivity, tunable hydrophobicity, and controllable gas solubility. In particular, protic ionic liquids (PILs) have received considerable interest as proton transfer media in proton exchange membrane fuel cells (PEMFC).

Previous researchers have shown that PILs with high $O_2$ solubility enhance the oxygen reduction reaction (ORR) activity[1-2]; therefore, the goal of this research is to design PILs with even higher $O_2$ uptake in order to achieve better PEMFC performance. A series of PILs with different fluoroalkyl chain lengths (shown below), all paired with the $[MTBD]^+$ cation, were investigated. Here we present the $O_2$ solubility and diffusivity in these ILs, measured using a gravimetric microbalance. Other physical properties, such as thermal stability, density, and viscosity, are also reported.

Further, we compare the $O_2$ solubility in the PILs to ILs we have designed for other applications[3]. These ILs include [PP13][$Tf_2N$], [PP12O1][$Tf_2N$], [Pyrro12O1][$Tf_2N$], [DEME][$Tf_2N$], [P2228][$Tf_2N$], whose structures are shown below. Additionally, the temperature dependence of $O_2$ dissolution in the ILs is examined, to provide information on both the interaction energy between the IL and $O_2$, and the underlying driving force for $O_2$ dissolution.

Figure 1. Structures of Investigated Ionic Liquids

References:

1. Snyder, J.; Livi, K.; Erlebacher, J., Oxygen Reduction Reaction Performance of [MTBD][beti]_Encapsulated Nanoporous NiPt Alloy Nanoparticles. *Advanced Functional Materials* **2013,** *23* (44), 5494-5501.

2. Snyder, J.; Fujita, T.; Chen, M.; Erlebacher, J., Oxygen reduction in nanoporous metalÐionic liquid composite electrocatalysts. *Nature materials* **2010,** *9* (11), 904-907.

3. Seo, S.; Song, T. M.; Mizuno, F.; Takechi, K.; Brennecke, J. F., Influence of Ion Diffusivity and Gas Solubility in Ionic Liquids on Li-Air Battery Performance. *ECS Transactions* **2016,** *75* (15), 147-160.

# 495608 The Formation and Regulation Mechanism of Meso-Structure of Ionic Liquids

**Abstract:**

Ionic liquids (ILs) have been studied extensively as green solvents, catalysts and electrolytes due to their unique properties and great potentials in industry applications. However, the lack of knowledge on the formation and regulation mechanism of meso-structure in ionic liquid systems may lead to possible failure and difficulties in the new reactor and process design. Focusing on the meso-science of ILs, the following research has been conducted. We firstly discovered that H-bond in ILs showed the co-valent feature by density functional theory method and far-IR spectra. Based on the special features in molecular structure, interaction energy, electronic, dynamics and spectroscopic, the special H-bond was defined as a new concept: Z-bond.
model has been developed. The ionic clusters in ILs or IL mixtures at nano-level have been proved by molecular dynamic simulation and experimental characteristic. Angstrom Aggregates (AA) in IL systems with the specific size at angstrom scale might possess unique structure-activity relationship, which may benefit in the area of chemical reaction, separation, and electrochemistry. Focusing on the AA in IL systems, we discovered that Z-bond is one of the most important interactions for the formation of AA in ILs.[2] As important properties for scaling up a process, the gas bubble behavior in IL system, such as bubble size distribution, gas holdup, Sauter diameter were investigated, and then the mass transfer properties were measured and predicted. The methodology of IL-based process design was proposed and applied to develop several novel technologies[3], such as the clean technology of dimethyl carbonate, the efficiently catalytic alkylation process, $SO_2$ recovery, and a green spinning process from wool keratin.

# 496455 Structural Characterization and Catalytic Valorization of Co-Solvent Enhanced Lignocellulosic Fractionation Pretreated Lignin

**Abstract:**

A novel pretreatment named Co-solvent Enhanced Lignocellulosic Fractionation (CELF) using THF with water and dilute acid to reduce biomass recalcitrance was recently developed. During CELF pretreatment, more than 90% of lignin could be extracted and dissolved into the liquid hydrolysate. The extracted lignin, known as CELF lignin, can be then isolated from the hydrolysate by rotatory evaporation. Physicochemical characteristics of CELF lignin including molecular weights, monolignol composition, and hydroxyl groups content were measured by various analytical techniques such as GPC, $^{13}C$-$^{1}H$ HSQC NMR, $^{31}P$ NMR. GPC results indicated a dramatic decrease in MW of lignin after CELF pretreatment of poplar. HSQC NMR revealed that lignin Î²-O-4 linkages were significantly decreased after CELF pretreatment, and in fact, almost all the interlinkages were disappeared for pretreatment done at 180 $^{o}C$ using $FeCl_3$ as the catalyst. $^{31}P$ NMR showed that CELF pretreatment resulted in significant decrease of aliphatic OH group, possibly due to the oxidation of lignin side chains. On the other hand, the content of total phenolic hydroxyl groups was significantly increased, suggesting the drastic cleavage of interunit linkages in CELF lignin as confirmed by 2D NMR. A catalytic hydrodeoxygenation of CELF lignin into fuels was also achieved in supercritical methanol using Cu-doped porous metal oxide as the catalyst.

# 496921 Group IB Metal Activated Anatase $TiO_2$ for Selective Catalytic Deoxygenation of Lignin Fragments

**Abstract:**

Lignin is a major component of lignocellulosic biomass. The primary chemical structure of lignin is composed of propyl benzene units linked by a complex network of aromatic C-O and aliphatic C-O bonds. Catalytic hydrodeoxygenation (HDO) has been recognized as an important process strategy to convert lignin fragments to useful products. Supported metal catalysts for HDO in the literature typically showed the following performance characteristics: (1) production of fully saturated products as alkanes, and (2) formation of mixed aromatics and phenolic products, often with saturated hydrocarbons. Catalyst deactivation due to coke formation on supports or metal sintering was often encountered. The supported noble metals or transition metals provide the active sites.

As an abundant and representative component of lignin, guaiacol as a model compound has been widely studied in evaluating the effectiveness of hydrodeoxygenation catalysts. We recently discovered that Group IB metals such as Au and Ag, known for poor hydrogenation activity are capable of activating anatase $TiO_2$ (TiO2-A) to generate unconventional hydrogenation sites for selective HDO of guaiacol by producing phenolics as the main products. While Au and Ag particles on several supporting materials, including rutile $TiO_2$, $ZrO_2$, $Al_2O_3$, $SiO_2$, and activated carbon, did not show distinguishable performance compared to that of the supports alone in the conversion of guaiacol, these metals on anatase $TiO_2$ support exhibited high activity with excellent stability and remarkable selectivity to phenolics in guaiacol hydrodeoxygenation. For example, at guaiacol conversion of 43% over a 0.4 wt% Au/$TiO_2$-A-40 nm catalyst, the selectivity to phenolics reached 87%. No aromatics and saturated hydrocarbons such as cyclohexane were formed. Gold particle size ranging from 2.7 to 41 nm was found to affect the catalyst activity but not the product selectivity. The activity of the catalysts did not markedly decrease in accordance with the increase of Au particle size or the decrease of total surface area of the gold particles in the catalysts, suggesting that activated $TiO_2$-A surface away from the interface between Au and the $TiO_2$-A also contributed to the HDO performance.Â The reaction rates of 0.26 and 0.91 (min$^{â''1}$ g-cat$^{â''1}$ cm$^3$) were determined for guaiacol hydrogenation and catechol hydrogenation, respectively. Bimolecular methylation was established as the dominant mechanism for methyl group transfer from guaiacol to phenolics. While silver and copper showed similar selectivity at subnanometer particle sizes, considerably less Ag and Cu were effective to achieve similar guaiacol conversion levels as with Au/ $TiO_2$-A. Temperature programmed reduction profiles of Ag/ $TiO_2$-A and Cu/ $TiO_2$-A showed that partial reduction of $TiO_2$-A was enhanced by the presence of Ag and Cu. Only when $TiO_2$-A reduction took place, the Group IB became active for the HDO, with phenolics as the only deoxygenation products.

Two major pathways of guaiacol hydrogenation to phenolics over anatase $TiO_2$supported Group IB metals catalysts are proposed: (1) direct hydrogenation of guaiacol to form phenol and methanol, (2) hydrodehydroxylation of catechol intermediate formed from transmethylation between guaiacol and phenolics.

The pyrolytic product of a virgin lignin extracted from corn-stalk was studied with the catalysts. The results are consistent with the proposed mechanistic pathways in producing phenolic compounds.

# 486632 Recovery of Coumaric and Ferulic Acids from Industrial Lignin Streams

**Abstract:**

Lignin, an aromatic renewable resource, is a low-cost byproduct from industrial streams such as cellulosic ethanol. Often these lignin streams are used for boiler fuel in the production of process heat for the plant. Byproduct lignin streams are a source of high value molecules that can be utilized as green building blocks. Cornstover, a herbaceous biomass, contains approximately 15-16% (w/w) lignin. Hydroxycinnamic acids (coumaric and ferulic) are phenolic acids found in herbaceous biomass which can be utilized to produce oxtinoxate, vanillin, green epoxides, as well as, other high value chemicals. Our research goal is to demonstrate cost-effective means of converting cornstover enzymatic hydrolysis lignin into an oil that can be utilized for the production of high-value chemicals. This research includes mild base extraction followed by low-temperature low-pressure hydrogenation (LTLP-H) to convert the lignin into oil. The base treatment is a 5 wt% solution of sodium hydroxide NaOH) in water refluxed for 6 h at 125°C and 145°C. The base treatment products were hydrogenated for 12 h at room temperature and 1 atmosphere absolute pressure, yielding 101 wt% dry basis (db) oil at 125°C and 95.1 wt% (db) oil at 145°C. The theoretical yield of coumaric acid in the cornstover enzymatic hydrolysis lignin were 2.42 ± 0.263 wt% (db), whereas, the theoretical yield for ferulic was 1.57±0.275 wt% (db). We obtained 1.83 wt% (db) coumaric acid and 0.764 wt% (db) ferulic in the 125°C treatment. This is 75.8 wt% (db) of theoretical yield for coumaric acid and 48.7 wt% (db) of theoretical yield for ferulic acid. For the 145°C treatment we obtained 1.98 wt% (db) coumaric acid and 0.789 wt% (db) ferulic acid, 81.7 wt% (db) theoretical yield and 50.2 wt% (db) theoretical yield, respectively.

# 494880 Enabling Bioconversion of Biorefinery Wastes to Lipids with Oleaginous *Rhodococci*

**Abstract:**

Conversion of cellulosic ethanol biorefinery wastes, which contain mainly lignin and some carbohydrates, to value-added bioproducts will greatly increase the competiveness of biomass refinery. Wild and engineered *Rhodococcus* strains (*R. opacus* PD630, *R. jostii* RHA1 and *R. jostii* RHA1 VanA$^{-}$) with lignin degradation and/or lipid biosynthesis capacities were used to co-ferment lignin in order to establish a fundamental understanding of the pathways and functional modules necessary to enable a platform for biological conversion of biomass-derived lignin to lipids. In this study, genomics, transcriptomics, and proteomics analysis were carried out to help identify catabolic pathways in co-fermentation for funneling biosynthesis of TAGs from both of lignin and carboxylates. About 50% of loaded lignin was converted by the microbes. Some intermediates (e.g. vanillin, vanillic acid, and benzoic acid) that could further form acetyl-CoA for lipid production were detected in the co-fermentation broths. In addition, NMR analysis indicated that monomers derived from lignin depolymerization could be instantly consumed by these strains as carbon sources to accumulate lipids. Accumulation of fatty acids (C14-C18) in cells, especially palmitic acid (C16:0, 35.8%) and oleic acid (C18:1, 47.9%), was detected. In addition, systems biology approach revealed the carbon utilization pattern from both lignin and carbohydrates in unique coordinative pathways during co-fermentation.

# 497320 Understanding Trends in Hydrodeoxygenation of Biomass-Derived Phenolics from C-C Versus C-OÂ Scission Reaction in Ethanol on Stepped Surfaces

**Abstract:**

Hydrodeoxygenation (HDO) reactions constitute an essential class of biomass processing reactions employed for the catalytic deoxygenation of platform molecules. The C-C versus C-O bond cleavage plays an important role in these reactions. To gain insights into the HDO of biomass derived molecules, ethanol has been used as a model molecule. Density Functional Theory and microkinetic modelling using CatMAP[1] have been utilized to determine the catalytic activity and selectivity trends of HDO reaction over stepped sites of transition state metals. For ethanol hydrogenation, various routes for decomposition have been studied: a-H abstraction, ß-H abstraction, C-C, C-O, and O-H bond scission. Methane, carbon monoxide and ethane have been found to be the major products. Methane and carbon monoxide are formed upon the C-C scission and ethane upon the C-O scission in the ethanol HDO reaction. The activity trend for ethanol HDO has been observed to be: Rh > Ru > Ni > Ir > Co > Pd > Re > Pt > Fe > Cu > Ag > Au for methane production and Co > Ir > Ni > Rh > Ru > Fe > Pt > Pd > Cu > Ag > Re > Au for ethane production (Figure 1(a)). Lee et al. have experimentally determined the trend for HDO product of guaiacol to be Rh > Pt ~ Ru > Pd which is similar to the activity trend observed for ethane formation[2]. Rh and Ru cleave both C-C as well as C-O bonds, however, the C-C scission products have a higher turnover than the ones produced upon C-O bond breaking. At the reaction conditions, Co, Ir, Pt and noble metals are the metals selective for ethane formation. To better understand the HDO reactions, the change in catalytic activity has been studied as a function of temperature across the length of the reactor. The activity trend for the metal catalysts was observed to be the same throughout the reactor; however, at inlet conditions (373K), the activity is much smaller than that of the reaction conditions (523K). At 373K, most of the transition metal catalysts viz. Pd, Pt, Rh, Ir, Ru, Fe, Ni, Co have been found selective towards ethane production. In order to explore alloys that give a high turnover and selectivity towards desired products, a number of bimetallic have been studied (Figure 1(b)). $Co_3Ni$ and $Ni_3Fe$ have been found be the metals with the highest turnover of ethane ($10^{-3}s^{-1}$) and selectivity greater than 0.98 towards ethane. $Ni_3Rh$ has shown the highest production rate of methane, followed by $Ni_3Pt$, $Ni_3Cu$, and $Co_3Ru$. The selectivity for methane on these alloys is greater than 0.9.

1. Medford, A. J., Shi, C., Hoffmann, M. J., Lausche, A. Fitzgibbon, S., Bligaard, T., Nørskov, J. K. Catalysis Letters, 145(3), 2015, 794-807

2. Lee, C., Sun, J., Suh, Y., Choi, J. & Ha, J. Catalysis Communication,17 (2012), 54-58

# 497366 Lignin Conversion and Upgrading to High Octane Number Gasoline

**Abstract:**

The current vision of biofuels production undervalues lignin's potential to address the nation's high quality liquid fuel requirements. Upgrading lignin to high octane number gasoline would significantly improve the total carbon use in biomass, enhance revenues, make biomass conversion more economically competitive, and support our country's mission of national energy security. In this study, hydrodeoxygenation of lignin model compounds and Kraft lignin over the combinations of Pd/alumina, Ni/alumina, and Ni-Mo/alumina in water-alcohol mixed solution was investigated. Results showed that 82% lignin was converted to products with up to 96% product selectivity that had the octane number over 100. In addition, the reactivity of lignin and its interactions with the chemical catalysis systems were studied to reveal the principles of the reaction mechanism, its control, and applications.

# 495895 Effect of Carbonaceous Susceptors on Selective Production of Phenolics and Syngas from Lignin Via Microwave Pyrolysis

**Abstract:**

At present, global energy supply and production of chemicals are largely dependent on fossil fuels such as coal, petroleum and natural gas. However, the utilization of these resources is limited owing to their non-renewable nature and its environmental adversity. Lignocellulosic biomass is projected as the only source of renewable carbon on earth. The primary structural building blocks of biomass are cellulose, hemicellulose, and lignin; of these components, contributes to 30% and 40% by weight and energy of the total biomass, respectively. Energy, syngas, fuels and chemicals can be obtained from lignin through thermochemical processing techniques like combustion, gasification and pyrolysis. Fast pyrolysis is a promising process to produce bio-oil and chemicals in high yields. High yields of aromatic hydrocarbons and organics can be obtained due to the aromatic functionality in the lignin structure. In order to extract resources from lignin on a large scale by fast pyrolysis technology, challenges such as high yield of condensable products, product selectivity and energy efficiency are to be improved. Achieving fast pyrolysis conditions with improved efficiency is possible by microwave assisted heating. Microwave assisted fast pyrolysis technology can be used to overcome the techno-economic barriers by reducing the processing time, improving the heat distribution, enhancing the product selectivity, and aiding in better heating efficiency.

The intention of the current study is to enhance the yield of bio-oil and its quality from lignin by microwave assisted pyrolysis. The effects of (i) lignin (L) to activated carbon (AC) ratio , (ii) particle size of activated carbon (0.45 mm to 3 mm), and (iii) carbonaceous susceptors (AC, graphite and charcoal) on heating rate, product yields (oil, gas and char fractions), product composition and process efficiency were investigated at a constant microwave power of 480 W. It was found that, owing to the high amount of physically bound moisture in the mesoporous capillaries of AC (1mm), high bio-oil yield of 66 wt. % at optimum conditions of 10: 90 g/g of L: AC was observed. In L:AC blend of 100 g, increase in mass of AC from 10 g to 90 g decreased the heating rate significantly from 128 °C/min to 32 °C/min, correspondingly a decrease in energy efficacy of the system from 95% to 34% was observed. From the oil composition analyses, it was found that bio-oil comprises of phenolics (simple phenols, guaiacols and syringols), aromatic hydrocarbons, naphthalenes, indene derivatives and furans. The highest yield of total phenolics (64.2%) was obtained at 10: 90 g/g of L: AC (1mm). Initial lignin sample mass of 9 to 40 wt. % was converted into simple phenolics and its composition increased with increase in L: AC ratio.

The carbonaceous materials, such as activated carbon, graphite, charcoal and bio-char are found to be the efficient microwave absorbers because of their high microwave coupling nature even at ambient temperatures **[1,2]**. For this reason, we have chosen activated carbon, graphite and charcoal as the susceptors to investigate their effect on lignin pyrolysis. The effect of susceptors on the heating rate, product mass and energy yields were investigated. The yield (wt. %) of bio-oil is in line with the heating rates and varies in the following order: charcoal (36 wt. %) > graphite (34.5 wt. %) > AC (0.45mm) (29.2 wt. %). Phenolic derivatives are dominant in oil fraction, 97% of oil obtained with AC (0.45mm) contains phenolics followed by charcoal (79%) and graphite (72%). Selectivity (%) of simple phenolics

is varying in the following order: 60.1 (AC (0.45mm)) > 55.8 (charcoal) > 40.9 (graphite). A significant enhancement in heating value was noticed both in char (28-30 MJ/kg) and oil (28-33 MJ/kg) fractions compared to raw lignin (25 MJ/kg). Tiny micro plasma spots generated during the pyrolysis process increased tiny cylindrical and spherical micro-pores within the residual char. Yield of $CO_2$ decreased and that of $H_2$ increased with increase in pyrolysis temperature and its evolution profiles were influenced by the nature of susceptor and lignin to susceptor mass ratio.

**References:**

1. T. Ohra-aho, J. Linnekoski. Catalytic pyrolysis of lignin using analytical pyrolysis-GC-MS. J. Anal. Appl. Pyrol. 113 (2015) 186-92.
2. S. Adhikari, V. Srinivasan, O. Fasina. Catalytic pyrolysis of raw and thermally treated lignin using different acidic zeolites. Energy & Fuels 28 (2014) 4532-38.

# 497990 Catalytic Conversion of Lignin in Ionic Liquids *Via* Catalysis and Biocatalysis

**Abstract:**

Lignin makes up 20-30% of the plant biomass and is the most abundant aromatic polymers in nature. Converting lignin into high-value chemicals adds revenues for a biorefinery thus improves the economic viability of biofuel production. Ionic liquids (ILs) have received increasing interest because of their potentials in fractionating and pretreating lignocellulosic biomass. Given the unique properties of aqueous ILs that provide for lignin solubility and biocompatibility to enzymes and microbes, we foresee a great opportunity to develop a new strategy for lignin extraction and depolymerization *via* catalysis and biocatalysis in aqueous ILs. The stability and activity of commercial fungal and plant laccases and a recombinant bacterial laccase were evaluated in difference concentration of ILs. Results suggest that certain ILs were fully compatible with laccases; while some appeared as non-competitive inhibitors. Immobilization of laccase on solid matrix, i.e. membrane surface helped to retain the enzyme activity and overcome the inhibition effects. We also explored catalytic oxidation route in ILs by screening a range of metal catalysts and oxidizing agents. Molecular weight distribution of lignin fraction was followed using gel permeation chromatography while the oxidation products such as vanillic acid, syringaldehyde, and syringic acid were identified and quantified using a GC-MS. Cleavage of the lignin structures was further investigated by NMR to understand the plausible oxidation mechanisms. Results from this study provide insights into the oxidative degradation pathway of lignin in the presence of IL and the selectivity of catalyst towards product formation. In situ lignin conversion in IL will also help to tackle the challenges associated with IL recycle and recovery.

# 498243 Depolymerization of Lignin Using Peracetic Acid Under Mild Conditions

**Abstract:**

Lignin is the largest non-carbohydrate component in lignocellulosic biomass, so utilization of lignin can significantly improve total biomass utilization. Depolymerization of lignin is an important step for effective utilization of lignin. Effective and selective depolymerization of lignin into mono-aromatics and/or dicarboxylic acids does not only convert lignin into value-added products, but they also enhance the total biomass utilization by valorizing the co-products from biomass. Diverse lignin depolymerization strategies have been discussed; however, there are few commercially feasible lignin conversion processes because of low conversion yield, poor product selectivity, expensive solvents and catalysts, and severe reaction conditions. PAA removed and/or decomposed lignin. In this study, Peracetic acid (PAA) was applied to decompose lignin under mild reaction temperature with minimal side reactions. Structural changes on lignin were evaluated by diverse analytical methods such as GPC and NMR. In addition, quantitative analysis with depolymerization products was conducted by GC/MS and HPLC.

# 498934 Low-Energy Catalytic Electrolysis for Simultaneous Hydrogen Evolution and Lignin Depolymerization

**Abstract:**

Here, a new proton-exchange-membrane electrolysis is presented, in which lignin was used as the hydrogen source at the anode for hydrogen production. Either polyoxometalate (POM) or $FeCl_3$ was used as the catalyst and charge-transfer agent at the anode. Over 90% Faraday efficiency was achieved. In a thermal-insulation reactor, the heat energy could be maintained at a very low level for continuous operation. Compared to the best alkaline--water electrolysis reported in literature, the electrical-energy consumption could be 40% lower with lignin electrolysis. At the anode, the Kraft lignin (KL) was oxidized to aromatic chemicals by POM or $FeCl_3$, and reduced POM or Fe ions were regenerated during the electrolysis. Structure analysis of the residual KL indicated a reduction of the amount of hydroxyl groups and the cleavage of ether bonds. The results suggest that POM- or $FeCl_3$-mediated electrolysis can significantly reduce the electrolysis energy consumption in hydrogen production and, simultaneously, depolymerize lignin to low-molecular-weight value-added aromatic chemicals.

Green Chemistry

PAPER

View Article Online
View Journal | View Issue

Check for updates

Cite this: *Green Chem.*, 2017, **19**, 4939

# Synergistic maximization of the carbohydrate output and lignin processability by combinatorial pretreatment†

Zhi-Hua Liu,[a,b,c] Michelle L. Olson,[d] Somnath Shinde,[e] Xin Wang,[a,b,c] Naijia Hao,[e] Chang Geun Yoo,[f] Samarthya Bhagia,[e] John R. Dunlap,[g] Yunqiao Pu,[f] Katy C. Kao,[d] Arthur J. Ragauskas,[e,f,h] Mingjie Jin[i,j] and Joshua S. Yuan *[a,b,c]

Lignocellulosic biorefineries have gained much attention worldwide as a potential solution to the challenges of energy demand and global climate change. However, the industrial implementation of biorefineries has been hindered by low fermentable sugar yields and low lignin processability. Combinatorial pretreatments with a low holding temperature were investigated in an effort to synergistically improve the carbohydrate output and lignin processability from corn stover. Upon combinatorial pretreatment with 1% $H_2SO_4$ for 30 min followed by 1% NaOH for 60 min at 120 °C, glucan and xylan conversion increased by 11.2% and 8.3% respectively relative to single pretreatment. This combinational pretreatment removed the amorphous portion, disrupted the rigid structure, and increased the water holding capacity of corn stover, thus increasing the hydrolysis performance. With whole fractionation by combinatorial pretreatment, glucose and xylose yields were 88.4% and 72.6%, respectively, representing increases of 10.0% and 8.1%. The lignin yield was 19.7% in the solid residue and 77.6% in the liquid stream, which increased by 33.4%. When grown in fed-batch fermentation mode, a record level of polyhydroxyalkanoate (PHA) concentration (1.0 g $l^{-1}$) was obtained using lignin as a carbon source by *Pseudomonas putida* KT2440. Lignin characterization results showed that combinatorial pretreatment increased the G- and H-lignin content, reduced the β–β and β-O-4 groups, and fractionated more aromatic monomers, thus facilitating lignin processability into PHA. These results highlighted the use of combinational pretreatment at a low holding temperature as a means to synergistically maximize the carbohydrate output and lignin processability, which provides a unique set of features to improve the biorefining performance.

Received 7th July 2017, Accepted 6th September 2017

DOI: 10.1039/c7gc02057k

rsc.li/greenchem

[a]*Synthetic and Systems Biology Innovation Hub (SSBiH), Texas A&M University, College Station, TX, 77843, USA. E-mail: syuan@tamu.edu*
[b]*Department of Plant Pathology and Microbiology, Texas A&M University, College Station, TX 77843, USA*
[c]*Institute for Plant Genomics and Biotechnology, Texas A&M University, College Station, TX 77843, USA*
[d]*Department of Chemical Engineering, Texas A&M University, College Station, TX 77843, USA*
[e]*Department of Chemical & Biomolecular Engineering, University of Tennessee, Knoxville, Tennessee 37996, USA*
[f]*Biosciences Division, Oak Ridge National Laboratory, Oak Ridge, Tennessee 37831, USA*
[g]*Advanced Microscopy and Imaging Center, University of Tennessee, Knoxville, TN 37996, USA*
[h]*Center for Renewable Carbon, Department of Forestry, Wildlife, and Fisheries, University of Tennessee Institute of Agriculture, Knoxville, TN 37996, USA*
[i]*School of Environmental and Biological Engineering, Nanjing University of Science and Technology, Nanjing 210094, China*
[j]*Guangdong Cleamol LTD., Foshan 528225, China*
† Electronic supplementary information (ESI) available. See DOI: 10.1039/c7gc02057k

## 1 Introduction

Lignocellulosic biorefineries offer a potential solution to the current energy and environmental challenges facing our generation.[1–3] Despite their potential, using biorefineries to produce biofuels and other bio-based products from lignocellulosic biomass (LCB) has yet to be fully commercialized.[4,5] Several major issues hinder the industrial implementation of biorefineries, including unsatisfactory pretreatment performance, low fermentable sugar yield, and lack of lignin utilization for fungible products.

LCB biorefining involves several technologies and processes to fractionate LCB into its primary building blocks: cellulose, hemicellulose, and lignin, which can be further converted to biofuels, chemicals, and materials. The critical challenges in biorefineries are full utilization of these three main components of the plant cell wall.[6] Generally, cellulose and hemicellulose are converted into fermentable sugars, which are then used to produce ethanol in cellulosic ethanol projects. In contrast, lignin is viewed as residual waste, or it is burned to produce energy with low-value utilization. Lignin (15–30%, dry weight) is the second most abundant polymer after cellulose, yet it is significantly underutilized in biorefineries and in the pulp and paper industry.[7–10] Recently, these industries have recognized that valorization of the lignin-containing residue is essential in enhancing the overall biorefinery competitiveness.[11–15] However, to achieve an appreciable product yield, all current lignin bioconversion platforms require biological or chemical fractionations of the lignin-containing biorefinery residue, which are costly and time-consuming.[11,14,16,17] The development of pretreatment technologies to both achieve higher carbohydrate output and derive more processable lignin stream thus has the potential to deliver more cost-effective and sustainable biorefineries.

Generally, most of the integrated biorefineries consist of three major unit operations: pretreatment, hydrolysis, and fermentation.[5,18] The goal of the pretreatment is to remove the structural and compositional impediments that hinder the hydrolysis efficiency and thus to increase the sugar productivity and yield lignin-rich streams. Hydrolysis is used to convert cellulose to glucose and hemicellulose to xylose, arabinose, *etc.*[7,19] During fermentation, monosaccharides and fractionated lignin can be converted into bioproducts *via* microorganisms.

Various pretreatments have been developed to deconstruct LCB and increase the hydrolysis efficiency. The pretreatment conditions mainly include catalyst types, holding temperature, and residence time.[20–24] Pretreatments vary in severity, which is ranked using a pretreatment severity (PS) score. For example, dilute acid pretreatment dissolves most of the hemicelluloses into monosaccharides, which in turn enhances the hydrolysis of cellulose. However, dilute acid pretreatment with a high PS causes irreversible degradation of hemicellulose and formation of inhibitors.[25] Acidic pretreatment may also form "pseudo-lignin" at a high holding temperature, which can hinder the hydrolysis performance and high-value utilization of LCB.[26] Liquid hot water uses only water to deconstruct LCB by the auto-hydrolysis effect at a high holding temperature. However, it usually results in incomplete LCB matrix opening and low hydrolysis efficiency. Alkaline pretreatment alters the lignin fraction to increase the hydrolysis efficiency.[27,28] Unfortunately, alkaline pretreatment degrades hemicellulose and releases monomeric lignin compounds, generating enzyme inhibitors. When one considers all these advantages and disadvantages, it is unlikely that a single pretreatment can become the universal method to maximize the yield of fermentable sugars and lignin as well as the lignin processability at the same time.

To overcome these challenges, we aimed to develop a novel pretreatment technology that would not only achieve a high carbohydrate and lignin output, but also enhance the processability of lignin containing the biorefinery residue. We hypothesized that the application of combinatorial pretreatment may enhance both carbohydrate and lignin yields and improve the lignin processability. To test this hypothesis, we exploited several combinatorial pretreatment technologies to deconstruct corn stover (CS). Pretreatment efficiency was assessed by measuring the enzymatic hydrolysis of pretreated CS. We then characterized the pretreated CS and lignin fractions to determine the effects of each pretreatment on the properties of different solids. Lignin processability was assessed by measuring the direct bioconversion of the biorefinery residue to polyhydroxyalkanoate (PHA). These analyses helped us to get a deep understanding of how combinatorial pretreatment improves the fermentable sugar output as well as lignin processability.

## 2 Materials and methods

### 2.1 Combinatorial pretreatment strategies of corn stover

The composition analysis of corn stover (CS) was conducted following the Laboratory Analysis Protocol (LAP) of the National Renewable Energy Laboratory (NREL), Golden, CO, USA. To study the effects of pre-washing on the composition transformation, CS was washed 20 times with distilled water based on its dry weight (dw). The compositions of CS are given in ESI 1.†

Seven pretreatment strategies were used, alone or in combination: dilute sulfuric acid, liquid hot water, sodium hydroxide, and ethanol pretreatment (Table 1). All pretreatments were conducted in a 1.0 L screw bottle heated by using an Amsco® LG 250 Laboratory Steam Sterilizer (Steris, USA). The

Table 1 Combinatorial pretreatment strategies of corn stover

| | Step 1 | | | PS | Step 2 | | | PS |
|---|---|---|---|---|---|---|---|---|
| Case | Chemicals | Conditions | SL (w/w) | $\log R''_0$ | Chemicals | Conditions | SL (w/w) | $\log R''_0$ |
| 1 | 1% NaOH | 120 °C, 60 min | 10% | 8.4 | | | | |
| 2 | 50% Ethanol | 120 °C, 60 min | 10% | 3.2 | | | | |
| 3 | 50% Ethanol + 1% NaOH | 120 °C, 60 min | 10% | 8.2 | | | | |
| 4 | Liquid hot water | 120 °C, 30 min | 10% | 2.8 | 1% NaOH | 120 °C, 60 min | 10% | 8.4 |
| 5 | Liquid hot water | 120 °C, 30 min | 10% | 2.8 | 50% Ethanol + 1% NaOH | 120 °C, 60 min | 10% | 8.2 |
| 6 | 1% $H_2SO_4$ | 120 °C, 30 min | 10% | 7.3 | 1% NaOH | 120 °C, 60 min | 10% | 8.4 |
| 7 | 1% $H_2SO_4$ | 120 °C, 30 min | 10% | 7.3 | 50% Ethanol + 1% NaOH | 120 °C, 60 min | 10% | 8.2 |

SL, solids loading; S1 stands for Step 1; S2 stands for Step 2; PS, pretreatment severity; % was calculated based on the weight percent, w/w.

holding temperature and pressure profiles of each pretreatment strategy are shown in ESI 2.† Fig. 1 provides a flow diagram of the fractionation process for single and combinatorial pretreatments, describing the points where fermentable sugars and lignin were obtained.

During single pretreatment (Cases 1, 2 and 3), 50 g CS (dw) were loaded into the screw bottle at 10% (w/w) solid loading for pretreatment. After pretreatment, the pretreated slurry was filtered by vacuum filtration to separate pretreated solids from the liquid stream. The pretreated solids were post-washed 10 times with distilled water. The liquid stream and washing stream were collected for further analysis. For combinatorial pretreatments (Cases 4, 5, 6, and 7), CS (loaded as described above) was pretreated first with liquid hot water or dilute sulfuric acid in Step 1, followed by EtOH + NaOH in Step 2. The morphological properties of the pretreated solids exposed to each of the pretreatments are shown in ESI 3.†

Pretreatment severity ($\log R_0$) is calculated as follows:[29]

$$\log R_0 = t \exp[(T - T_b)/\omega] \quad (1)$$

where $t$ is the residence time, min; $T$ is the pretreatment temperature, °C; $T_b$ is the base temperature, 100 °C; $\omega$ is the fitted value based on the activation energy, 14.75.

The severity factor ($\log R''_0$) was used to compare the severity of combinatorial pretreatment, where the effects of pH value are taken into account (Table 1).[20]

$$\log R''_0 = \log R_0 + |\text{pH-7}| \quad (2)$$

## 2.2 Enzymatic hydrolysis of pretreated corn stover

The pretreated CS was hydrolyzed using the cellulase preparations Cellic CTec2 and HTec2. Filter paper activity (FPU) of Cellic CTec2 is 96 FPU ml$^{-1}$, while the cellobiase activity of β-glucosidase is 1270 CBU ml$^{-1}$. The protein content of Cellic CTec2 and HTec 2 determined by the Bradford assay is 178 ± 19.9 mg ml$^{-1}$ and 103 ± 9.6 mg ml$^{-1}$, respectively.

Enzymatic hydrolysis assays were conducted at 3% (w/v) solid loading in citrate buffer solution (50 mM, pH 4.8) with an enzyme loading of 10 FPU g$^{-1}$ solid in a 250 mL

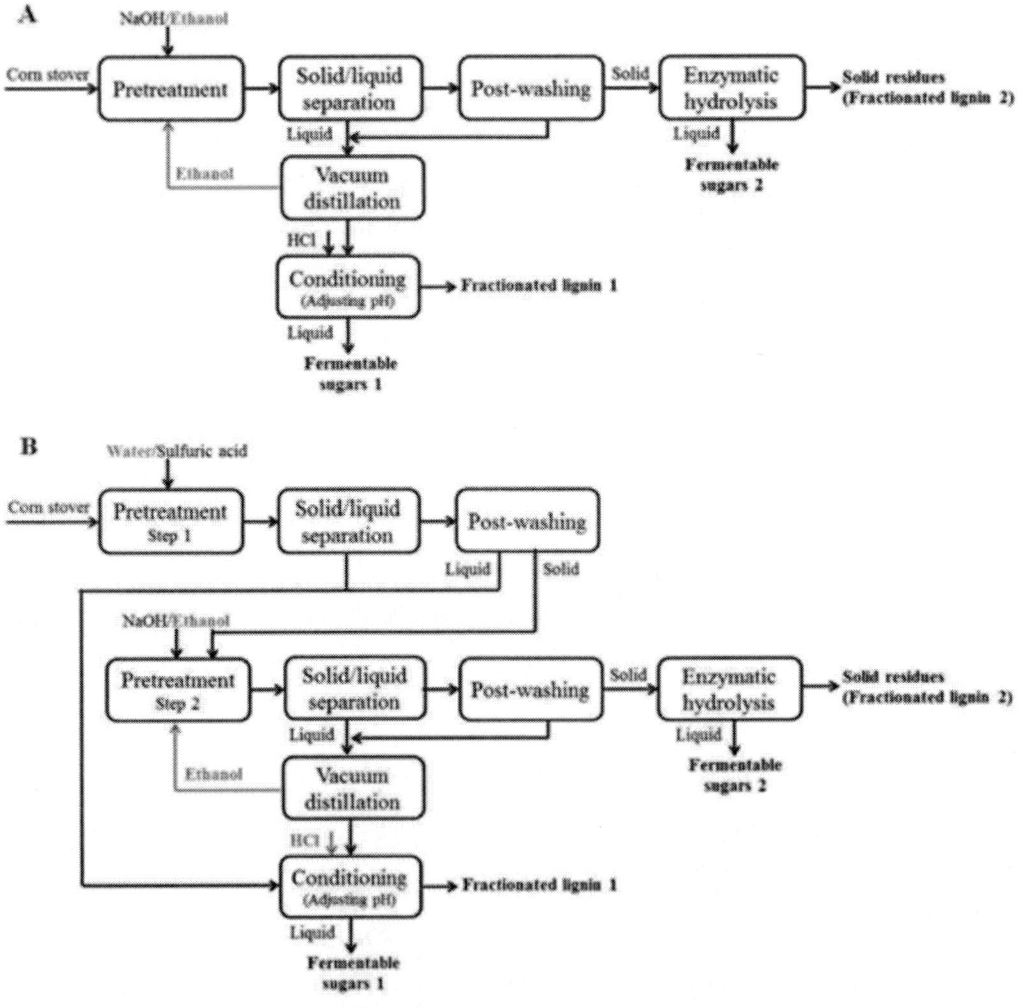


Fig. 1 The fractionation process flow diagram of fermentable sugars and lignin from corn stover by (A) single pretreatments (Cases 1–3) and (B) combinatorial pretreatments (Cases 4–7).

Erlenmeyer flask. The volumetric ratio of CTec2 : HTec2 was 10 : 1 used in hydrolysis assays. The mixture was stirred with a shaker at 50 °C and 200 rpm for 168 h. A 1.0 ml sampling of supernatant was collected at 6, 12, 24, 48, 96 and 168 h for sugar analysis. Sugar conversions are calculated as follows:

$$\text{Glucan conversion } (\%) = (\text{Glucose in hydrolysis} \times 162/180)/\text{Glucan in pretreated corn stover} \times 100\% \tag{3}$$

$$\text{Xylan conversion } (\%) = (\text{Xylose in hydrolysis} \times 132/150)/\text{Xylan in pretreated corn stover} \times 100\% \tag{4}$$

### 2.3 Characterization of untreated and pretreated corn stover

The biomass crystallinity of the untreated corn stover (UCS) and pretreated CS was measured using a D8 Fucos X-ray diffractometer (Bruker AXS Co., Germany). The samples with a particle size less than 125 μm were placed in a quartz sample holder using a Rigaku Miniflex diffractometer in conjunction with a CuKα radiation source ($k$ = 0.154 nm), and then scanned at a speed of 1° $min^{-1}$ ranging from $2\theta$ = 5–40°, and with a step size of 0.02°. The crystallinity index (CrI) is calculated as follows:

$$\text{CrI} = [(I_{002} - I_{\text{amorphous}})/I_{002}] \times 100 \tag{5}$$

where $I_{002}$ is the intensity for the crystalline portion of LCB (*i.e.*, cellulose) at about $2\theta$ = 22.1°, and $I_{\text{amorphous}}$ is the peak for the amorphous portion (*i.e.*, cellulose, hemicellulose, and lignin) at about $2\theta$ = 18.2°.

The water holding capacity (WHC) of CS solids before and after pretreatment was determined by the centrifuge separation method.[30] 1.0 g sample (dw) was mixed well with 10 ml water in a 50 mL beaker for 1 h, and then centrifuged for 5 min at a certain separation factor ($F$ = 1000). The separation factor ($F$) was defined as the ratio of rotational acceleration and gravity acceleration:

$$F = \omega^2 R/g = (2\pi Rn/60)^2/Rg = Rn^2/895 \tag{6}$$

where $\omega$ is angular velocity, rad $s^{-1}$; $R$ is the distance of the rotator and the axis of rotation, m; $n$ is rotating speed, rpm; $g$ is gravity acceleration, 9.8 m $s^{-2}$. WHC was defined as the adsorbed water weight per gram of dry sample.

### 2.4 PHA fermentation of lignin stream

The liquid stream from pretreatment, which contains mainly lignin, was used for polyhydroxyalkanoate (PHA) fermentation. The fermentation process and PHA analysis were conducted following a previous report.[31] The lignin stream was adjusted using 1.0 M HCl to pH 7.0, and then sterilized by using an autoclave at 121 °C for 20 min. For medium preparation, the lignin stream was used as a carbon source, dissolved to a certain soluble substrate concentration (SSC) with $ddH_2O$, and then transferred into 250 mL Erlenmeyer flasks in a working volume of 100 ml. *Pseudomonas putida* KT2440 cell pellets for inoculation were obtained by centrifuging the seed culture at 5000 rpm for 10 min. Fermentation was carried out at pH 7.0, 28 °C, and 200 rpm for 18 h.

### 2.5 Characterization of fractionated lignin from different pretreatments

**2.5.1 2D HSQC NMR analysis.** For lignin recovery from different pretreatments, the liquid stream was filtered with a 0.22 μm filter membrane and then acidified to a pH less than 2.0 with concentrated HCl. The lignin precipitates were collected in tared centrifuge tubes by centrifugation at 10 000 for 10 min. The lignin solids after acidification were washed with $ddH_2O$ and then freeze-dried by using a lyophilizer at −55 °C for 24 h.

Native lignin from corn stover (cellulolytic enzyme lignin, CEL) was isolated according to procedures described in the previous study.[32] In brief, the extractive-free corn stover was air-dried and ball-milled using a planetary ball mill (Retsch PM 100) at 600 rpm with zirconium dioxide vessels (50 ml) containing $ZrO_2$ ball bearings (10 mm × 10) for 2 h (5 min grinding and 5 min break). The ball-milled CS was then hydrolyzed using Cellic CTec2 (0.1 ml $g^{-1}$ biomass) and HTec2 (0.1 ml $g^{-1}$ biomass) at 50 °C for 24 h. The supernatants were removed by centrifugation, and the solid residues were hydrolyzed again under the same conditions with fresh enzyme mixtures. The residue was then extracted with 96% dioxane at ambient temperature for 48 h. The CS native lignin was recovered using a rotary evaporator, freeze-dried, and vacuum dried for further analysis.

For the 2D HSQC NMR analysis, the lignin samples (30–50 mg) were dissolved in DMSO-$d_6$ (0.6 ml). Two-dimensional (2D) $^1H$–$^{13}C$ heteronuclear single quantum coherence (HSQC) nuclear magnetic resonance (NMR) experiment was performed using a Bruker Avance-III 400 MHz spectrometer equipped with a 5 mm Broadband Observe probe (5 mm BBO 400 MHz W1 with $Z$-gradient probe, Bruker) and a Bruker standard pulse sequence ('hsqcetgpsi2'). The spectra were obtained with a spectral width of 11 ppm in F2 ($^1H$, 2048 data points) and 190 ppm in F1 ($^{13}C$, 256 data points). Sixty-four scans with a 1 s delay were used for each sample.

**2.5.2 $^{31}P$ NMR analysis.** $^{31}P$ NMR spectra were acquired according to published methods with some modifications.[33] In detail, the lignin sample was weighed into a vial with a PTFE cap. The stock solution of pyridine/$CDCl_3$ (0.7 ml, v/v = 1.6/1), including 1.25 mg $ml^{-1}$ Cr(acac)$_3$ and 2.5 mg $ml^{-1}$ internal standard *endo-N*-hydroxy-5-norbene-2,3-dicarboxylic acid imide (NHND) was then added to the vial. The vial was shaken until the lignin was dissolved completely. Prior to the $^{31}P$ NMR experiment, 70 μl of the phosphorylating reagent 2-chloro-4,4,5,5-tetramethyl-1,3,2-dioxaphospholane (TMDP) was added to the vial. Quantitative $^{31}P$ NMR spectra were acquired on a Varian 500 MHz spectrometer using an inverse-gated decoupling pulse sequence, 90° pulse angle, 1.2 s acquisition time, 25 s pulse delay, and 64 scans. The phase was adjusted automatically and the baseline was corrected using a Bernstein polynomial fit with a parameter of 6. The quantitative calculation of the hydroxyl groups was based on the amount of the internal standard.

**2.5.3 Gel-permeation chromatography (GPC).** The lignin samples (dried under vacuum at 40 °C overnight) were acetylated with acetic anhydride/pyridine (1/1, v/v) at ambient temperature for 24 h in a sealed flask under an inert atmosphere. The lignin concentration in the solution was approximately 2 mg $ml^{-1}$. After 24 h, the solution was diluted with approximately 20 ml of ethanol and stirred for an additional 30 min, after which the solvents were removed with a rotary evaporator, followed by drying in a vacuum oven at 40 °C. Prior to GPC analysis, the acetylated lignin samples were dissolved in tetrahydrofuran (1.0 mg $ml^{-1}$), filtered through a 0.45 μm filter, and placed into a 1 ml auto-sampler vial. The molecular weight distributions of the acetylated lignin samples were analyzed on an Agilent GPC SECurity 1200 system equipped with four Waters Styragel columns (HR1, HR2, HR4, HR6), an Agilent refractive index (RI) detector, and an Agilent UV detector (270 nm), using tetrahydrofuran (THF) as the mobile phase (1.0 ml $min^{-1}$), with an injection volume of 20.0 μl. A standard polystyrene sample was used for calibration. The number-average molecular weight ($M_n$) and weight-average molecular weight ($M_w$) were determined by GPC.

**2.5.4 Fourier transform infrared (FTIR) analysis.** FTIR spectroscopy experiments of lignin samples were carried out using a Spectrum One FTIR Spectrometer from PerkinElmer (USA). The resolution was set at 4 $cm^{-1}$, 16 scans were recorded for each analysis and the scanning range was from 600 $cm^{-1}$ to 4000 $cm^{-1}$. Five analyses were performed at five locations per sample. For calculating ratios, absorption intensities were considered and normalization was performed on the band at 1511 $cm^{-1}$ (C=C stretching vibrations of the aromatic skeletal of lignin).

### 2.6 Scanning electron microscopy (SEM) of untreated and pretreated corn stover

The samples for SEM analysis were mounted onto stubs with carbon tape and sputter-coated with gold. SEM was then carried out on a Zeiss EVO MA15 at an accelerating voltage of 20 kV with a back scatter detector at 100 to 1000 times magnification. Raw images were adjusted for brightness and contrast in ImageJ software.[34] Images were merged using Adobe Photoshop CC v. 2017.

### 2.7 Derivatives analysis by gas chromatography-mass spectrometry (GC-MS)

The liquid streams produced from different pretreatments were filtered with a 0.22 μm filter membrane and then acidified to pH 1–2 with concentrated HCl. 3.0 ml of acidified supernatant were mixed well with 1.0 ml butanedioic acid-$d_6$ (0.04 mg $ml^{-1}$) as an internal standard, and then extracted with three volumes of methyl *tert*-butyl ether (MTBE) at 4500 rpm for 30 min. The organic layer was collected and dried under a stream of nitrogen gas. Prior to GC-MS analysis, 1.0 ml MTBE was added to dissolve the sample.

GC-MS was performed on a GCMS-QP2010SE (Shimadzu Scientific Instruments, Inc.). The analytical column used was a Shimadzu SH-Rxi-5Sil column (30 m × 250 μm × 0.25 μm). The eluted sample was analyzed using helium as carrier gas at a flow rate of 1.0 ml $min^{-1}$. The temperature profile of the GC-method was 3 min at 50 °C, and then it was increased to 290 °C at 8 °C $min^{-1}$. Ions were generated with a 70 eV electron beam at an ionization current of 40 μA. Mass spectral peak quantification was performed using GCMSsolution software Ver. 2.6. The acquired peak height was normalized against that of an internal standard.

### 2.8 Analysis methods and mass balance

The sugars were analyzed using HPLC in an HPLC 1260 Infinity (Agilent Technologies, CA) equipped with an Aminex HPX-87P carbohydrate analysis column (Bio-Rad Laboratories, CA) and a refractive index detector at 85 °C with HPLC grade water as the mobile phase at a flow rate of 0.6 ml $min^{-1}$. The sugar yield in the whole fractionation process (pretreatment + hydrolysis) was calculated as follows:

$$\text{Glucose yield (\%)} = \left[\text{Glucose}_{\text{Liquid of pretreatment}} + \text{GO}_{\text{Liquid of pretreatment}} \times (180/162) + \text{Glucose}_{\text{Hydrolysis}}\right] / \left[\text{Glucan}_{\text{Feedstock}} \times (180/162)\right] \quad (7)$$

$$\text{Xylose yield (\%)} = \left[\text{Xylose}_{\text{Liquid of pretreatment}} + \text{XO}_{\text{Liquid of pretreatment}} \times (150/132) + \text{Xylose}_{\text{Hydrolysis}}\right] / \left[\text{Xylan}_{\text{Feedstock}} \times (150/132)\right] \quad (8)$$

where GO represents the glucose oligomers and XO represents the xylose oligomers. Error bars in the figures represent the standard deviation of the replicates. Significance was determined using the Student *t*-test. A *p*-value less than 0.05 was considered significant.

## 3 Results and discussion

### 3.1 Effects of pre-washing on the composition transformation of untreated corn stover

Untreated corn stover (UCS) was analyzed to establish the baseline composition. The total sugar content in UCS was 56.4%, and the total lignin content was 22.0% (Table 2). For the biorefinery purpose, feedstock with a high sugar content has high potential to increase the biorefining efficiency.[35] Interestingly, water and ethanol extractives in UCS were 10.1% and 3.2%, respectively, and the free glucose in the water extractives was only 0.7% (ESI 1†).

Pre-washing was evaluated for an impact on the components of UCS (ESI 1†). The results showed that pre-washing increased the glucan, xylan, and lignin content. The total sugars in washed UCS were 58.8%, which increased by 4.3% compared to those in UCS. The water extractives were decreased by 50% after pre-washing. Notably, free glucose was removed almost completely, and the ash content decreased more than 36% after pre-washing, as confirmed by the

**Table 2** Components of untreated corn stover and pretreated corn stover with each pretreatment

| Components/% | Glucan | Xylan | Araban | Galactan | Lignin (AIL) | Lignin (ASL) | Ash | Total |
|---|---|---|---|---|---|---|---|---|
| UCS | 33.1 (1.3) | 17.5 (0.6) | 0.9 (0.3) | 4.4 (0.3) | 19.7 (0.1) | 2.3 (0.2) | 3.6 (0.2) | 81.5 |
| Case 1 | 44.9 (1.3) | 21.1 (0.5) | 0.8 (0.2) | 4.9 (0.5) | 10.8 (1.6) | 1.2 (0.1) | 2.4 (0.2) | 86.1 |
| Case 2 | 34.4 (0.7) | 19.1 (1.6) | 1.4 (0.4) | 4.9 (0.4) | 19.4 (1.9) | 1.9 (0.1) | 2.6 (0.2) | 83.8 |
| Case 3 | 44.5 (2.1) | 21.5 (1.3) | 1.4 (0.5) | 4.8 (0.1) | 11.0 (1.3) | 1.4 (0.2) | 3.0 (0.1) | 87.6 |
| Case 4 S1 | 35.8 (1.3) | 19.3 (1.1) | 1.0 (0.3) | 5.0 (0.1) | 21.0 (0.6) | 2.0 (0.1) | 2.0 (0.2) | 86.1 |
| Case 4 S2 | 49.5 (0.5) | 22.1 (0.1) | 0.9 (0.1) | 4.5 (0.0) | 10.4 (0.7) | 1.4 (0.1) | 1.8 (0.3) | 90.7 |
| Case 5 S1 | 35.8 (1.3) | 19.3 (1.1) | 1.0 (0.1) | 5.0 (0.1) | 21.0 (0.6) | 2.0 (0.2) | 2.0 (0.2) | 86.1 |
| Case 5 S2 | 45.6 (2.3) | 23.8 (0.1) | 1.4 (0.6) | 4.5 (0.2) | 10.3 (1.4) | 1.4 (0.1) | 1.9 (0.3) | 88.9 |
| Case 6 S1 | 43.9 (0.1) | 9.0 (0.0) | — | 4.2 (0.2) | 28.8 (1.3) | 1.3 (0.0) | 2.2 (0.4) | 89.3 |
| Case 6 S2 | 74.4 (1.9) | 8.1 (0.5) | — | 3.8 (0.5) | 11.0 (2.4) | 0.9 (0.3) | 1.6 (0.3) | 99.9 |
| Case 7 S1 | 43.9 (0.1) | 9.0 (0.0) | — | 4.2 (0.2) | 28.8 (1.3) | 1.3 (0.1) | 2.2 (0.4) | 89.3 |
| Case 7 S2 | 65.9 (0.9) | 8.8 (0.3) | — | 3.6 (0.3) | 14.3 (1.6) | 0.7 (0.1) | 1.9 (0.3) | 95.3 |

UCS, untreated corn stover; S1 stands for Step 1; S2 stands for Step 2; AIL, acid-insoluble lignin; ASL, acid-soluble lignin. Water and ethanol extractives are not included in the total content of UCS presented in this table. All data in the table are mean values of duplicate experiments; standard deviations are shown in parentheses.

washing stream analysis. Previous studies have confirmed that free sugars are easily degraded in the pretreatment to generate inhibitors of enzymes and microbes.[36,37] The high ash content may impede the performance of hydrolysis and fermentation.[38,39] Thus, lower free sugar and ash content should improve the bioconversion of LCB. The results indicate that pre-washing removes the non-structural components and increases the content of structural components, both of which may improve the biorefinery performance.

### 3.2 Combinatorial pretreatment increases the carbohydrate content

Considering that non-structural components exist in UCS, pretreatment conditions should be designed to remove them to avoid the degradation of non-structural components and improve the carbohydrate output. The hypothesis of this study was that combinatorial pretreatment with mild severity may reduce inhibitor generation by removing non-structural components first, maximize the carbohydrate release, and improve lignin processability for more cost-effective and sustainable biorefineries. In order to validate the hypothesis, four types of combinatorial pretreatments were designed and evaluated (Table 1), where liquid hot water (Cases 4 and 5) and dilute sulfuric acid (Cases 6 and 7) in Step 1 were used to remove non-structural components with a low holding temperature of 120 °C. Following Step 1, NaOH (Cases 4 and 6) or EtOH + NaOH (Cases 5 and 7) in Step 2 was used to further deconstruct CS and fractionate carbohydrate and lignin. Single pretreatment was used as a control with NaOH (Case 1), EtOH (Case 2), and EtOH + NaOH (Case 3).

The results showed that the composition of pretreated CS varies under different pretreatments (Table 2). Single pretreatment of 50% EtOH (Case 2) and liquid hot water (Cases 4S1 and 5S1) led to the lowest glucan content (about 35%). Combinatorial pretreatments with 1% $H_2SO_4$ followed by 1% NaOH (Case 6S2) and 50% EtOH + 1% NaOH (Case 7S2) resulted in the highest glucan content at 74.4% and 65.9%, respectively. These glucan contents were 2.3 and 2.0 times as that of UCS. Interestingly, pretreatments with 1% NaOH (Cases 1, 4S2 and 6S2) had a slightly higher glucan content than those with 50% EtOH + 1% NaOH (Cases 3, 5S2, and 7S2). Importantly, combinational pretreatments (Cases 5–7S2) resulted in a higher glucan content than single pretreatments (Cases 1–3). The increased glucan content generally indicates the removal of extractives, ash, hemicellulose, and/or lignin.[36,40] These results highlight the effectiveness of combinational pretreatments to increase the glucan content.

Pretreatments in Cases 1–5 resulted in the xylan content ranging from 19.1–23.8% while pretreatments in Cases 6–7 produced the xylan content at lower than 9.0%. The reason was that pretreatments in Cases 1–5 were more effective at dissolving the soluble components, such as extractives and ash, and thus increased the xylan content. However, pretreatment of 1% $H_2SO_4$ (Cases 6S1 and 7S1) dissolved most of the hemicellulose (mainly xylan) into the liquid stream, leading to the decreased xylan content.[21,41] Pretreatments with NaOH (Cases 1, 4S2, and 6S2) produced less xylan content than those with EtOH + NaOH (Cases 3, 5S2 and 7S2). These results, taken together, suggest an inverse relationship between the xylan and glucan contents. Araban and galactan contents were also evaluated for each pretreatment. Araban in pretreated CS produced from Cases 6 and 7 was not detected because pretreatment with $H_2SO_4$ dissolves araban. The galactan content in each pretreatment was increased by 5–20% over that in UCS.

In support of the hypothesis, the total sugars in pretreated CS reached 77.1% for Case 4, 75.3% for Case 5, 86.3% for Case 6 and 78.3% for Case 7, which were greater than those for Cases 1–3. The results suggest that combinatorial pretreatment increases the total sugar contents of UCS. Previous studies have confirmed that the high-value utilization of LCB is hindered by the high ash content.[38,39,42,43] Pretreatment step 1 in combinatorial pretreatments (Cases 4–7S1) removed most of the ash, and thus could facilitate the high-value utilization of three main components.

Different combinatorial pretreatments resulted in significant differences of sugar recovery (Fig. 2). Glucan recovery was more than 90% for pretreatment Cases 1, 2, 4–7S1, and 4–7S2.

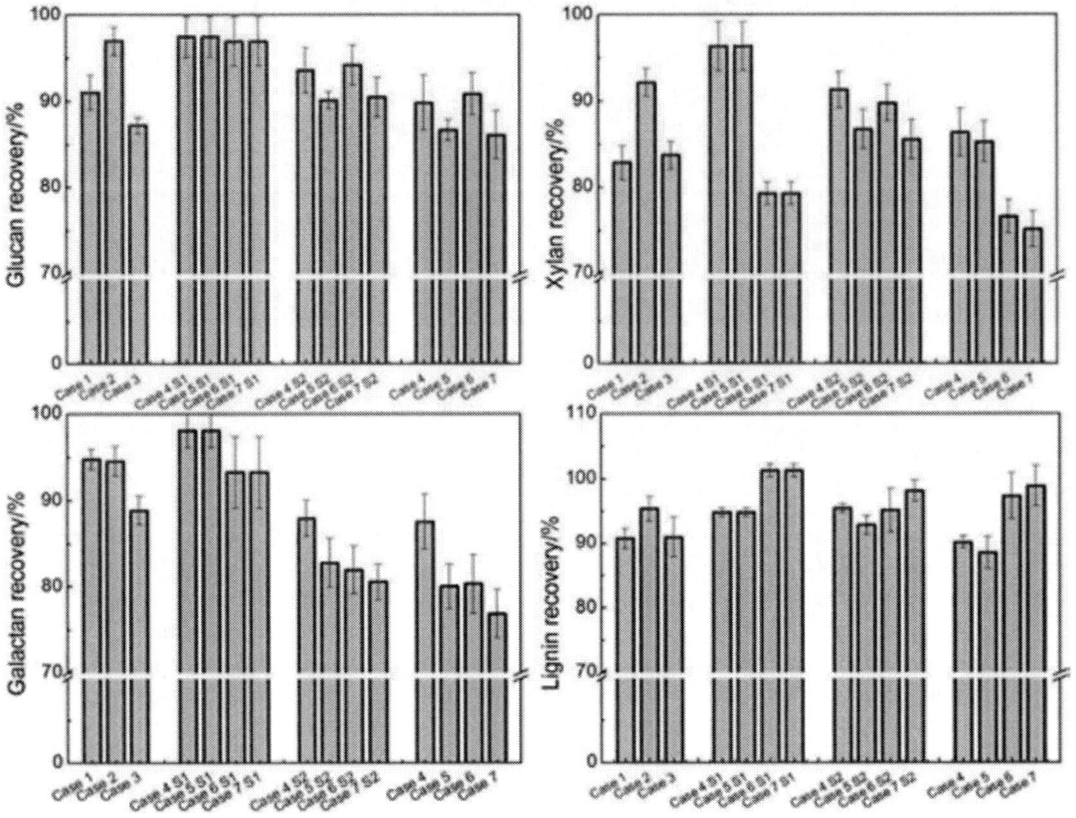


Fig. 2 The recovery of glucan, xylan, galactan, and lignin by each pretreatment of corn stover. S1 stands for Step 1; S2 stands for Step 2. The conditions of each pretreatment strategy are shown in Table 1.

Xylan recovery was more than 80% for pretreatment Cases 1–5 and 6–7S2, while it was 75–77% for Cases 6–7S1 and 6–7. This result was due to the degraded xylan in Cases 6–7S1, which was supported by the xylan content analysis. Pretreatment Cases 1, 4S2, 4, 6S2, and 6 (NaOH) obtained 3.6–4.3% more glucan, 1.2–5.2% more xylan, and 1.1–9.4% more galactan than pretreatment Cases 3, 5S2, 5, 7S2 and 7 (EtOH + NaOH). Overall, combinatorial pretreatment Cases 4–7 resulted in comparable glucan recovery compared to single pretreatments. Xylan recovery results were mixed: in Cases 6 and 7, combinatorial pretreatments recovered less xylan than single pretreatments, whereas in Cases 4 and 5, single and combinatorial pretreatments had comparable xylan recovery.

### 3.3 Combinatorial pretreatment dissolves more of the lignin fraction from corn stover

In biorefinery design, the pretreatment efficiency should also be a factor in lignin transformation and processability.[11,12,14] As shown in Table 2, pretreatments with EtOH (Case 2), liquid hot water (Cases 4–5S1) and $H_2SO_4$ (Cases 6–7S1) resulted in 19.4%, 21% and 28.8% lignin content, respectively, which was slightly higher than that in UCS. However, pretreatment in Cases 1, 3, and 4–7S2 produced the lignin content at lower than 14.3%, suggesting that the combinatorial pretreatment with NaOH was more effective at lignin fractionation into the liquid stream. The dissolution of lignin also contributed to the increased sugar contents, consisting with the above results of sugar analysis. In addition, the efficient lignin dissolution by combinatorial pretreatment will enable the bioconversion of lignin-containing waste to by-pass the additional fractionation step.

Lignin recovery was greater than 90% with all pretreatments and depended on the combinatorial pretreatments (Fig. 2). Pretreatment in Cases 6–7S1 (which included 1% $H_2SO_4$) led to the highest lignin recovery, while pretreatment in Cases 1 and 3 led to the lowest one. Pretreatment Cases 4S2 and 4 led to higher lignin recovery than Cases 5S2 and 5, respectively. Pretreatment Cases 6S2 and 6 resulted in lower lignin recovery than Cases 7S2 and 7. Lignin recovery by combinatorial pretreatments (Cases 4–7) was similar to or higher than that by single pretreatments (Cases 1–3). These results from combinatorial pretreatments help to recover the lignin fraction from corn stover.

### 3.4 Combinatorial pretreatment improves carbohydrate conversion

Pretreated CS produced from different pretreatments exhibited a range of enzymatic hydrolysis (Fig. 3). The initial glucan conversion, measured before 12 h, was the greatest in Case 7S2 (74.2%), followed by Case 5S2 (69.1%), Case 1 (65.4%), Case 3

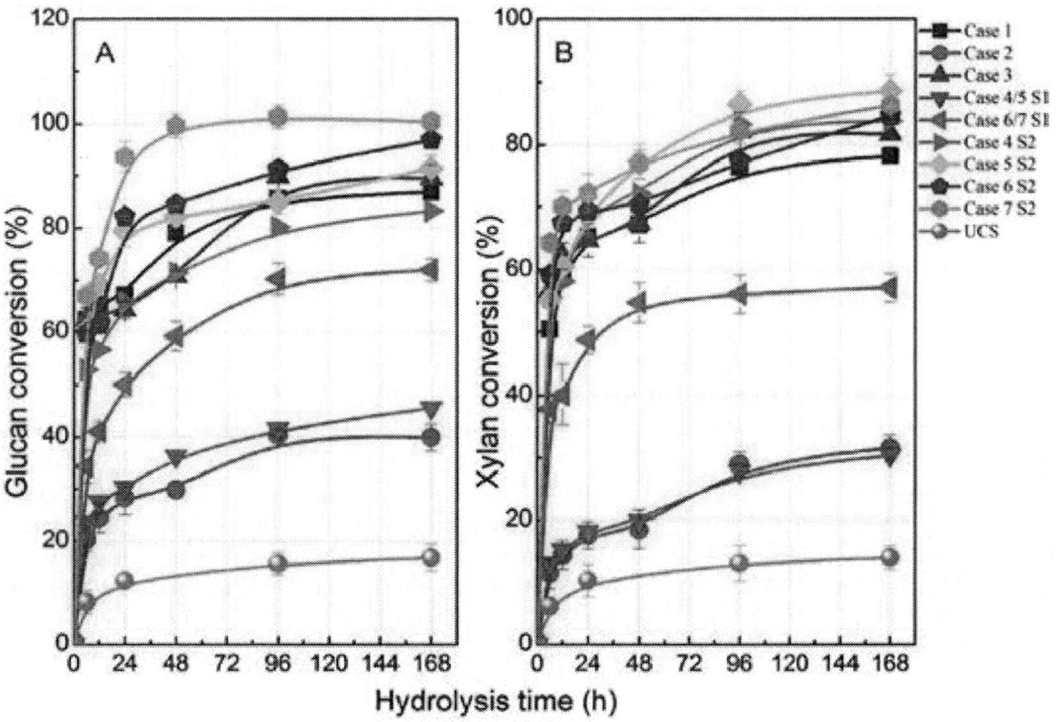


**Fig. 3** Enzymatic hydrolysis of untreated corn stover (UCS) and pretreated corn stover by different pretreatments. The initial sugar conversion is calculated before 12 h hydrolysis. S1 stands for Step 1; S2 stands for Step 2. The conditions of each pretreatment strategy are shown in Table 1.

(63.2%), Case 6S2 (61.3%), Case 4S2 (54.0%), Cases 6/7S1 (41.1%), Cases 4/5S1 (27.5%), and Case 2S (24.5%). Most of the pretreatments with NaOH (Cases 1 and 6) or EtOH + NaOH (Cases 3, 5, and 7) led to more than 60% initial glucan conversion. The final glucan conversion, measured at 168 h, was the greatest in Case 7S2 (100.4%), followed by Case 6S2 (96.9%), Case 5S2 (91.4%), Case 3 (89.6%), Case 1 (86.8%), Case 4S2 (78.1%), Cases 6/7S1 (72.0%), Cases 4/5S1 (45.6%), and Case 2 (40.0%). Glucan conversion was 4.3–6.0 times higher in pretreated CS from Cases 1, 3, 6–7S1 and 4–7S2 than that in UCS. Most of the pretreatments of NaOH (Case 1 and 4) or EtOH + NaOH (Case 3, 5, and 7) resulted in more than 90% glucan conversion. Furthermore, pretreatment Cases 3, 5S2, and 7S2 had higher glucan conversion than those in Cases 1, 4S2, and 6S2, respectively. Combinational pretreatments in Cases 5–7S2 resulted in more glucan conversion than single pretreatments (Cases 1–3). Xylan conversion data had shown a similar trend. In all cases, combinational pretreatments (Cases 4–7S2) produced more than 81% xylan conversion, a value higher than that of any single pretreatments (Cases 1–3). These results highlight that combinatorial pretreatments (Cases 4–7) outperform single pretreatment on the carbohydrate conversion and improve the hydrolysis efficiency of CS.

The enzymatic hydrolysis of a LCB substrate depends on its structural property.[44,45] The biomass crystallinity result showed that the CrI value of pretreated CS increased over UCS by 7.6–18.8% for single pretreatments (Cases 1–3 and 4–7S1) and by 21.1–33.8% for combinatorial pretreatments (Cases 4–7S2) (ESI 4†). Combinational pretreatments (Cases 5–7S2) increased the CrI values by 2.4–15.2% over single pretreatments (Cases 1–3). Pretreatments in Cases 3, 5S2 and 7S2 resulted in higher CrI values than those in Cases 1, 4S2 and 6S2, respectively. The same trends were observed between CrI and glucan conversion. An increase in CrI value generally indicates the removal of the amorphous portion (*i.e.*, cellulose, hemicellulose, and lignin). The biomass crystallinity results correlate well with the composition analysis, in that combinatorial pretreatment can better remove the amorphous portion and increase the cellulose crystallinity.

Combinatorial pretreatments might have also altered the physicochemical properties of CS to increase the accessible surface area of carbohydrates to enzymes. Water holding capacity (WHC) is a measure of this property and represents the adsorption of water on/in the carbohydrates.[30,46] The WHC of the pretreated CS increased by 14.7–56.9% with single (Cases 1–3 and 4–7S1) and by 53.3–105.5% with combinatorial pretreatments (Cases 4–7S2) (ESI 5†). Interestingly, combinational pretreatments (Cases 5–7S2) showed a 14.3–46.3% higher WHC than single pretreatments (Cases 1–3). Pretreatments in Cases 3, 5S2, and 7S2 led to higher WHC than those in Cases 1, 4S2, and 6S2, correlating with the crystallinity data. The WHC trends were consistent with glucan conversion.

The ultrastructure of UCS and pretreated CS as observed through SEM imaging indicated a large variation in biomass morphology (ESI 6†). The variations divided into roughly three shapes: long fiber-like strands, long or wide cuboid-like bulky structures with near 90° edges, and ellipsoid-like bulk structures with curved edges (EIS 6C1). The results showed that there were discernible changes in the ultrastructure of the pretreated CS compared to those of UCS. All pretreatments resulted in an overall decrease in the size of CS. Ellipsoid shaped particles had the largest changes in morphology with pretreatment. These particles suggested the distorted structure of CS. The cuboid-like bulky structures indicated that the pretreatments led to delamination, seen as appearance of fibrillary surfaces. The thin fiber strands indicated disentangling of the fiber bundle at the ends. These results suggested that all pretreatment led to the deconstruction of CS and increased the accessible surface area of CS to enzymes. One interesting feature was the absence of lignin droplets that are often produced after liquid hot water and dilute acid pretreatments (Cases 4–7) due to the movement of lignin from the inner regions to the outer surface.[26,47] This was likely due to the use of alkali and/or organosolv pretreatments in Step 2 that solubilized the lignin molecules on the surface of these particles into liquid streams. This result should be helpful to increase enzymatic hydrolysis and bioconversion performance of the lignin stream.

Overall, these results suggested that combinational pretreatments removed the amorphous portion, altered the structural and compositional properties of CS, increased the accessible surface area of cellulose, and thus improved the hydrolysis performance.

### 3.5 Combinatorial pretreatment improves the sugar and lignin yields significantly

Techno-economic analyses have emphasized the importance of the sugar yield on the biorefinery optimization.[41,48] As shown in Fig. 4, the glucose yield from the whole fractionation process (pretreatment + hydrolysis) was the highest for Case 7 (91.4%), followed by Case 6S2 (91.3%), Case 7S2 (90.9%), Case 6 (88.4%), Case 5S2 (82.9%), Case 5 (82.8%), and Case 1 (80.5%). For all other cases, the glucose yield was less than 80%. Pretreatments with 50% EtOH + 1% NaOH (Cases 5S2, 5, 7S2, and 7) led to higher glucose yields than those with 1% NaOH (Cases 4S2, 4, 6S2, and 6), which was consistent with glucan conversion. The combinatorial pretreatment of Cases 5–7S2 and 5–7 increased glucose yields by 3.0–16.7% over single pretreatment of Cases 1 and 3. Similar trends were observed for xylose yields. The xylose yield increased by 7.2–20.1% in pretreatment Cases 4, 4S2, 6, and 6S2 over that for Case 1 with NaOH, and it increased by 4.0–11.4% for Cases

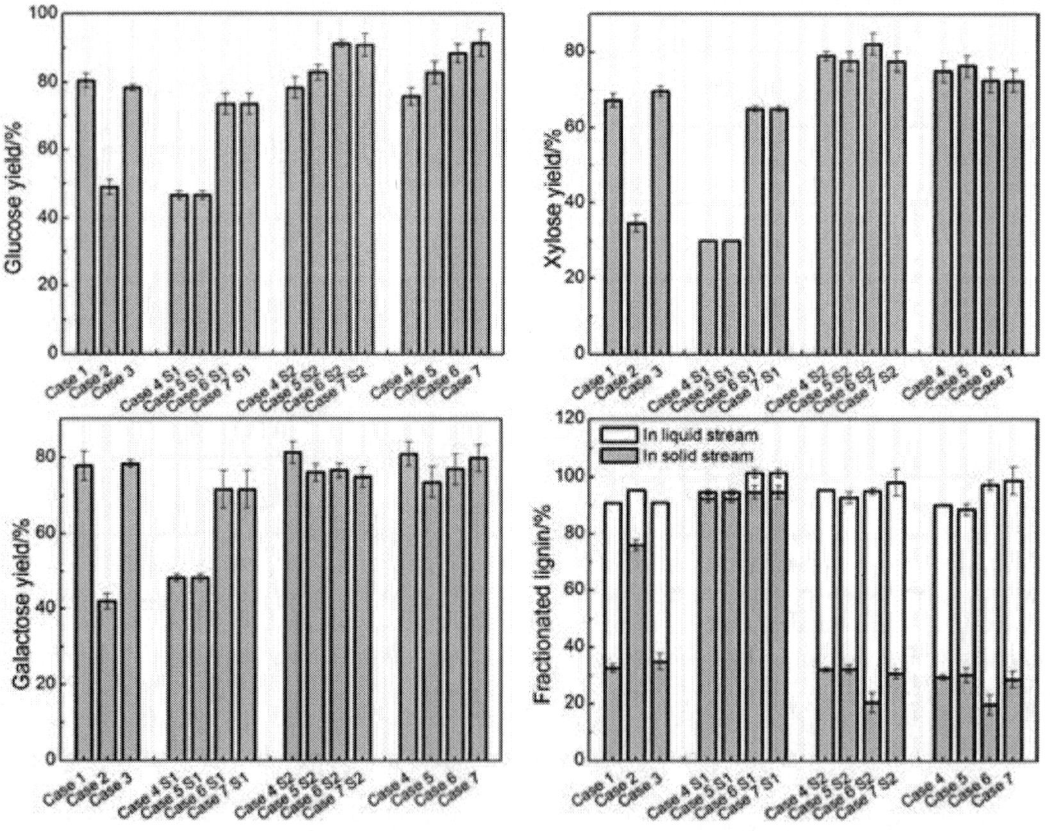


Fig. 4 Yields of fermentable sugar and fractionated lignin in the whole fractionation process (pretreatment + hydrolysis) of corn stover by different pretreatment strategies. S1 stands for Step 1; S2 stands for Step 2. The conditions of each pretreatment strategy are shown in Table 1.

5, 5S2, 7, and 7S2 over that for Case 3 with EtOH + NaOH. The total xylose yields from CS exposed to combinatorial pretreatments were 77.6–80.5% for Cases 4–7S2 and 72.1–76.3% for Cases 4–7. Thus, these results show that combinatorial pretreatments improved the fermentable sugar yields.

Lignin was measured in the solid fraction and liquid stream produced after each pretreatment. More than 90% of the lignin was retained in the residue solids after hydrolysis with pretreatment Cases 4–7S1, and 75% of the lignin was retained with pretreatment Case 2 (Fig. 4). However, only 20% lignin was retained in the residue solids after pretreatment Cases 6S2 and 6, and about 30% lignin was retained by other pretreatments. Correspondingly, more than 60% of the lignin was fractionated and dissolved into the liquid stream by combinatorial pretreatments (Cases 4–7S2 and 4–7). The highest lignin yield in the liquid stream was 77.8% produced from combinatorial pretreatment Case 6. Pretreatment in Cases 1, 4S2, 4, 6S2, and 6 (NaOH) led to 4.0–11.1% more lignin in the liquid stream than that in Cases 3, 5S2, 5, 7S2, and 7 (EtOH + NaOH). These results paralleled the trend observed for sugar recovery, but were opposite to those trends seen in the glucose yield. Compared with single pretreatment Cases 1 and 3, combinatorial pretreatments in Cases 4S2, 4, 5S2, and 5 increased the lignin yield in the liquid stream by 4.2–8.8%, while Cases 6S2, 6, 7S2, and 7 increased this yield by 20.4–33.5%. Overall, these results indicate that combinatorial pretreatments, especially Cases 6, improve the lignin output from CS, thus enhancing the biorefining performance.

### 3.6 Combinatorial pretreatment improves lignin valorization to polyhydroxyalkanoate (PHA) without additional fractionation

Besides the carbohydrate and lignin output, an important yet unexplored aspect of pretreatment is the processability of the lignin stream. Recent studies indicated that lignin can be further processed into high value products like PHA *via* bioconversion. Despite the progress, the product yield is still low and an additional fractionation step or enzymatic depolymerization is often needed to promote lignin degradation and conversion. Combinatorial pretreatment may improve the lignin processability and bioconversion due to two reasons. First, the combination of different conditions will maximize the lignin fractionation and solubilization, which further provides suitable carbon sources for bioconversion. Second, the low holding temperature will reduce the inhibitors to improve the fermentation performance. The hypothesis was validated by both fermentation and structure characterization of lignin.

The lignin fractionated by different pretreatments (Lignins 1–7 as labeled by their corresponding pretreatment Cases 1–7) was used to generate PHA through fermentation with *P. putida* KT2440. Working with a soluble substrate concentration (SSC) of 20 g $l^{-1}$ and an initial OD of 0.5, the highest cell dry weight was 2.53 g $l^{-1}$, fermented from lignin 6 (pretreatment with 1% $H_2SO_4$ followed by 1% NaOH) using engineered *P. putida* KT2440 (Fig. 5A). The PHA concentration from lignins 4, 5, and 6 produced by combinatorial pretreatments was 0.66 to 1.38 times higher than that from lignin 1 (a single pretreatment) (Fig. 5B). The highest PHA concentration was 0.48 g $l^{-1}$, which was also fermented from lignin 6 using engineered *P. putida* KT2440. The trends for PHA content were similar to those for PHA yield (Fig. 5C and D). PHA contents from lignins 4, 5, and 6 were 0.56–1.0 times higher than that from lignin 1. The highest PHA content was 0.19 g $g^{-1}$ dried cell, produced by fermenting lignin 6 using engineered *P. putida* KT2440. The PHA yields from lignin 4, 5, and 6 were 0.21–0.58 times higher than from lignin 1 using engineered *P. putida* KT2440. The highest PHA yield was 14.1% from lignin 6 by engineered *P. putida* KT2440. As a result, combinatorial pretreatments improved the lignin processability, as suggested by the increase in PHA fermentation performance with engineered *P. putida* KT2440.

To produce a high PHA concentration from the lignin stream, fed-batch fermentation was conducted using engineered *P. putida* KT2440 with the addition of 20 g $l^{-1}$ SCC at 0 h plus 20 g $l^{-1}$ SSC at 18 h (Fig. 6). The highest cell dry weight was 5.3 g $l^{-1}$ from lignin 6 produced from the pretreatment of 1% $H_2SO_4$ followed by 1% NaOH. The PHA concentration pro-

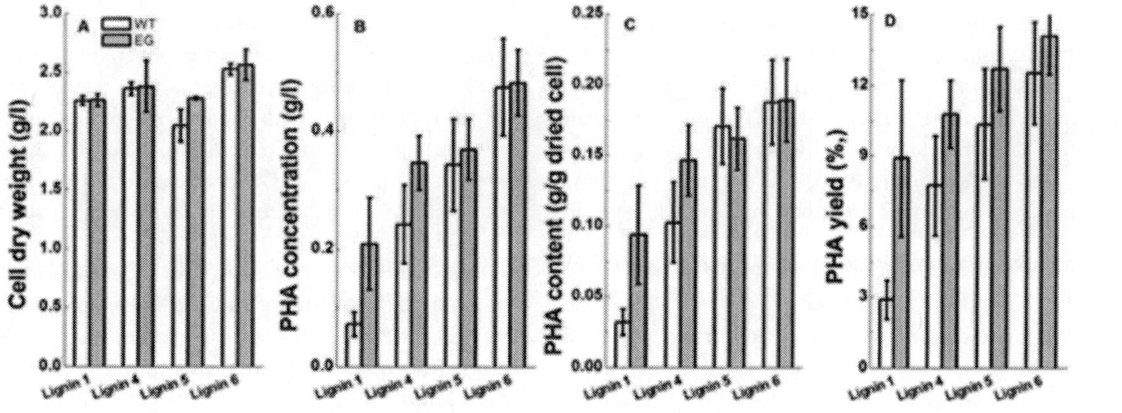


Fig. 5 PHA fermentation from the lignin stream produced by different CS pretreatments using wild type (WT) and engineered (EG) *Pseudomonas putida* KT2440. The fermentation conditions were: 20 g $l^{-1}$ soluble substrate concentration (SSC), OD 0.5, pH 7.0, 28 °C, 200 rpm, and 18 h. "Lignin 1" represents the fractionated lignin produced by pretreatment Case 1 described in Table 1.

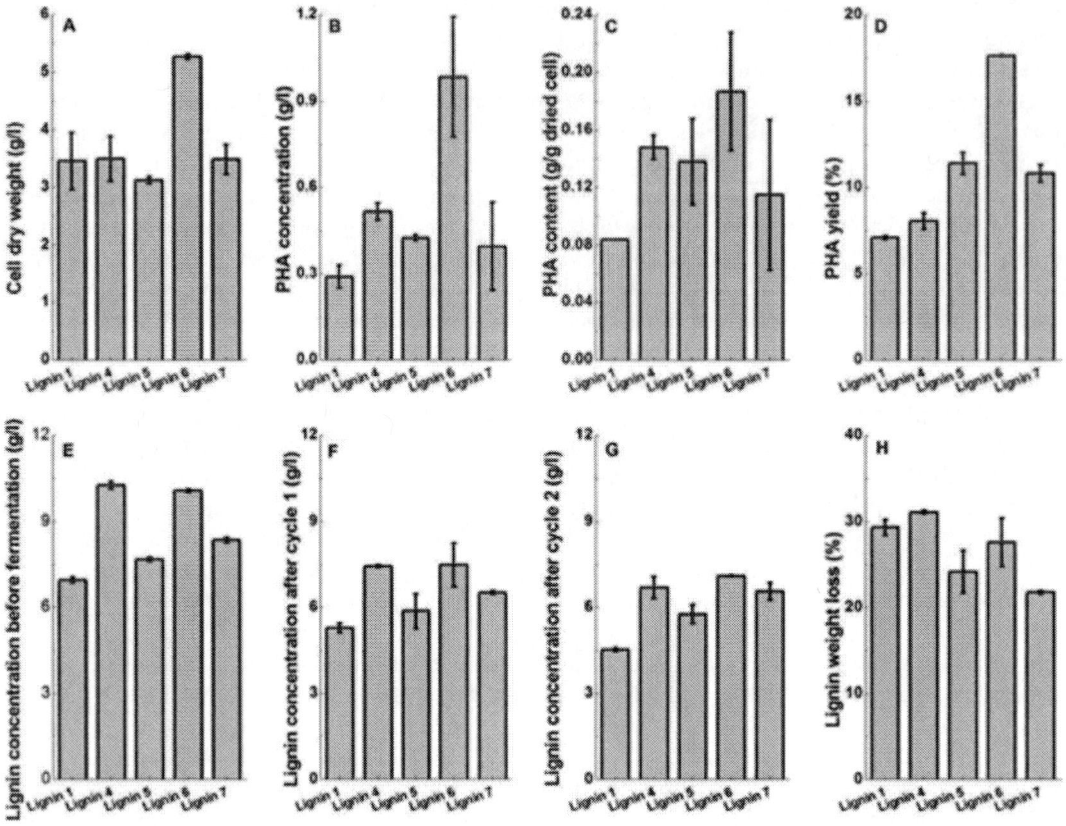

Fig. 6 PHA fermentation from the lignin streams produced by different pretreatments using engineered (EG) *Pseudomonas putida* KT2440 grown in fed-batch fermentation mode. The fermentation conditions were: 20 g $l^{-1}$ soluble substrate concentration (SSC) in cycle 1 and 20 g $l^{-1}$ SSC in cycle 2, OD 2.0, pH 7.0, 28 °C, 200 rpm, and 18 h for cycle 1 and 18 h for cycle 2. "Lignin 1" represents the fractionated lignin produced by pretreatment Case 1 described in Table 1.

duced from lignins 4, 5, 6, and 7 was 0.77, 0.46, 2.39, and 0.36 times higher than that from lignin 1, respectively. The highest PHA concentration was 1.0 g $l^{-1}$, generated from lignin 6. This result represents the highest PHA concentration reported using lignin as a carbon source for fermentation. From this sample, the PHA content was 0.19 g $g^{-1}$ dried cell and the PHA yield was 17.6%. Notably, the PHA content and the yield from lignins 4–7 were 0.36–1.22 and 0.61–1.48 times higher than that from lignin 1, respectively. The lignin weight loss was around 30% for lignins 1, 4, and 6, which supports the results described above (Fig. 6E–H). These results highlight that combinatorial pretreatments lead to a higher cell dry weight as well as increased PHA concentration, content, and yield, thus improving the lignin bioconversion performance.

### 3.7 Lignin characterization using 2D- and $^{31}$P-NMR revealed the mechanisms for improved lignin processability

The lignin fractionated from each pretreatment was characterized to reveal the molecular mechanisms for improved lignin processability by HSQC-NMR, quantitative $^{31}$P NMR, and gel-permeation chromatography (GPC), and FTIR. Fig. 7 presents the aromatic regions of the HSQC NMR spectra, including lignin subunits and hydroxycinnamates.[33] Previous studies have confirmed that H- and G-lignin are more readily degraded and utilized by a lignin-degradable microbe than S-lignin.[11,49–51] Fig. 8 shows that fractionated lignin by a single pretreatment (Case 1) had the highest lignin S/G ratio (1.22), implying the least readily degradable lignin composition. Combinatorial pretreatments in Cases 4–7 increased the (G + H)/(G + H + S) ratios by 9.0–35% over single pretreatment in Case 1. The increases in G- and H-lignin may be the underlying cause for the improved bioconversion of lignin to PHA.

The selection of pretreatments also affected the lignin composition. Compared with the lignin fractionated using a NaOH pretreatment in Step 2 (*i.e.*, lignin 4 and 6), the lignin fractionated using EtOH + NaOH pretreatment in Step 2 (*i.e.*, lignin 5 and 7) had more hydroxycinnamates such as FA and *p*CA. Most importantly, the lignin fractioned from combinatorial pretreatments in Cases 5, 6, and 7 had more *p*CA than that from single pretreatment in Case 1. Previous studies have confirmed that FA and pCA are primary aromatic compounds produced from alkaline pretreatment, and are readily consumed by *P. putida* KT2440 to produce cell biomass and PHA.[51,52] These results demonstrate that combinatorial pretreatments increase the hydroxycinnamate content, and thus facilitate the bioconversion of lignin to PHA.

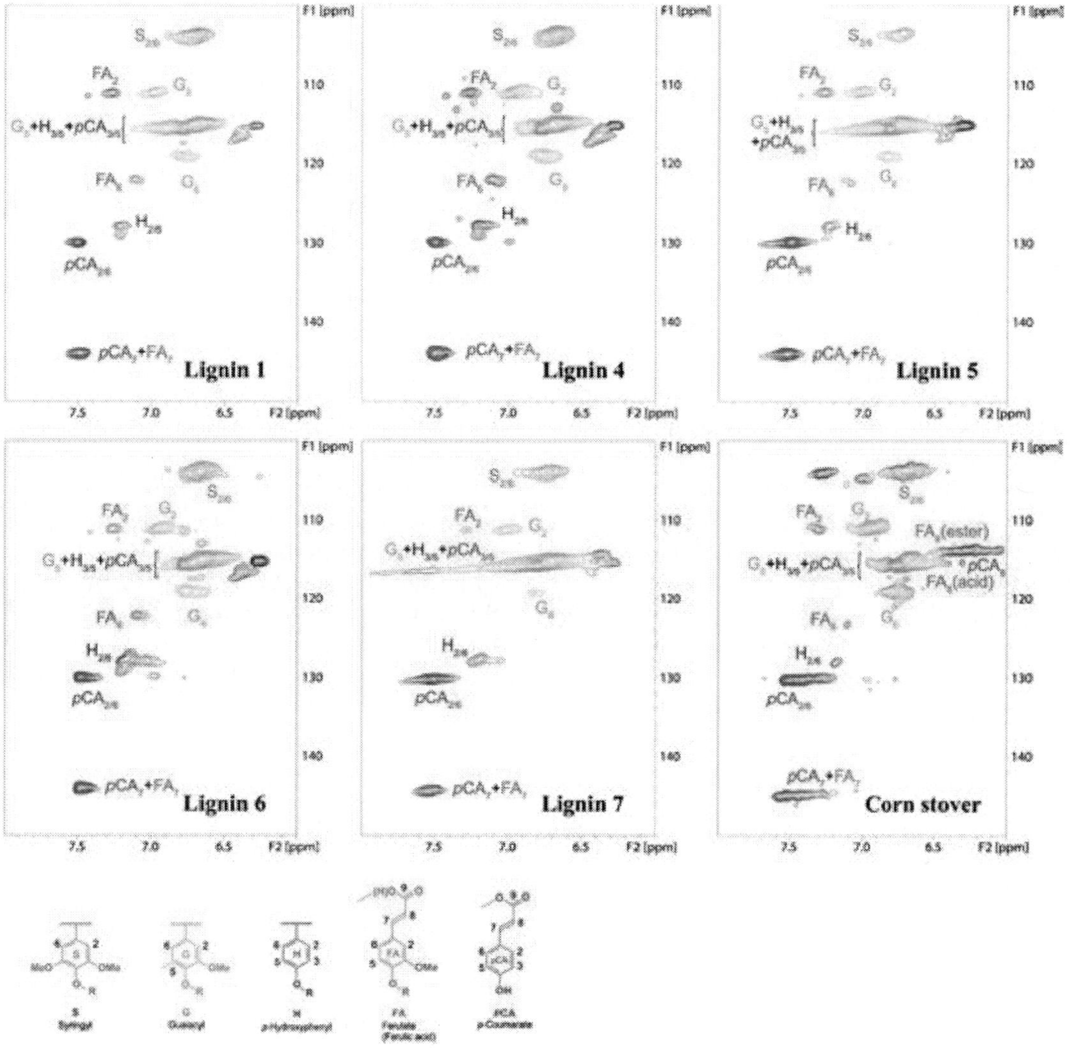

Fig. 7 Aromatic regions of 2D HSQC NMR spectra from corn stover native lignin and the fractionated lignin produced after each pretreatment. "Lignin 1" represents the fractionated lignin produced by pretreatment Case 1 described in Table 1.

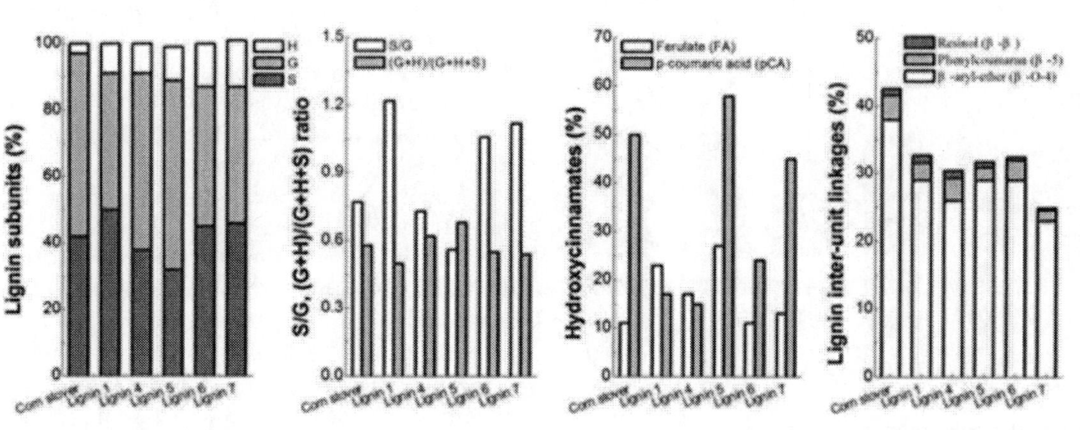

Fig. 8 Semi-quantitative information of fractionated lignin produced after each pretreatment as detected using 2D HSQC NMR. Contents were expressed as a percentage relative to the total lignin subunits (G+H+S). "Lignin 1" represents the fractionated lignin produced by pretreatment Case 1 as described in Table 1.

The occurrence of three major lignin interunit linkages (β-aryl ether, phenylcoumaran, and resinols) was estimated by measuring the correlation of the α position of β-aryl ether (β-O-4), phenylcoumaran (β-5), and resinols (β–β), respectively (Fig. 8). All pretreatments significantly decreased the amount of β-O-4 in fractionated lignin compared to that in CS native lignin. Combinatorial pretreatments led to similar amounts of β-O-4 to the single pretreatment in Case 1, or a bit less. The amounts of β-5 in the lignin produced by pretreatments in Cases 5 and 7 were less than those from pretreatments in Cases 1, 4, and 6. These results indicate that combinatorial pretreatment with EtOH + NaOH is more effective at cleaving the β-5 linkage than NaOH alone. The selection of pretreatments also affected the amount of β–β. For instance, the lignin produced by combinatorial pretreatments (Cases 4–7) decreased the amount of β–β by 18–73% compared to single pretreatment (Case 1). The lignin produced by pretreatments in Cases 6 and 7 showed less β–β than Cases 4 and 5. These results suggest that combinatorial pretreatment reduces the amounts of β–β and β-O-4 present in the lignin, indicating more degradation of the lignin polymer to facilitate the bioconversion of lignin to PHA.

The hydroxyl groups including the aliphatic, C5 substituted phenolic, guaiacyl, *p*-hydroxy phenyl, and acid hydroxyl groups in fractionated lignin and native lignin from CS were determined using $^{31}$P NMR. As shown in Fig. 9, all pretreatment strategies significantly reduced the aliphatic hydroxyl groups in fractionated lignin compared with those in CS native lignin. Combinatorial pretreatments in Cases 5, 6, and 7 decreased aliphatic hydroxyl groups by 29.5–38.1% over single pretreatment in Case 1. The lignin produced by pretreatment in Cases 6 and 7 showed higher C5 substituted phenolic hydroxyl group abundance (0.52 mmol $g^{-1}$). It is possible that this result is caused by the condensation of G units because both syringyl and C5 condensed hydroxyl groups are counted as the C5 substituted phenolic hydroxyl groups. The guaiacyl and *p*-hydroxy phenyl phenolic hydroxyl groups from combinatorial pretreatments in Cases 5, 6, and 7 were 1.4–1.7 and 1.9–2.9 times as that from single pretreatment in Case 1, respectively. The lignin fractionated by pretreatments in Cases 6 and 7 contained 2.2 and 2.3 times more phenolic hydroxyl groups than those fractionated by pretreatment in Case 1, respectively.

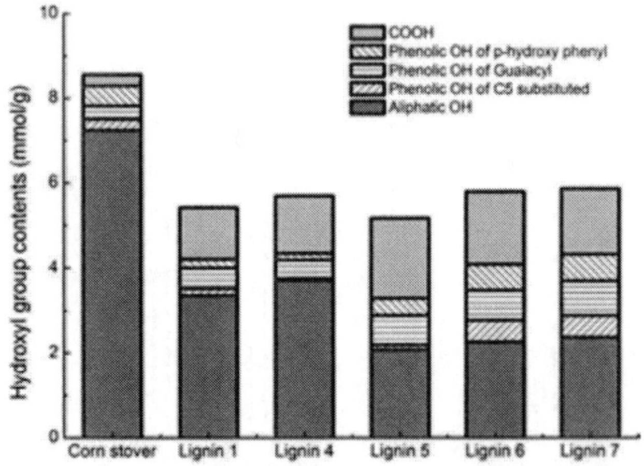


Fig. 9 Contents of hydroxyl groups in fractionated lignin produced after each pretreatment as detected using $^{31}$P NMR. "Lignin 1" represents the fractionated lignin produced by pretreatment Case 1 described in Table 1.

All the pretreatments significantly increased the abundance of carboxyl groups in the fractionated lignin compared to that in the CS native lignin. The lignin fractionated by combinatorial pretreatments in Cases 4–7 contained 10.1–54.5% more carboxyl groups than that fractionated by pretreatment in Case 1. The lignin fractionated by combinatorial pretreatments in Cases 5 and 6 produced most carboxyl groups: 1.87 and 1.70 mmol $g^{-1}$, respectively. During biodegradation and bioconversion of lignin, fewer or shorter side chain groups containing aromatic rings could help to increase the lignin utilization efficiency by lignin-degradable microbes.[50,51] Combinatorial pretreatments led to fewer aliphatic hydroxyl groups, more phenolic hydroxyl groups and more carboxyl groups, resulting in more breakage of the lignin linkage, and thus facilitating the lignin bioconversion to PHA.

The relative intensities of the infrared spectra of the lignin from different pretreatments showed the decrease in functional groups corresponding to the removal of glucan and xylan, which was consistent with the composition analysis results (ESI 7†). Lignins 3 and 5 showed similar functionalities, while lignins 2 and 4 showed similar functionalities in the FTIR spectra. The ester bond signal at 1730 $cm^{-1}$ was weaker in the lignin samples from all the pretreatments, suggesting that ester linkages between lignin and carbohydrates were cleaved. The band at 1640 $cm^{-1}$ assigned to conjugated carbonyl groups of lignin showed slightly less relative adsorption in the lignin fractioned by pretreatment than that in CS native lignin. The absorption of unconjugated carbonyl groups at 1708 $cm^{-1}$ was increased for all the fractioned lignin compared to CS native lignin. The results indicated the presence of unconjugated ketone, aldehyde and/or acid groups in all the fractioned lignin. These results were consistent with the above composition analysis and NMR results.

The molecular weight distribution of lignin fractions produced from different pretreatments was analyzed. The different pretreatment also showed an effect on the fragmentation of lignin. As shown in Fig. 10, the number-average molecular weight ($M_n$) of fractionated lignin ranged from 700 to 900 g $mol^{-1}$ and the weight-average molecular weight ($M_w$) ranged from 1350 to 1981 g $mol^{-1}$, while CS native lignin showed a $M_n$ of 1497 g $mol^{-1}$ and a $M_w$ of 2962 g $mol^{-1}$. All pretreatments significantly decreased the molecular weight of lignin. Combinatorial pretreatment in Cases 4 and 6 exhibited lower $M_n$ and $M_w$ of lignin than Cases 5 and 7. Generally, low molecular weight lignin is easy to convert by lignin-degradable microbes.[53,54] The lignin produced from pretreatment in Cases 4 and 6 had higher polydispersity index (PDI) values than those in Cases 5 and 7. A higher polydispersity index (PDI) value may also facilitate lignin utilization by lignin-degradable

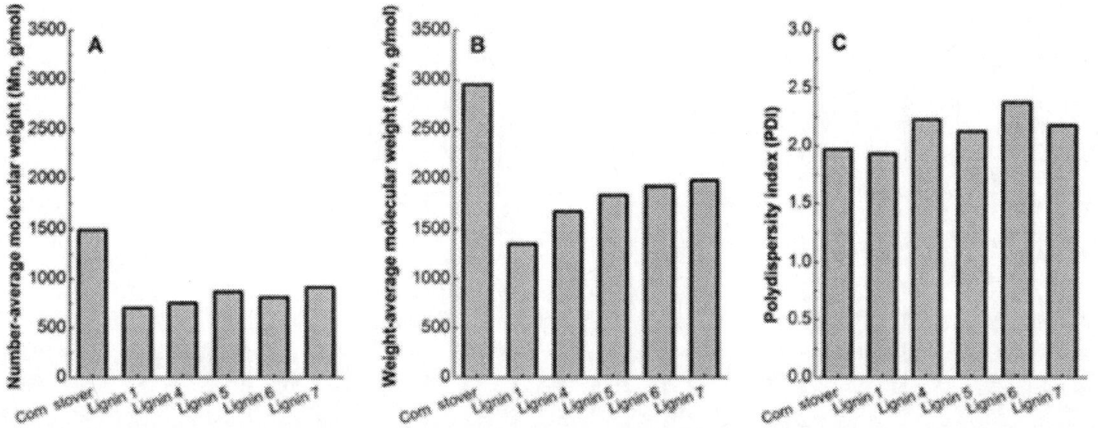


**Fig. 10** Molecular weight distributions of corn stover native lignin and fractionated lignin produced after each pretreatment. "Lignin 1" represents the fractionated lignin produced by pretreatment Case 1 described in Table 1.

microbes. These results highlighted that combinatorial pretreatment reduced the molecular weight of lignin and increased the PDI value, facilitating its bioconversion to PHA. Overall, the study suggested that the bioconversion performance of lignin could be impacted by the molecular weight, PDI, the functional groups and the type of lignin.

### 3.8 Derivatives of carbohydrates and lignin with different pretreatments

The pretreatment of LCB normally generates derivatives, such as weak acids, furans, and phenolic compounds.[37,55,56] In particular, aromatic derivatives can be more readily processed by lignin-degraded microbes, and the amount of aromatic derivatives will impact the processability and fermentation performance of the lignin stream. The results showed that the relative abundance and the type of derivative varied with the pretreatment option (ESI 8†). Malic acid, hydroxybutyric acid and valeric acid were the major non-volatile carboxylic acid derivatives formed, accounting for 60–95% of the quantifiable carboxylic acid detected. The pretreatments at a low holding temperature produced few furan aldehydes, which are usually not stable under alkaline conditions.[55] Combinatorial pretreatments in Cases 4–7S2 led to higher concentrations of carboxylic acids and furan aldehydes than those in Cases 1 and 3, likely due to their higher pretreatment severity. However, the concentrations of weak acids and furans produced after combinatorial pretreatments were obviously lower compared to those in previous reports.[36]

In total, 35 different aromatic monomers were quantified in the liquid stream from the different pretreatments. *p*-Coumaric acid, ferulic acid, 4-hydroxy-3-methyl-acetophenone, 4-vinylguaiacol, coumarone, acetoxycinnamic acid, ethyl ferulate, vanillin, guaiacol, and syringaldehyde were the major derivatives that formed from the lignin polymer, which accounted for about 67–81% of the quantifiable derivatives detected. The aromatic monomers produced from pretreatments in Cases 3, 5S2, and 7S2 were 6.3, 1.6, and 1.7 times higher than those in Cases 1, 4S2, and 6S2, respectively. Notably, the evaporation and concentration effects caused by ethanol distillation after pretreatment (Cases 3, 5, and 7) should contribute to this result (Fig. 1). Combinatorial pretreatments in Cases 4S2 and 6S2 produced 4.2 and 3.1 times more aromatic monomers than single pretreatment in Case 1. Aromatic monomers are more easily converted by *P. putida* KT2440 compared to the high molecular weight lignin polymer.[31,51,57] Therefore, combinatorial pretreatment produced more aromatic monomers, and thus facilitated lignin bioconversion to PHA.

### 3.9 Combinatorial pretreatment holds promise for improving biorefining efficiency and cost-effectiveness

Overall, combinatorial pretreatments offered several significant advantages over single pretreatment. Firstly, combinatorial pretreatment increased the carbohydrate output. The mass balance of the whole fractionation process (pretreatment + hydrolysis) showed that combinatorial pretreatment in Case 6 led to 96.6% glucan, 84.6% xylan, and 89.2% galactan conversions, which increased by 11.2%, 8.3%, and 21.0%, respectively, over single pretreatment in Case 1 (Fig. 11). In the whole fractionation process, glucose and xylose yields were 88.4% and 72.6%, respectively, representing increases of 10.0% and 8.1% over single pretreatment in Case 1. On one hand, pretreatment Step 1 removed degradable components, water extractives, ash, and other impurities, which reduced the formation of potential inhibitors. On the other hand, pretreatment Step 2 further deconstructed CS and improved hydrolysis performance as indicated by the high WHC and the SEM image.

Secondly, combinatorial pretreatment enhanced lignin processability and bioconversion performance for PHA. The lignin yield from combinatorial pretreatment in Case 6 was 77.6% in the liquid stream, which represents an increase of 33.4% over single pretreatment. The lignin stream was used by *P. putida* KT2440 to produce PHA. In the fed-batch fermentation mode,

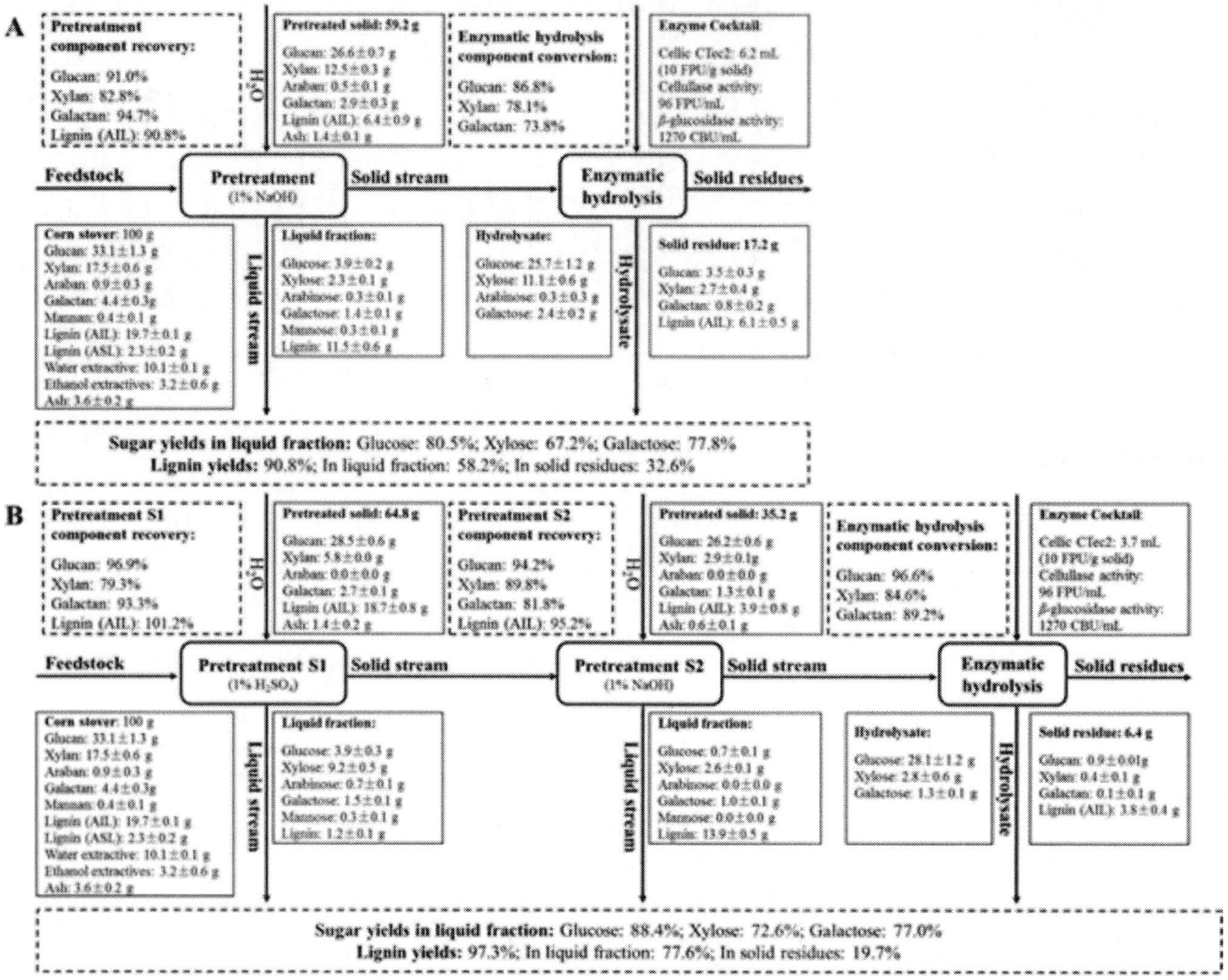


Fig. 11 Mass balance around the whole fractionation process of corn stover by single (A, Case 1) and combinatorial pretreatment (B, Case 6) strategies. S1 stands for Step 1; S2 stands for Step 2.

a record PHA concentration (1.0 g $l^{-1}$) was produced. Such an increase of lignin processability could be due to the high G- and H-type lignin content, the improved cleavage of β–β and β-O-4 linkages, the broader distribution of the molecular weight, the reduced aliphatic hydroxyl groups, and the increased phenolic hydroxyl and carboxyl groups in the lignin. Additionally, the reduced inhibitor and non-structural components in the lignin stream further improved the lignin processability compared to traditional pretreatment, contributing the record PHA yield for biorefinery waste utilization.

Thirdly, the low holding temperature of combinatorial pretreatment offers a range of advantages. Conventional hydrothermal pretreatment is carried out between 160–240 °C holding temperature.[5,15,18] Such a high holding temperature leads to increased pretreatment severity and improved hydrolysis performance, but it also leads to sugar loss, high-energy consumption and inhibitor formation.[15,18] Combinatorial pretreatment here was conducted at a low holding temperature (120 °C), which reduced sugar degradation, inhibitor generation, and the need for energy consumption (Fig. 1 and ESI 2†).

Fourth, conventional acidic or alkaline pretreatments require conditioning of the liquid stream before fermentation. Combinatorial pretreatments using 1% $H_2SO_4$ in Step 1 and 1% NaOH in Step 2 helped to eliminate the need for conditioning. Instead, the liquid streams from Step 1 and Step 2 were mixed, which balanced the pH (Fig. 1). In addition, the use of acid pretreatment followed by base pretreatment allowed the maximum dissolving of lignin and avoided the lignin condensation and precipitation. Together with the low holding temperature, the combination of acid and base pretreatments also avoids the formation of pseudo-lignin, a major cause of low hydrolysis efficiency in acid pretreatment.[26,36] Fifth, the ethanol can be recycled by distillation and reused, reducing water consumption.[58] The use of ethanol as a model organic solvent proves the concept that other solvents more readily for recovery can also be used in the process. Altogether, these advantages highlighted that combinatorial pretreatment synergistically increased the carbohydrate output and enhanced the lignin processability, which would open new avenues for optimizing biorefining performance, cost-effectiveness, and sustainability. The research thus proved an important concept for synergistic maximization of carbohydrate output and lignin processability *via* combinatorial pretreatment at a low holding temperature. Further optimization can be carried out by evaluating the combination of different chemical concentrations, holding temperatures, and residence time.

## 4 Conclusions

Combinatorial pretreatments with a low holding temperature were developed to synergistically increase both the carbohydrate output and the lignin processability. Using combinatorial pretreatments with 1% $H_2SO_4$ followed by 1% NaOH, glucose and xylose yields were 88.4% and 72.6%, respectively, representing an increase of 10.0% and 8.1% over single NaOH pretreatment. The lignin yield from this combinatorial pretreatment was 77.6% in the liquid stream, representing an increase of 33.4%. Using *P. putida* KT2440, the highest observed PHA concentration was 1.0 g $l^{-1}$, with a 17.6% yield. Combinational pretreatments maximized the output of both carbohydrate and lignin and enhanced the lignin processability with a record level of PHA concentration. In summary, combinational pretreatments may improve energy production from CS and maximize its carbon utilization in biorefinery applications.

## Author contributions

ZHL and JSY designed the study. ZHL carried out the experiments, performed the statistical analysis, and drafted the manuscript. MLO and KK determined the sugar samples. XW determined the derivative samples. SS, NH, CY, YP, and AJR determined the lignin characterization. SB and JRD conducted SEM analysis. MJ and JSY revised the manuscript. All authors provided critical inputs to the manuscript and read and approved the final draft.

## Conflicts of interest

There are no conflicts to declare.

## Acknowledgements

The work was supported financially by the U.S. DOE (Department of Energy) EERE (Energy Efficiency and Renewable Energy) BETO (Bioenergy Technology Office) (DE-EE0007104 and DE-EE0006112).

## References

1 M. E. Himmel, S.-Y. Ding, D. K. Johnson, W. S. Adney, M. R. Nimlos, J. W. Brady and T. D. Foust, *Science*, 2007, **315**, 804–807.
2 S. Chu and A. Majumdar, *Nature*, 2012, **488**, 294–303.
3 D. M. Alonso, S. G. Wettstein, M. A. Mellmer, E. I. Gurbuz and J. A. Dumesic, *Energy Environ. Sci.*, 2013, **6**, 76–80.
4 M. J. Jin, C. Gunawan, N. Uppugundla, V. Balan and B. E. Dale, *Energy Environ. Sci.*, 2012, **5**, 7168–7175.
5 J. J. Bozell and G. R. Petersen, *Green Chem.*, 2010, **12**, 539–554.
6 P. C. A. Bruijnincx, R. Rinaldi and B. M. Weckhuysen, *Green Chem.*, 2015, **17**, 4860–4861.
7 P. Azadi, O. R. Inderwildi, R. Farnood and D. A. King, *Renewable Sustainable Energy Rev.*, 2013, **21**, 506–523.
8 D. S. Argyropoulos and C. Crestini, *ACS Sustainable Chem. Eng.*, 2016, **4**, 5089–5089.
9 L. D. Sousa, M. J. Jin, S. P. S. Chundawat, V. Bokade, X. Y. Tang, A. Azarpira, F. C. Lu, U. Avci, J. Humpula, N. Uppugundla, C. Gunawan, S. Pattathil, A. M. Cheh, N. Kothari, R. Kumar, J. Ralph, M. G. Hahn, C. E. Wyman, S. Singh, B. A. Simmons, B. E. Dale and V. Balan, *Energy Environ. Sci.*, 2016, **9**, 1215–1223.
10 K. H. Kim, B. A. Simmons and S. Singh, *Green Chem.*, 2017, **19**, 215–224.
11 G. T. Beckham, C. W. Johnson, E. M. Karp, D. Salvachúa and D. R. Vardon, *Curr. Opin. Biotechnol.*, 2016, **42**, 40–53.
12 A. J. Ragauskas, G. T. Beckham, M. J. Biddy, R. Chandra, F. Chen, M. F. Davis, B. H. Davison, R. A. Dixon, P. Gilna and M. Keller, *Science*, 2014, **344**, 1246843.
13 S. Xie, A. J. Ragauskas and J. S. Yuan, *Ind. Biotechnol.*, 2016, **12**, 161–167.
14 M. De Bruyn, J. Fan, V. L. Budarin, D. J. Macquarrie, L. D. Gomez, R. Simister, T. J. Farmer, W. D. Raverty, S. J. McQueen-Mason and J. H. Clark, *Energy Environ. Sci.*, 2016, **9**, 2571–2574.
15 D. Salvachua, E. M. Karp, C. T. Nimlos, D. R. Vardon and G. T. Beckham, *Green Chem.*, 2015, **17**, 4951–4967.
16 Z. Chen and C. Wan, *Renewable Sustainable Energy Rev.*, 2017, **73**, 610–621.
17 S. X. Xie, Q. Li, P. Karki, F. J. Zhou and J. S. Yuan, *ACS Sustainable Chem. Eng.*, 2017, **5**, 2817–2823.
18 Z. Y. Zhang, M. D. Harrison, D. W. Rackemann, W. O. S. Doherty and I. M. O'Hara, *Green Chem.*, 2016, **18**, 360–381.
19 S. Roth and A. C. Spiess, *Bioprocess Biosyst. Eng.*, 2015, **38**, 2285–2313.
20 M. Pedersen and A. S. Meyer, *New Biotechnol.*, 2010, **27**, 739–750.
21 N. Mosier, C. Wyman, B. Dale, R. Elander, Y. Y. Lee, M. Holtzapple and M. Ladisch, *Bioresour. Technol.*, 2005, **96**, 673–686.
22 Z. H. Liu, L. Qin, B. Z. Li and Y. J. Yuan, *ACS Sustainable Chem. Eng.*, 2015, **3**, 140–146.

23 X. Z. Meng, T. Wells, Q. N. Sun, F. Huang and A. Ragauskas, *Green Chem.*, 2015, **17**, 4239–4246.
24 A. Mittal, T. B. Vinzant, R. Brunecky, S. K. Black, H. M. Pilath, M. E. Himmel and D. K. Johnson, *Green Chem.*, 2015, **17**, 1546–1558.
25 Y. Q. He, J. Zhang and J. Bao, *Biotechnol. Biofuels*, 2016, **9**, 19.
26 P. Sannigrahi, D. H. Kim, S. Jung and A. Ragauskas, *Energy Environ. Sci.*, 2011, **4**, 1306–1310.
27 Y. Chen, M. A. Stevens, Y. M. Zhu, J. Holmes and H. Xu, *Biotechnol. Biofuels*, 2013, **6**, 8.
28 R. J. Stoklosa, A. del Pilar Orjuela, L. da Costa Sousa, N. Uppugundla, D. L. Williams, B. E. Dale, D. B. Hodge and V. Balan, *Bioresour. Technol.*, 2017, **226**, 9–17.
29 R. P. Overend and E. Chornet, *Philos. Trans. R. Soc. London, Ser. A*, 1987, **321**, 523–536.
30 Z. H. Liu and H. Z. Chen, *Bioresour. Technol.*, 2015, **193**, 345–356.
31 L. Lin, Y. Cheng, Y. Pu, S. Sun, X. Li, M. Jin, E. A. Pierson, D. C. Gross, B. E. Dale, S. Y. Dai, A. J. Ragauskas and J. S. Yuan, *Green Chem.*, 2016, **18**, 5536–5547.
32 C. G. Yoo, M. Li, X. Z. Meng, Y. Q. Pu and A. J. Ragauskas, *Green Chem.*, 2017, **19**, 2006–2016.
33 P. Yunqiao, C. Shilin and R. J. Arthur, *Energy Environ. Sci.*, 2011, **4**, 3154–3166.
34 C. A. Schneider, W. S. Rasband and K. W. Eliceiri, *Nat. Methods*, 2012, **9**, 671–675.
35 C. Krishnan, L. D. Sousa, M. J. Jin, L. P. Chang, B. E. Dale and V. Balan, *Biotechnol. Bioeng.*, 2010, **107**, 441–450.
36 Z. H. Liu, L. Qin, M. J. Jin, F. Pang, B. Z. Li, Y. Kang, B. E. Dale and Y. J. Yuan, *Bioresour. Technol.*, 2013, **132**, 5–15.
37 L. J. Jonsson, B. Alriksson and N. O. Nilvebrant, *Biotechnol. Biofuels*, 2013, **6**, 16.
38 C. A. Mullen, A. A. Boateng, R. B. Dadson and F. M. Hashem, *Energy Fuels*, 2014, **28**, 7014–7024.
39 B. Yu and H. Z. Chen, *Bioresour. Technol.*, 2010, **101**, 9114–9119.
40 D. Szczerbowski, A. P. Pitarelo, A. Zandoná Filho and L. P. Ramos, *Carbohydr. Polym.*, 2014, **114**, 95–101.
41 M. Kapoor, S. Soam, R. Agrawal, R. P. Gupta, D. K. Tuli and R. Kumar, *Bioresour. Technol.*, 2017, **224**, 688–693.
42 H. Z. Chen and Z. H. Liu, *Biotechnol. Biofuels*, 2014, **7**, 53.
43 S. V. Vassilev, D. Baxter, L. K. Andersen and C. G. Vassileva, *Fuel*, 2013, **105**, 40–76.
44 S. Park, J. O. Baker, M. E. Himmel, P. A. Parilla and D. K. Johnson, *Biotechnol. Biofuels*, 2010, **3**, 10.
45 H. Z. Chen and Z. H. Liu, *Biotechnol. J.*, 2015, **10**, 866–885.
46 Z. H. Liu and H. Z. Chen, *ACS Sustainable Chem. Eng.*, 2016, **4**, 1274–1285.
47 M. J. Selig, S. Viamajala, S. R. Decker, M. P. Tucker, M. E. Himmel and T. B. Vinzant, *Biotechnol. Prog.*, 2007, **23**, 1333–1339.
48 Z. H. Liu, L. Qin, J. Q. Zhu, B. Z. Li and Y. J. Yuan, *Biotechnol. Biofuels*, 2014, **7**, 167.
49 R. K. Le, T. Wells, P. Das, X. Z. Meng, R. J. Stoklosa, A. Bhalla, D. B. Hodge, J. S. Yuan and A. J. Ragauskas, *RSC Adv.*, 2017, **7**, 4108–4115.
50 T. Wells and A. J. Ragauskas, *Trends Biotechnol.*, 2012, **30**, 627–637.
51 J. G. Linger, D. R. Vardon, M. T. Guarnieri, E. M. Karp, G. B. Hunsinger, M. A. Franden, C. W. Johnson, G. Chupka, T. J. Strathmann and P. T. Pienkos, *Proc. Natl. Acad. Sci. U. S. A.*, 2014, **111**, 12013–12018.
52 A. Rodriguez, D. Salvachúa, R. Katahira, B. A. Black, N. S. Cleveland, M. Reed, H. Smith, E. E. K. Baidoo, J. D. Keasling, B. A. Simmons, G. T. Beckham and J. M. Gladden, *ACS Sustainable Chem. Eng.*, 2017, **5**, 8171–8180.
53 C. Zhao, S. Xie, Y. Pu, R. Zhang, F. Huang, A. J. Ragauskas and J. S. Yuan, *Green Chem.*, 2016, **18**, 1306–1312.
54 Z. Wei, G. Zeng, F. Huang, M. Kosa, D. Huang and A. J. Ragauskas, *Green Chem.*, 2015, **17**, 2784–2789.
55 H. B. Klinke, B. K. Ahring, A. S. Schmidt and A. B. Thomsen, *Bioresour. Technol.*, 2002, **82**, 15–26.
56 E. Palmqvist and B. Hahn-Hagerdal, *Bioresour. Technol.*, 2000, **74**, 17–24.
57 M. E. Brown and M. C. Chang, *Curr. Opin. Chem. Biol.*, 2014, **19**, 1–7.
58 X. B. Zhao, S. M. Li, R. C. Wu and D. H. Liu, *Biofuels, Bioprod. Biorefin.*, 2017, **11**, 567–590.

# 501845 Cleavage of β-*O*-4 Ether Bonds in Acidic Lithium Bromide Trihydrate for Lignin Depolymerization

**Abstract:**

Depolymerization of lignin is essential in pulping and biorefining processes. It was observed in our previous studies that lignin could be extensively depolymerized in acidic lithium bromide trihydrate media. In this study, the mechanism of lignin depolymerization in acidic lithium bromide trihydrate was investigated. First, it was demonstrated that the native lignin in lignocellulosic biomass was easily depolymerized under mild conditions (e.g. in 60 % LiBr with 40 mM HCl at 110 °C). Six types of biomass, including hardwood, softwood, and herbaceous biomass, were treated in acidic lithium bromide trihydrate, and the isolated lignin fractions were characterized with GPC and NMR. The result indicated that the β-*O*-4 ether linkages of the native lignin were selectively cleaved, leading to lignin depolymerization (M*w*: 2000-4000). Second, lignin model compound (guaiacylglycerol-β-guaiacyl ether) was used to elucidate the cleavage mechanism of the β-*O*-4 ether bonds. It was found that the ether bonds were cleaved via the enol ether pathway initiated by the a-C catioinzation, and both LiBr and HCl catalyzed the cleavage. In addition, new C-C bonds were formed, resulting from the nucleophilic attack of the benzyl carbocations by the electron-rich lignin aromatic rings. Third, this lignin depolymerization strategy was applied to technical lignins isolated from pulping and biorefining processes. It was found that the residual β-*O*-4 ether linkages of the lignin could be selectively cleaved as well.

# 501979 Oxidative Functionalization of Lignins for the Synthesis of Polyamides and Polyesters

**Abstract:**

Lignin, a renewable resource available from lignocellulosic biomass, has the potential to replace petrochemical feedstock in the production of synthetic polymers. Valorization of lignin coproduct from enzymatic conversion offers opportunity to improve the economic feasibility of lignocellulosic sugar production. Applications of lignin as a polymer material, however, is hindered by brittleness and thermal instability of the material. Unless blended with other polymers (e.g. PLA, PP, PVA, PET) or added as a filler in composites, technical lignins alone are not suited for material applications. The complexity in lignin structure and condensation chemistry also complicates the use of lignin as a source of polymer building blocks.

We here present a new process that converts a wide variety of lignins to high molecular weight (>10,000 Da) polymer that contains over 80% lignin by weight. In this process, enzymatic lignin and organosolv lignin are functionalized via Cu-catalyzed oxidation to form oligomers with acyl chloride groups. Using these oligomers and difunctional linkers, we have successfully synthesized block copolymers of lignin that contain amide and ester linkages.

Characterization of the synthesized polymer demonstrated the preservation of native beta-aryl ether linkages in lignin, while FTIR data confirms the formation of amide and ester linkages. Our studies on lignin polymerization chemistry revealed the association between polymerization parameters and mechanical properties. This new polymer material has potential in coating, adhesives, and drug delivery applications.

# 489038 Propane Adsorption on ZIF-8

**Abstract:**

We present the results of an adsorption isotherm study of Propane on the metal-organic framework ZIF-8. This material undergoes a structural transition (â€œgate-openingâ€) as a function of increasing pressure and sorbent loading for many sorbates. Gate-opening manifests itself in the isotherm data as a quasi-vertical substep, corresponding to an increase in the amount that can adsorb in the ZIF-8. We measured multiple isotherms over a temperature range of nearly 100 K between the triple point and critical point temperatures for bulk Propane. In the range studied, we did not find in any of the isotherms the characteristic substep indicative of the gate-opening. It is possible that the temperature range over which this transition would manifest in the data simply has not been explored in our measurements. The adsorption isotherm data was used to determine the isosteric heat of adsorption of Propane on this sorbent as a function of sorbent loading. The resulting isosteric heat plot contains a dramatic peak which might be associated with the gate opening transition. At sorbent loadings below that which corresponds to the peak, isosteric heat values are less than the latent heat of condensation for Propane. We have studied the kinetics of adsorption for this system using the linear driving force model to obtain concise summaries of kinetic data for multiple temperatures. Times to reach fractional uptakes of 0.999 for mass loading data corresponding to individual points along the isotherms will be presented. Additionally, effective specific surface area measurements obtained from the Point-B and BET methods will be presented.

# 499630 Conformal Zr-Based Metal-Organic Framework Thin Films on Nanofibers for Ultra-Fast Degradation of Chemical Warfare Agents

**Abstract:**

The use of nerve agent in recent chemical attacks in Syria has urged the development of new materials and equipment to protect first-responders, soldiers and the general public. With large porosity and abundant Lewis-acidic catalytic sites, metal-organic frameworks (MOFs) are promising candidates for adsorption and catalytic degradation of chemical warfare agents (CWAs). However, it is still challenging to integrate MOFs into functional textiles for protective gas filters, masks and suits. Here, we present a series of MOF-coated nylon nanofibers enabling fast detoxification of CWAs. We found that a 5-nm-thick $TiO_2$ layer deposited onto electrospun nylon nanofibers via atomic layer deposition (ALD) promotes the heterogeneous nucleation of Zr-based MOFs including UiO-66, UiO-66-$NH_2$ and UiO-67. SEM and cross-sectional TEM images reveal that conformal MOF coatings localized on the nanofiber surface form unique MOF-nanofiber kebab structures. These MOF-functionalized nanofibers were tested for catalytic hydrolysis of a CWA simulant Dimethyl 4-Nitrophenyl Phosphate (DMNP) and a nerve agent *O*-Pinacolyl Methylphosphonofluoridate (GD, or Soman). The half-lives of DMNP are as short as 7.3 min, while the half-lives of GD are all less than 4 min for the three types of Zr-based MOF thin films on nanofibers. The results demonstrate the excellent performance of our MOF-coated nanofibers for CWA destruction, and also provide new opportunities for the development of air filters, wearable sensors, and smart multifunctional textile materials.

# Engineering Copper Carboxylate Functionalities on Water Stable Metal-Organic Frameworks for Enhancement of Ammonia Removal Capacities

**Jayraj N. Joshi**, Erika Garcia-Gutierrez, Colton M. Moran, Jacob I. Deneff, Krista S. Walton

AIChE 2017
November 2nd, 2017

Georgia Institute of Technology
School of Chemical & Biomolecular Engineering

Georgia Institute of Technology • School of Chemical & Biomolecular Engineering
311 Ferst Drive, N.W. • Atlanta, GA 30332-0100
404-894-1838 • www.chbe.gatech.edu

---

Nanomaterials & Adsorption Lab
WaltonGroup

2

## Walton Research Group

ChBE Department, Georgia Institute of Technology

- PI: Dr. Krista Walton
- Walton Research Group

Acknowledgements

- Funding Sources
  - ARO and DTRA Contracts W911NF-10-1-0076
  - UNCAGE-ME Energy Frontier Research Center (EFRC) Award # DE-SC0012577
- K-edge XAS experiments at APS, sector 12-BM

UNCAGE ME

U.S. DEPARTMENT OF ENERGY | Office of Science

Nanomaterials & Adsorption Lab
WaltonGroup

# Novel Design of Adsorbent Materials

## Metal-Organic Frameworks (MOFs)

3

- MOFs are crystalline, nanoporous materials typically formed via solvothermal synthesis
- Promising candidates for air purification applications
  - High surface area and porosity
  - Facile structural tuning
  - **Possible active open metal sites (OMS)**

Georgia Institute of Technology
School of Chemical & Biomolecular Engineering

---

Nanomaterials & Adsorption Lab
WaltonGroup

# Novel Design of Adsorbent Materials

## Copper Carboxylates (-COOCu) for Ammonia Capture

4

- Ammonia ($NH_3$)
  - Toxic industrial chemical (TIC), chemical warfare agent (CWA)
  - **6 mmol $NH_3$ $g^{-1}$ target**
- Analogous copper OMS in several MOFs (e.g. Cu-BTC) afford high ammonia uptake capacities[1,2]
  - OMS MOFs often experience poor stability[3]

| functional group | $NH_3$ | $H_2O$ | lowest binding energy of $H_2O$ − lowest binding energy of $NH_3$ |
|---|---|---|---|
| R−COOCu | −161.2 | −97.8 | 63.4 |
| R−COOAg | −84.9 | −51.5 | 33.4 |
| R−$HSO_4$ | −65.9 | −46.9 | 19.0 |
| R−COOLi | −81.1 | −70.2 | 10.9 |
| R−OOH | −37.2 | −27.7 | 9.5 |
| R−$SO_3H$ | −52.1 | −42.7 | 9.4 |
| R−OP (=O) $(OH)_2$ | −54.2 | −44.9 | 9.3 |
| R−P (=O) $(OH)_2$ | −50.8 | −42.7 | 8.1 |

Computational screening by Snurr and coworkers[4]:
**Copper Carboxylates (-COOCu)**
Favorable adsorption energy for $NH_3$ (even in humidity)


1. Glover et al. *Chem Eng. Sci.* **2011**, *66*, 163-170
2. Petit et al. *Factors Affecting the Removal of Ammonia from Air on Carbonaceous Materials: Investigation of Reactive Adsorption Mechanism*. **2012**
3. Jiao et al. *Ind. Chem. Eng.* **2015**, *54*, 12408-12414
4. Kim, et al. *Langmuir* **2013**, *29*, 1446-1456

Nanomaterials & Adsorption Lab
Walton Group

# Novel Design of Adsorbent Materials

## Stability of Open Metal Site (OMS) MOFs

5

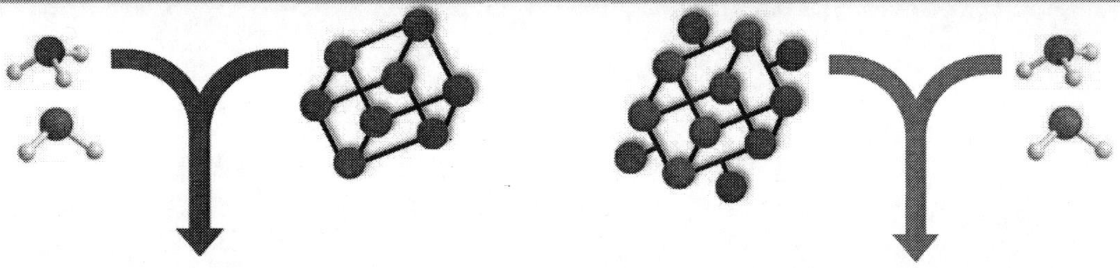

Reaction with **MOF Metal Center**

Framework Degradation

Let ● be reactive open metal sites (OMS)

Reaction with **Framework Supported Metal Sites**

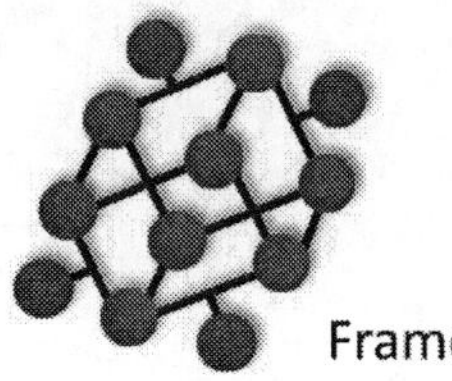

Framework Preservation

---

Nanomaterials & Adsorption Lab
Walton Group

# Novel Design of Adsorbent Materials

## UiO-66(Zr): A Stable Platform for Active Site Engineering

6

### UiO-66(Zr)

- Zirconium based framework
- High thermal and chemical stability[1,2]

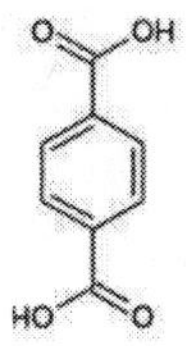


Terephthalic acid linker

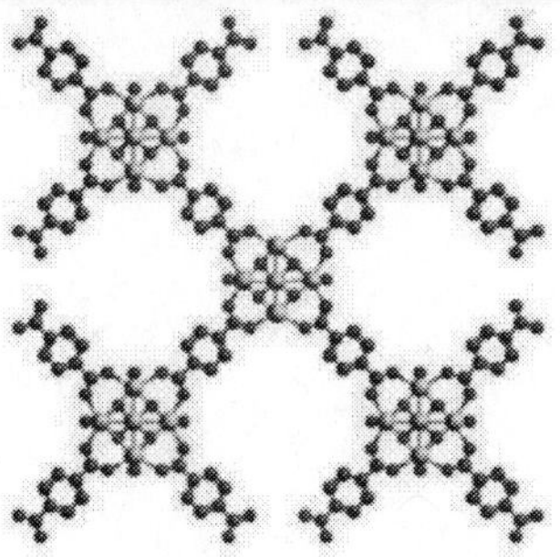

UiO-66 offers a **large degree of tailorability** through:

1. MOF linker based functional groups
2. Tunable framework defects[3,4]


1. Cavka et al. *J. Am. Chem. Soc.* **2008**, *130*, 13850-13851
2. Valenzano et al. *Chem. Mat.* **2011**, *23*, 1700-1718
3. Gutov et al. *Inorg. Chem.* **2015**, *54*, 8396-8400
4. Katz et al. *Chem. Commun.* **2013**, *49*, 9449-9451

# Novel Design of Adsorbent Materials

## UiO-66(Zr): A Stable Platform for Active Site Engineering



**Linker replacement**

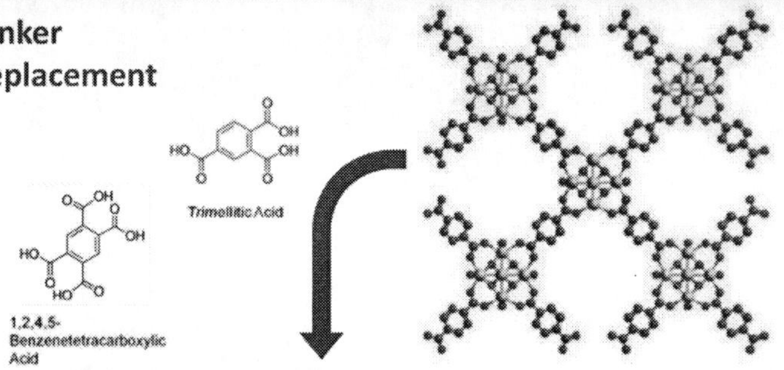


Carboxylic acid (-COOH) groups added to **MOF linker**

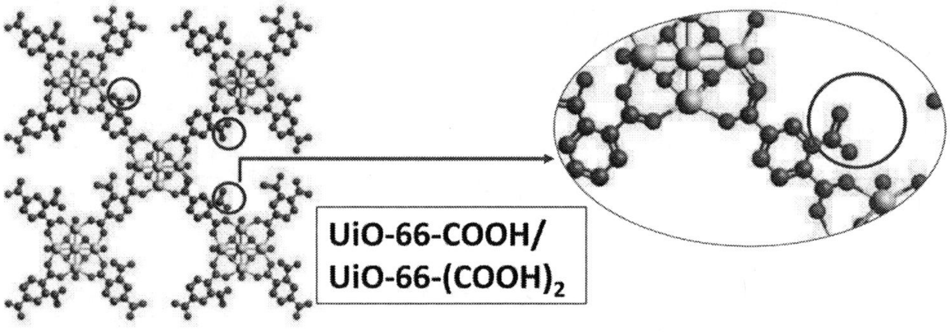




# Novel Design of Adsorbent Materials

## UiO-66(Zr): A Stable Platform for Active Site Engineering



**Defect Incorporation**

Carboxylic acid (-COOH) groups added to **missing linker vacancies** in framework[1,2]

- First requires creation of "defect-laden" framework: **UiO-66-vac**

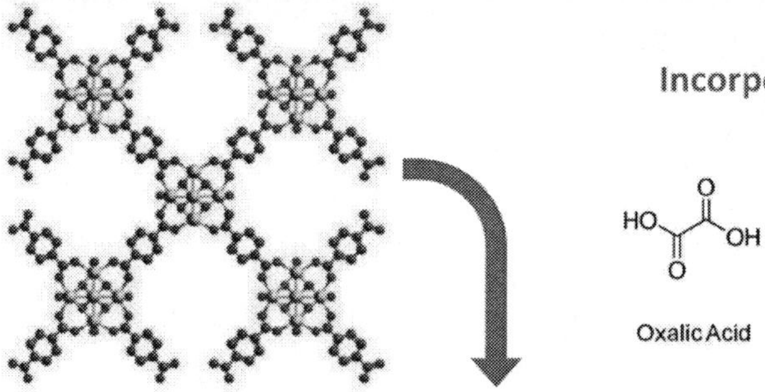


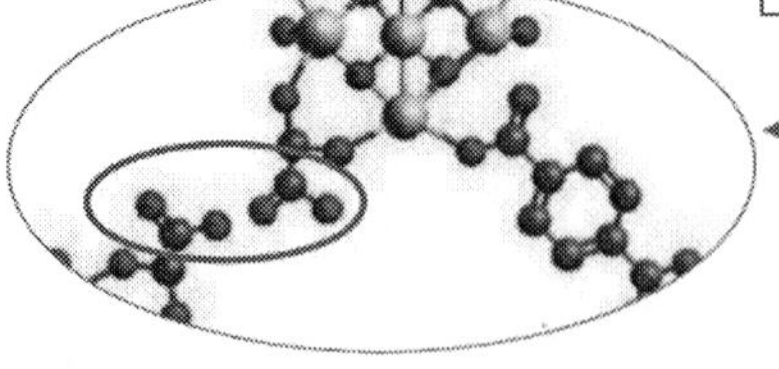

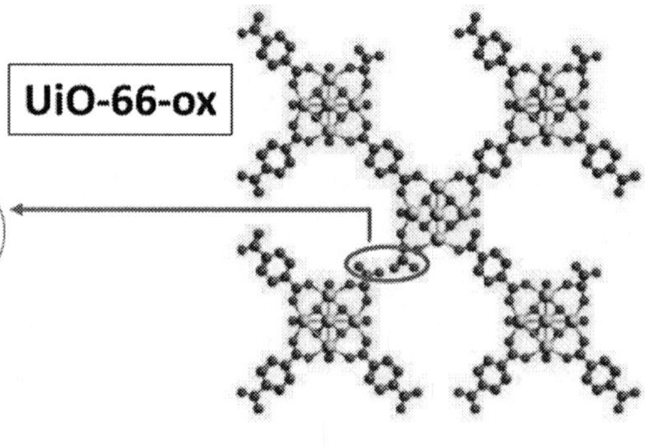


1. Katz et al. *Chem. Commun.* **2013**, *49*, 9449-9451
2. DeCoste et al. *New J. Chem.* **2015**, *39*, 2396-2399

Nanomaterials & Adsorption Lab
WaltonGroup

9

# Novel Design of Adsorbent Materials

## UiO-66(Zr): A Stable Platform for Active Site Engineering

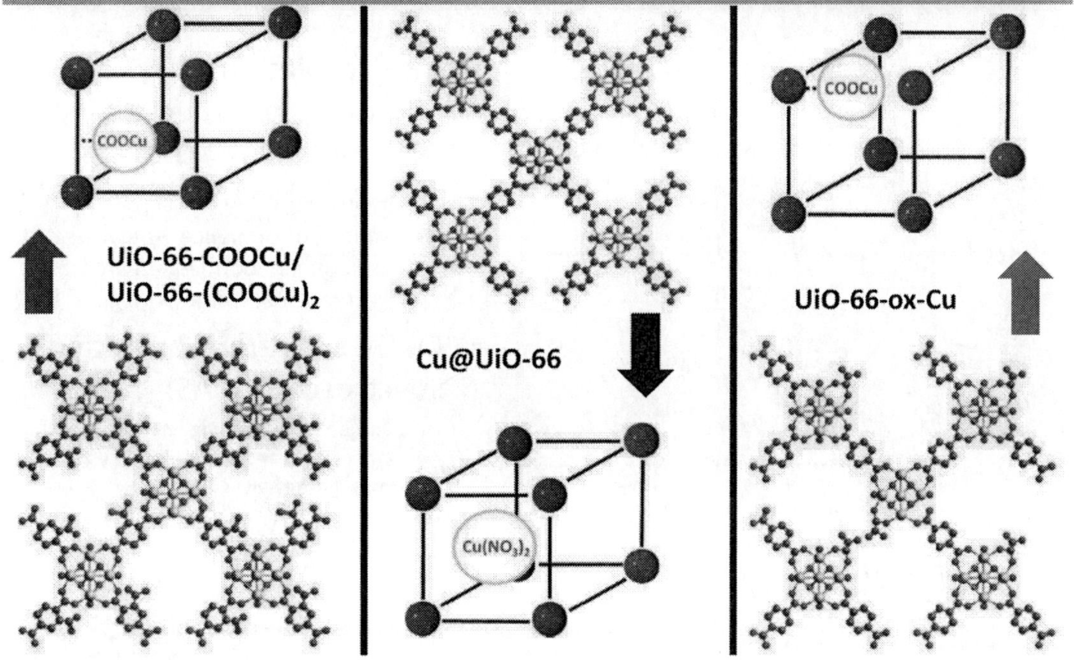


Nanomaterials & Adsorption Lab
WaltonGroup

10

# Copper Carboxylate MOFs

## Incorporating Copper into UiO-66(Zr) Analogues

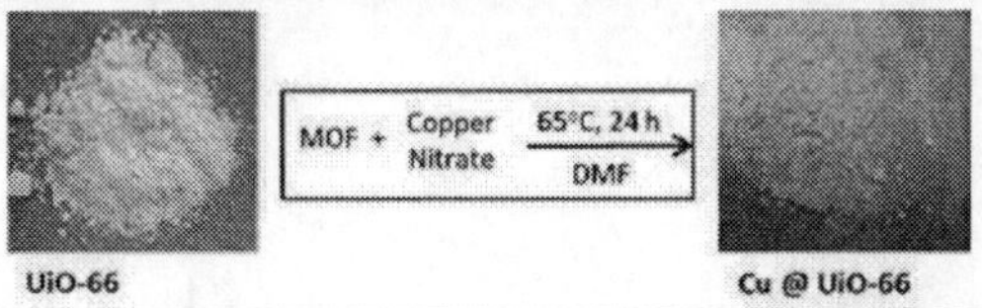


Solvothermal
Copper Incorporation

| MOF | Copper Mass Fraction (wt %) Exp. | Theoretical | Fractional Loading (-) | Surface Area ($m^2$ $g^{-1}$) | Total Pore Volume (cc $g^{-1}$) |
|---|---|---|---|---|---|
| UiO-66 | ---- | ---- | ---- | **1111** | **0.73** |
| [a]Cu @ UiO-66 | 4.51 | 0 | ---- | **781** | **0.46** |
| UiO-66-vac | ---- | ---- | ---- | **1534** | **0.80** |
| UiO-66-ox | ---- | ---- | ---- | **1158** | **0.59** |
| UiO-66-ox-Cu | 4.84 | 5.50 | 0.88 | **1116** | **0.56** |
| UiO-66-COOH | ---- | ---- | ---- | **658** | **0.28** |
| UiO-66-COOCu | 7.88 | 9.62 | 0.82 | **564** | **0.24** |
| UiO-66-$(COOH)_2$ | ---- | ---- | ---- | **364** | **0.22** |
| UiO-66-$(COOCu)_2$ | 3.15 | 14.52 | 0.22 | **357** | **0.18** |

[a]Cu@UiO-66 assumed to be free of open –COOH bonding sites for copper

Joshi et al. *J. Phys. Chem. C* **2017**, *121*, 3310-3319

Difference in porosity changes after Cu incorporation between:

1. MOFs **with** –COOH binding sites (UiO-66-ox-Cu, UiO-66-COOCu, UiO-66-$(COOCu)_2$)
2. MOFs **without** –COOH binding sites (Cu@UiO-66)

Georgia Institute of Technology
School of Chemical & Biomolecular Engineering

Nanomaterials & Adsorption Lab
WaltonGroup



# Copper Carboxylate MOFs

## Characterization of Incorporated Copper Sites

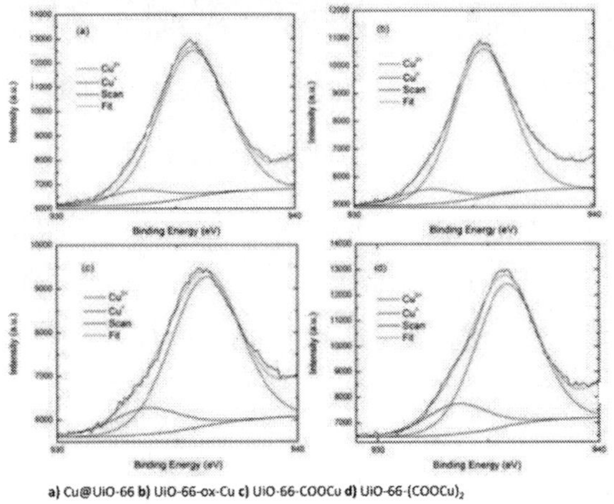


a) Cu@UiO-66 b) UiO-66-ox-Cu c) UiO-66-COOCu d) UiO-66-(COOCu)$_2$

- High resolution X-ray photoelectron spectroscopy (XPS) of Cu $2p_{3/2}$ core
  - Majority $Cu^{2+}$ state
  - Minor $Cu^{+}$ presence—may suggest some mixed bonding environment with residual framework guests
- Cu K-edge X-ray absorption spectroscopy (XAS)
  - Unique local bonding environment[1-3] for copper carboxylate MOFs vs. impregnate salt

1. Sarode et al. *Proc. Indian Acad. Sci. (Chem. Sci.).* **1983**, *92*, 527-542
2. Sibilia, J.P. *A Guide to Materials Characterization and Chemical Analysis.* **1988**
3. Trickett et al. *Agnew. Chemie Int. Ed.* **2015**, 11162-11167
4. Joshi et al. *J. Phys. Chem. C* **2017**, *121*, 3310-3319

Georgia Institute of Technology
School of Chemical & Biomolecular Engineering

Nanomaterials & Adsorption Lab
WaltonGroup



# Composite Adsorption & Stability

## Water Vapor Adsorption Experiments at 25°C

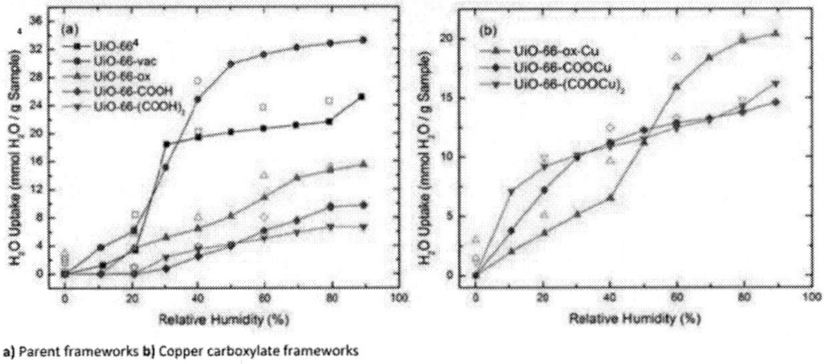


a) Parent frameworks b) Copper carboxylate frameworks

- Both synergistic and competitive adsorption interactions observed for humid ammonia adsorption[1-3]
- Likely pore blocking from –COOH functionalities
- Incorporation of –COOCu induces hydrophillicity

1. Moghadam et al. *J. Mater. Chem.* **2016**, *4*, 529-536
2. Petit et al. *Factors Affecting the Removal of Ammonia from Air on Carbonaceous Materials: Investigation of Reactive Adsorption Mechanism.* **2012**
3. Nijem et al. *J. Phys. Chem. C* **2015**, *119*, 24781-24788
4. Schoenecker, P.M. *High-Throughput Synthesis and Application Development of Water-Stable MOFs.* **2012**
5. Joshi et al. *J. Phys. Chem. C* **2017**, *121*, 3310-3319

Nanomaterials & Adsorption Lab
WaltonGroup



# Composite Adsorption & Stability

## Material Stability Post-Water Vapor Exposure

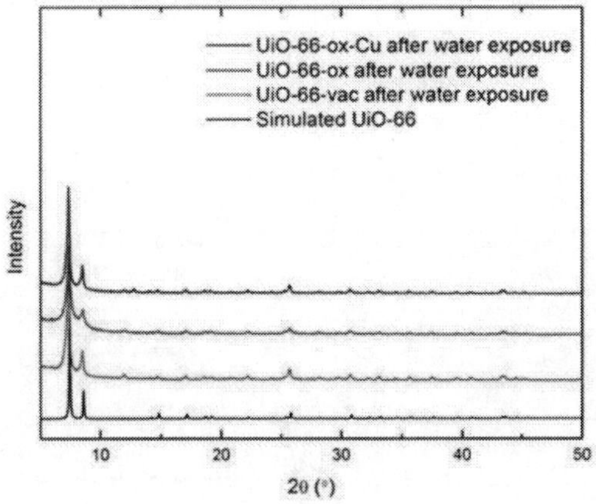


- Most materials retain characteristic water vapor stability of UiO-66(Zr)
  - **UiO-66-ox and UiO-66-ox-Cu:** 30% and 52% loss in BET surface area, respectively
  - No change in diffraction pattern—strongly suggests partial collapse of pore space

Joshi et al. *J. Phys. Chem. C* **2017**, *121*, 3310-3319

Georgia Institute of Technology
School of Chemical & Biomolecular Engineering

Nanomaterials & Adsorption Lab
WaltonGroup



# Composite Adsorption & Stability

## Improvement in Ammonia Adsorption Capacities

Dry (0% RH)

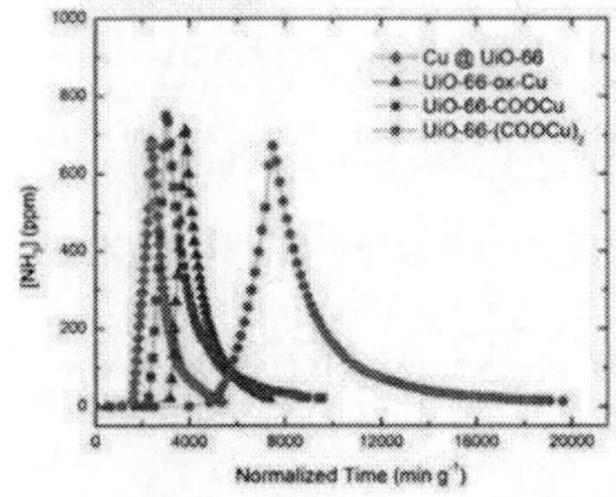


Humid (80% RH)

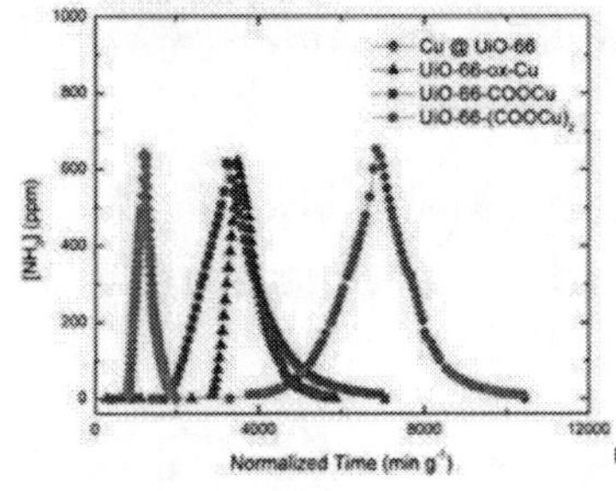


| MOF | $NH_3$ Dynamic Adsorption Capacity (mmol g$^{-1}$)[a] | |
|---|---|---|
| | Dry (0% RH) | Humid (80% RH) |
| UiO-66[b] | 1.79 | 2.75 |
| Cu@UiO-66 | 2.25 | 1.15 |
| UiO-66-ox | 2.50 | 2.30 |
| UiO-66-ox-Cu | 4.31 | 4.02 |
| UiO-66-COOH | 2.24 | 2.21 |
| UiO-66-COOCu | 3.04 | 3.10 |
| UiO-66-(COOH)$_2$ | 2.83 | 1.83 |
| UiO-66-(COOCu)$_2$ | **6.38** | **6.84** |

a) Capacities normalized by sample mass
b) Data from Jasuja et al.[3]

1. Kim et al. *Langmuir* **2013**, *29*, 1446-1456.
2. Joshi et al. *J. Phys. Chem. C* **2017**, *121*, 3310-3319
3. Jasuja et al. *Chem. Eng. Sci.* **2015**, *124*, 118-124

Georgia Institute of Technology
School of Chemical & Biomolecular Engineering

Nanomaterials & Adsorption Lab
Walton Group

15

# Composite Adsorption & Stability

## Stability in Dry and Humid Ammonia Environments

- Structural integrity preserved in dry exposures
- UiO-66-ox-Cu degradation after humid $NH_3$
  - Reflected in $N_2$ physisorption measurements
  - Instability not present in parent framework: UiO-66-ox

Degradation of UiO-66-ox-Cu

1. Enhanced accessibility in defect-laden framework
2. Strengthened adsorbate-adsorbent interactions from introduced –COOCu groups (e.g. cooperative binding[1,2])

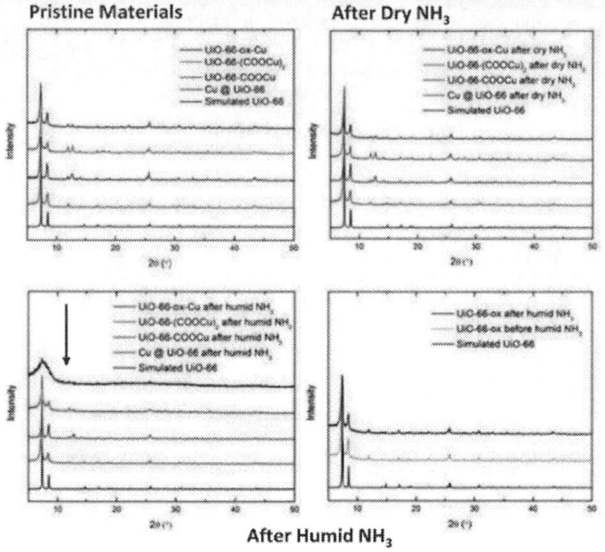


1. Nijem et al. *J. Phys. Chem. C* **2015**, *119*, 24781-24788
2. Watanabe et al. *J. Chem. Phys.* **2010**, *133*, 133
3. Joshi et al. *J. Phys. Chem. C* **2017**, *121*, 3310-3319

Georgia Institute of Technology
School of Chemical & Biomolecular Engineering

Nanomaterials & Adsorption Lab
Walton Group

16

# Conclusions

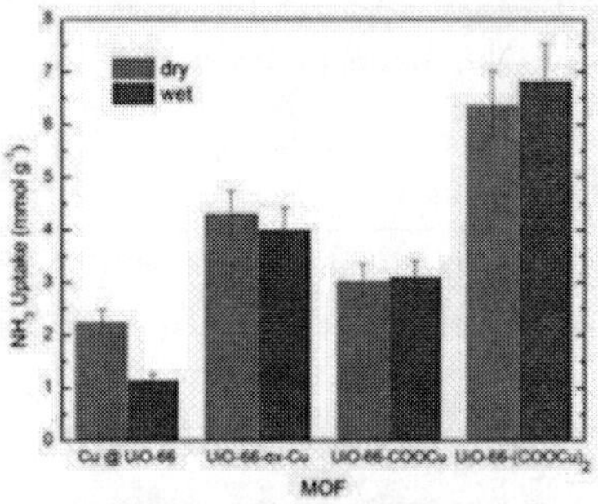


Extending the design motif towards new research:

1. Other MOF systems
2. Different coordination metals
3. Target adsorbates for separation

## Novel MOF Design Strategy

**Install reactive OMS without compromising framework integrity**

- Linker-bound –COOCu analogues of UiO-66(Zr)
  - Ammonia uptake in dry and humid conditions exceeded project target (6 mmol $NH_3$ $g^{-1}$)
  - New adsorbates exhibit chemical stability towards ammonia/water environments

Joshi et al. *J. Phys. Chem. C* **2017**, *121*, 3310-3319

Georgia Institute of Technology
School of Chemical & Biomolecular Engineering

# Engineering Copper Carboxylate Functionalities on Water Stable Metal-Organic Frameworks for Enhancement of Ammonia Removal Capacities

**Jayraj N. Joshi**, Erika Garcia-Gutierrez, Colton M. Moran, Jacob I. Deneff, Krista S. Walton

AIChE 2017
November 2nd, 2017

Georgia Institute of Technology
School of Chemical & Biomolecular Engineering

Georgia Institute of Technology • School of Chemical & Biomolecular Engineering
311 Ferst Drive, N.W. • Atlanta, GA 30332-0100
404-894-1838 • www.chbe.gatech.edu

# 488693 A Combined Molecular Simulation and Process Simulation Study of Benzene Removal from Vinyl Acetate in MOFs

**Abstract:**

Vinyl acetate (VAc) is a widely used kind of organic raw material which is mainly synthesized by the carbide acetylene method in China. However, the trace amounts of benzene (about 100 ppm) impurities in the VAc product have seriously limited its application in fine chemical industry. It is an urgent requirement to separate trace benzene from VAc down to 10 ppm. Pressure swing adsorption (PSA) is an efficient method to remove trace impurities and its main challenge is to find an effective adsorbent. Metal-organic frameworks (MOFs) possess many advantages compared with traditional materials and are believed to have perspective applications in this field, although few investigations are reported. In this work, multi-scale simulation method including molecular level and process simulation are used to screen an effective adsorbent which could be employed in the PSA process. This work would provide the guidance for the removal of trace benzene.

Grand Canonical Monte Carlo (GCMC) and configure bias Monte Carlo (CBMC) are combined to investigate the adsorption behavior of benzene and VAC in different MOFs: Cu-BTC, Zn-BDC, IRMOFs and M-DOBDC. The effects of exposed metal sites, organic chain length and different functional groups attached to organic chains on adsorption and separation behaviors were studied. The highest selectivity factor of benzene over VAc is as high as 12, showing the great ability to remove trace amount of benzene from VAc product.

After the molecular simulation screening work, the most effective MOF material is selected for the following PSA process to remove trace benzene in an annual output of 250,000 tons of VAc plant. In the designed PSA process, benzene concentration in the VAc product could be reduced less than 1 ppm (wt%) . The result indicate that MOF is an excellent adsorbent for removing trace benzene, and the purity is able to meet the requirement of high-quality VAc product.

# 496106 New Strategies of Enhancing Steam Stability of MOFs By Modification with Post-Synthesis and in-Situ Synthesis Methods

**Abstract:**

**New Strategies of Enhancing Steam Stability of MOFs by Modification with Post-Synthesis and In-situ Synthesis Methods**

Jing Xiao[1*], Zhedong Lin, Hao Li, Xin Zhou, and Zhong Li[2,*]

1 School of Chemistry and Chemical Engineering, South China University of Technology, Guangzhou, 510640, China

*2 The Key Laboratory of Enhanced Heat Transfer and Energy Conversation Ministry of Education, South China University of Technology, Guangzhou 510640, PR China*

* Presenting authors: cejingxiao@scut.edu.cn; cezhli@scut.edu.cn

**ABSTRACT**

MOFs are very potential materials for gas adsorption and separation due to their ultrahigh surface area, open crystalline structure and adjustable chemical functionality. Unfortunately, most of MOFs are unstable under humid circumstances, which greatly hinders their practical applications. Thus, enhancing water vapor stability of MOFs is a very challenging issue in research community. Here, we reported two strategies to enhance the steam stability of Cu-BTC (HKUST-1). Cu-BTCs were functionalized with some organic compounds to prepare TED@Cu-BTC and ACN@Cu-BTC separately by in-situ synthesis method and post-synthesis method, and then characterized. Water vapor stability and $CO_2$ adsorption capacity of the resulting samples were examined. Results show that the resulting samples possessed similar crystal structure and textural parameters as well as $CO_2$ capacity as the parent Cu-BTC. More interestingly, the water vapor stability of the resultant samples were significantly enhanced in comparison with Cu-BTC. Steam stability experiments proved that after being exposed to moist air with relative humidity of 80% for 20-30 days, the resulting samples preserved their main crystal structures and 80% of its initial $CO_2$ capacity, and in contrast, the parent Cu-BTC almost lost its crystallinity and its $CO_2$ adsorption capacity. The origination of the improved steam stabilities of the modified Cu-BTC were interpreted by DFT calculations. The strategy of incorporating TED and ACN into Cu-BTC is very efficient in improving its stability against moisture, which is worthy of further exploitation for enhancing stream stability of some MOFs.

# 494249 High Performance Gas Adsorption and Separation of Light Hydrocarbon in a Microporous Metal Organic Framework


**Abstract:**

The adsorption and separation of $CH_4$, $C_2H_6$, $C_3H_8$ and $C_4H_{10}$ on the $Zn(bdc)(ted)_{0.5}$ were studied. The samples were characterized by $N_2$ adsorption, X-ray diffraction (XRD), and Fourier Transform Infrared spectroscopy (FTIR). The adsorption isotherms of light hydrocarbon gases on the $Zn(bdc)(ted)_{0.5}$ were separately measured by volumetric method, and the $C_2H_6/CH_4$, $C_3H_8/CH_4$ and $C_4H_{10}/CH_4$ adsorption selectivities of the samples were calculated based on ideal adsorbed solution theory (IAST). Results showed that the isotherms of light hydrocarbon gases on the adsorbent were well fitted by Langmuir-Frendlich equation. $Zn(bdc)(ted)_{0.5}$ could take up a amount of $C_2H_6$ (4.9 mmol/g) $C_3H_8$ (6.6 mmol/g), and $C_4H_{10}$ (6.9 mmol/g), which were much higher than the amount of $CH_4$ (0.4 mmol/g) at 298 K and 1 atm. The $C_4H_{10}/CH_4$ selectivity was the highest and was in the range of 65â€“150. The $C_3H_8/CH_4$ selectivity is lower and the values were in the range of 8â€“16. The $C_2H_6/CH_4$ selectivity was the lowest and the values was in the range of 1.4â€“2.

# 492204 Emission of Sulphur and Nitrogen Based Pollutants in Commercial Scale Circulating Fluidized Bed Gasifiers

**Abstract:**

Coal gasification technology has attracted great attention and a number of commercial coal gasification plants have been in operation. Environmental concerns should been taken as well as gas heat value, cold gas efficiency and carbon conversion in coal gasification. In the next years, Institute of Engineering Thermophysics of Chinese Academy of Sciences have designed the circulating fluidized bed gasification system to produce industrial fuel gas, and about 30 suits of commercial coal gasification system have been built in coking and smelting industry. This paper studied wastewater of three different scale coal gasification system, and the release of minor contaminants (N, S compounds) was studied by sampling and analyzing the feedstock, ash, slag and the producer gas. Little tar and phenol was detectable in the three water samples. The main pollutants were $COD_{Cr}$, ammonia nitrogen and sulphate, the concentration of whom were 113mg/L to 890 mg/L, 270 mg/L to 903 mg/L, 14 mg/L to 60 mg/L, respectively detected in the gasification wastewater. There was low toxicity due to low level of complex organics. Ammonia nitrogen should been removed firstly because of the low C/N ratio. The $NH_3$ and $H_2S$ and COS concentrations in the producer gas were 300ppm, 530ppm and 80ppm. There was little HCN detected in producer gas. The elements analyze of fuel coal, ash and slag indicated that more than 96.61% of fuel-S transferred to $H_2S$ and COS. The conversion of fuel-N to ash and slag was less than 1%. A large portion of $NH_3$ in producer gas dissolved in water along the gas condensation which made ammonia nitrogen concentration in wastewater cannot reach the discharge standard.

# 494507 Insights of Biomass Recalcitrance in *Populus* for Biomass Valorization

**Abstract:**

For the effective utilization of biomass, understanding of its properties, in particular, biomass recalcitrance-related properties is essential. Diverse approaches are applicable to reveal the secret of biomass recalcitrance. Among many, physicochemical characteristics of biomass feedstock do not only elucidate the biomass recalcitrance, but also provide clues to overcome its physical and chemical barriers for biomass conversion. Transcriptome analysis also gives important overview about transcriptional regulation network of biomass synthesis related gene families. *Populus* has been investigated as a promising biomass feedstock for alternative fuels and chemicals. However, biomass recalcitrance-related properties and their causable variation of gene expression are not sufficiently understood. Here, recalcitrance of *Populus* natural variants was elucidated in three aspects: (1) fermentable sugar release, (2) physicochemical properties of biomass, and (3) relative variation of gene expression. Sugar release and physicochemical characteristics of *Populus* natural variants including chemical composition, cellulose and ligninâ€™s molecular weights/degree of polymerization, cellulose crystallinity, lignin composition, and cellulose accessibility were measured by diverse analytical methods such as HPLC, GPC, 2D $^{13}C$-$^{1}H$ HSQC NMR, solid-state NMR, and others. Differential Gene expression of each variant was also investigated by RNA-seq analysis.

# 499309 Evaluation of Industrial Hemp Varieties As Potential Biomass Feedstock for Biofuels and Bioproducts

**Abstract:**

Industrial hemp is an annual herbaceous crop historically grown for fiber in the U.S. and worldwide. Despite existing applications of industrial hemp based fiber and oil products, hemp-based biofuels and bioproducts represent a new potential application area. Several key questions still remain for the technical and economic feasibility of using hemp as a bioenergy crop. Furthermore, it is necessary to screen industrial hemp species and process conditions while aiming at maximized production of biofuel and bioproducts. Hence, this study takes combined agronomy, laboratory, and economic analysis approaches to evaluate industrial hemp in comparison with other biomass feedstocks e.g. kenaf, switchgrass and biomass sorghum. Several industrial hemp varieties, currently widely cultivated in North America, Europe and Asia, were evaluated. Experimental results show an ethanol yield of 85-100 gallons /dry ton hemp stems using dilute alkali or acid pretreatment, enzymatic hydrolysis and fermentation, which is comparable to the other three tested feedstocks. Agronomy data suggest that the per hectare yield of industrial hemp stem alone was at a similar level to the other energy crops such as switchgrass and sorghum; while the hemp plants may require reduced inputs. Field trial also showed that ~1200 kg/ha hemp grain can be harvested in addition to stems. A preliminary cost analysis indicates that industrial hemp could generate higher per hectare gross profit than the other crops if both hemp grains and biofuels from hemp stem were counted. These combined evaluation results demonstrate that industrial hemp has great potential to become a promising regional commodity crop for producing both biofuels and value-added products.

# 503939 Thermochemical Conversion of Blended Herbaceous and Woody Biomass Feedstocks

**Abstract:**

The United States has a vast potential supply of renewable lignocellulosic biomass, which can be utilized as feedstocks for the production of renewable fuels and chemicals via biomass conversion technologies. In order to support a large-scale bioeconomy, it will be required that multiple types of biomass be utilized such as dedicated bioenergy crops, forestry and agricultural residues, and municipal wastes. Blending several types of biomass to produce a uniform feedstock has the potential to control feedstock quality specifications. In order to exploit the potential benefits of biomass blending, we must gain a comprehensive understanding of its effects on thermochemical conversion. In this study, we target pyrolysis as the thermochemical conversion technology of interest as it is well suited to produce liquid fuels and chemicals through thermal decomposition and condensation of pyrolysis vapors into bio-oil. The chemical composition of the feedstock used has a significant impact of the yield and quality of the bio-oil produced. In this work, we first utilized pyrolysis gas chromatography mass spectroscopy (py-GC/MS) to measure the vapor-phase pyrolysis products produced from switchgrass, two pine residue feedstocks, and binary blends of these. Next, we used an intermediate-scale pyrolysis reactor to test the effect of blending on bio-oil yield and quality. Results indicate that the composition and content of alkali and alkaline earth metals in the feedstocks have a significant effect on the pyrolysis products, and that blending is a potential strategy to produce high quality and low variability feedstocks. Further results and discussion regarding the effects of blending on the yield, quality, and chemical composition of vapor-phase and liquid-phase pyrolysis products will be presented.

# 491952 Feedstock Preprocessing, Fractionation, and Blending to Improve Overall Cost, Supply, and Quality Considerations for Catalyzed and Uncatalyzed Fast Pyrolysis

**Abstract:**

The success of lignocellulosic biofuels and biochemical industries depends on an economical and reliable supply of high-quality biomass. As such, biomass cost, supply, and quality are critical parameters to consider when choosing feedstocks and locations for biorefineries. Biomass cost is dependent upon feedstock type, location, availability, transportation costs, and the quality specifications required by the conversion technology. Biomass quality depends on inherent feedstock properties, field and storage conditions, and any preprocessing methods used to improve quality. Biomass quantity depends on location, field conditions, annual weather, and collection and storage conditions. Because of these variables, it is necessary to develop strategies that enable supply chain improvements in biomass quality and consistency to improve lower overall conversion costs and improve yields, while also securing a less risky feedstock supply.

The presented work examines the interdependencies of these variables and how they affect the biomass blends required by biomass depots and/or biorefineries to achieve the lowest cost feedstock with sufficient quality at the quantities needed for biorefinery operation. Four biomass depots were proposed in South Carolina to each produce 200,000 tons of feedstock per year. These depots supply a centrally located 800,000 ton biorefinery that converts the feedstocks to bio-oil using either catalyzed or uncatalyzed fast pyrolysis. The four depots utilize biomass based upon availability, but the feedstock or feedstock blend still met the minimum quality requirements for the biorefinery. Costs were minimized by using waste biomass resources such as construction and demolition waste, logging residues, and forest residuals. Preprocessing methods such as air classification and water washing or acid leaching were used to upgrade biomass quality. Air classification was found to be an effective method for separating low- and high-ash fractions of woody residues. For example, a high-ash fraction of logging residues was generated that contained approximately 50 wt% of the total original ash content, but less than 8 wt% of the original organic content. Dilute-acid leaching was then effectively used to remove between 85% (25°C, 0.5 wt% acid) and 98% (90°C, 1.0 wt% acid) of the alkali and alkaline earth metals from the high-ash fraction, making it suitable for recombination with the low-ash fraction. Various combinations of such preprocessing methods were tested to produce an array of blends that were available for recombination to produce least-cost blendstocks that met various conversion specifications. For both uncatalyzed and catalyzed fast pyrolysis, all four depots produced blends that met quality and quantity specifications at a cost lower than using a single feedstock.

# 499592 New Approach to Ammonia Pretreatment Integrates Better Feedstock Logistics with Improved Sugar Conversion


**Abstract:**

The implementation of a large scale bioeconomy, where lignocellulosic residues can be used for production of commodity products (fuels, chemicals, materials, food additives, animal feed, etc.) highly depends on the ability of moving large amounts of biomass across long distances. To be able to effectively transport low density materials, such as agricultural residues, it is necessary to apply densification methods (e.g., pelletization). Though densification is expensive, it opens a range of new possibilities for feedstock logistics (e.g., ability to use various transportation methods, safer feedstock storage, etc.) and the ability to implement very large scale lignocellulosic biorefineries, as these require economic transportation of lignocellulosic feedstocks within a wide feedstock production radius. By increasing the size of the biorefinery, it is possible to take advantage of economies of scale and thus, reduce the capital cost per unit of processed biomass. To extract further value from using densified biomass, it is important to explore biomass conversion methods that benefit from having high density feedstock, which can generate higher sugar yields with lower energy inputs.

In this work, we have developed a new ammonia pretreatment strategy called Compacted Biomass with Reduced Ammonia (COBRA) pretreatment. In COBRA pretreatment, liquid ammonia is added to biomass pellets in ratios lower than 1:1, allowing native cellulose I to be converted to cellulose III at relatively low temperatures (~70 °C) and residence times of 3 to 4 hours. As pellets occupy less volume per unit mass relative loose biomass, lower ammonia volumes are required to completely submerge the solid biomass. Thus, as pellets contain low moisture content, the presence of liquid ammonia around the cellulose fibers will allow the formation of a highly digestible cellulose III. As a result, COBRA pretreated biomass can be effectively digested with 60% less enzyme relative to the traditional AFEX pretreatment. Furthermore, as COBRA pretreatment uses densified biomass, it also allows the reduction of reactor size compared to AFEX, even though it uses longer residence times. By using low ammonia loadings, small sized reactors, relatively low temperatures and pressures, low enzyme loading and taking advantage of economies of scale, COBRA pretreatment has the potential to reduce overall processing costs in the biorefinery. Here, we will present a complete optimization of COBRA pretreatment for corn stover, an enzyme cocktail optimization to maximize sugar yields and determination of sugar fermentability using *Saccharomyces cerevisiae* 424A (LNH-ST). In addition, a complete mass balance and process economics will be presented.

Danmarks Tekniske Universitet DTU

# Screening method for selection of biomass feedstock and location in biorefneries

**A. Jaksland,** M. Bertran, J.D. Woodley, R. Gani

2/11 AIChE Annual Meeting 2017

DTU Chemical Engineering
Department of Chemical and Biochemical Engineering

Danmarks Tekniske Universitet DTU

## Outline

- Motivation and Background
- Biomass selection
- One parameter screening method
- Case study
  - Biomass selection
  - Superstructure generation and solution
- Results
- Evaluation in three parameters
- Conclusion

DTU Chemical Engineering
Department of Chemical and Biochemical Engineering

Danmarks Tekniske Universitet

# Motivation

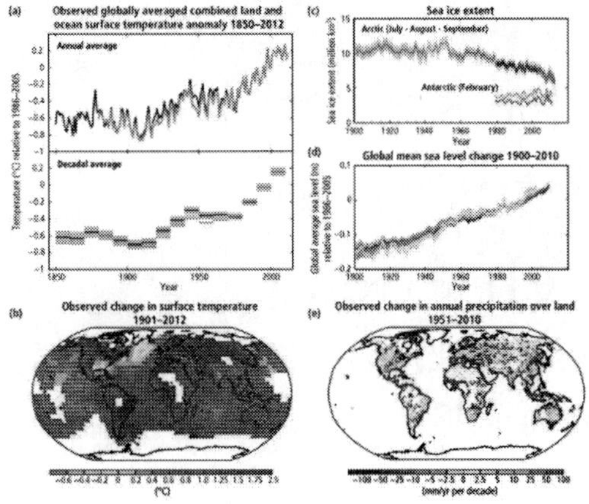


# Background

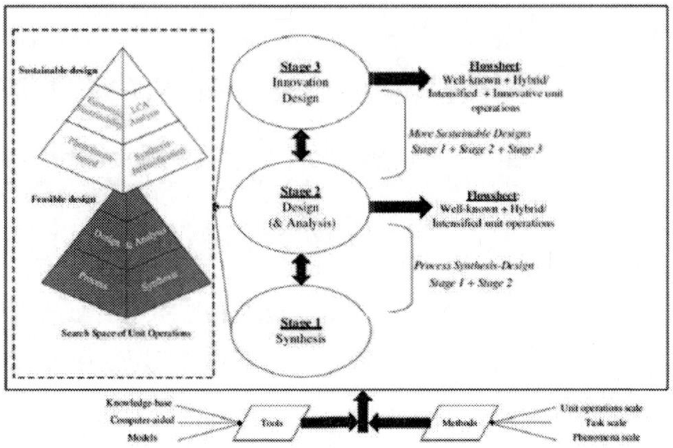


**DTU Chemical Engineering**
Department of Chemical and Biochemical Engineering

1) IPCC. Climate Change 2014: Synthesis Report. Contribution of Working Groups I, II and III to the Fifth Assessment Report of the intergovernmental Panel on Climate Change. 2014.
2) Deenesh K. Babi, Johannes Holtbruegge, Philip Lutze, Andrzej Gorak, John M. Woodley, and Raqul Gani. Sustainable process synthesis-intensication. Computers and Chemical Engineering, 81:218{244, 2015.

Danmarks Tekniske Universitet

# Superstructure Optimisation Framework

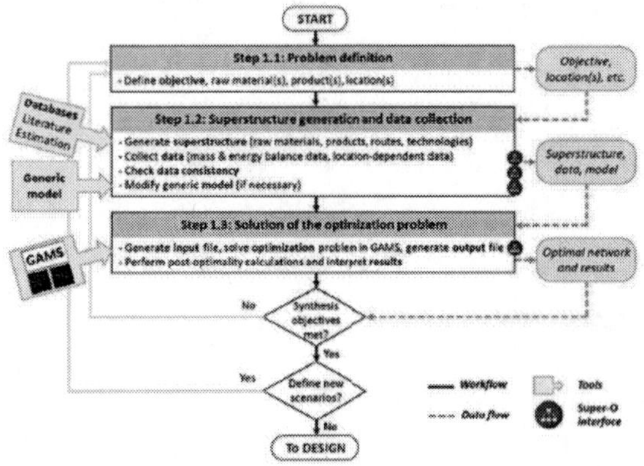


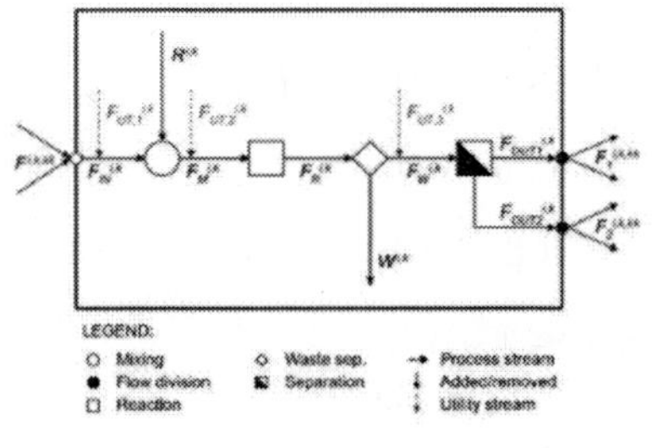


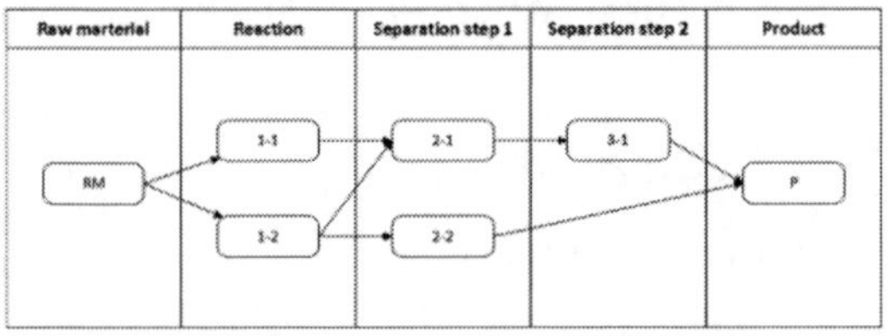


**DTU Chemical Engineering**
Department of Chemical and Biochemical Engineering

3) Maria Ona Bertran, Rebecca Frauzem, Lei Zhang, and Raqul Gani. A Generic Methodology for Superstructure Optimization of Different Processing Networks. Computer Aided Chemical Engineering, 38:685{690, 2016.

# Biomass as Feedstock

| Category | Potential biomass supply in 2050 (EJ/y) |
|---|---|
| (I) Biomass production on surplus agricultural land | 0-988 |
| (II) Biomass production on degraded lands | 8-110 |
| (III) Agricultural residues | 10-32 |
| (IV) Forest residues | 10-16 |
| (V) Animal Manure | 9-25 |
| (VI) Organic waste | 1-3 |
| (VII) Bio-materials | Minus 83-116 |
| Total | 33-1130 |

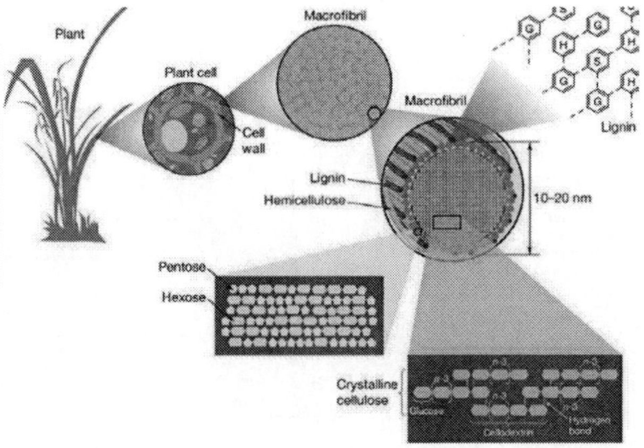




4) Moniqu Hoogwijk, Andre Faaij, Richard van den Broek, G??oran Berndes, Dolf Gielen, and Wim Turkenburg. Exploration of the ranges of the global potential of biomass for energy. Biomass and Bioenergy, 25:119{133, 2003.



# Biomass Selection

Location and specie dependent variables: Composition, density, harvest field yield, transportation cost, at-field price

Three parameters: Harvest field yield, at-field price and transportation cost

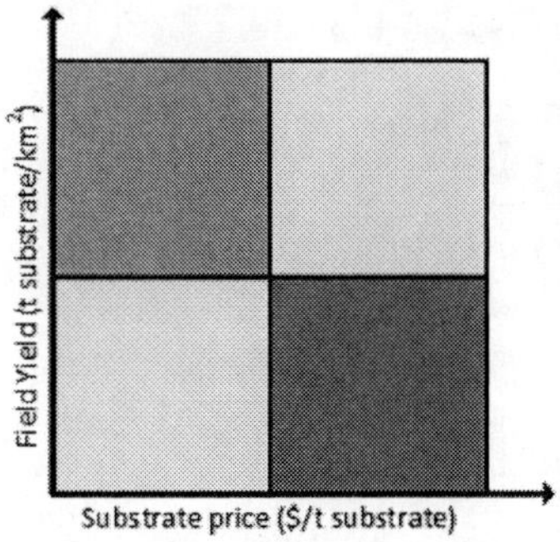


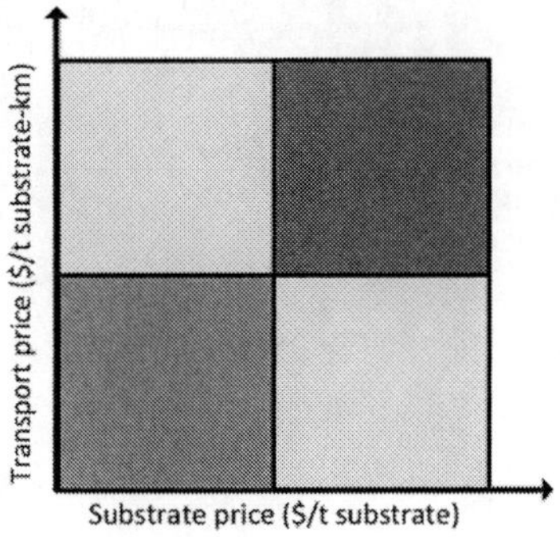


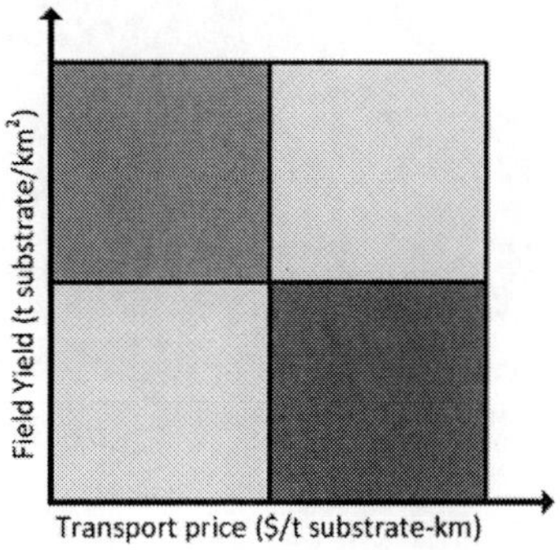

Danmarks Tekniske Universitet DTU

# Biomass Selection

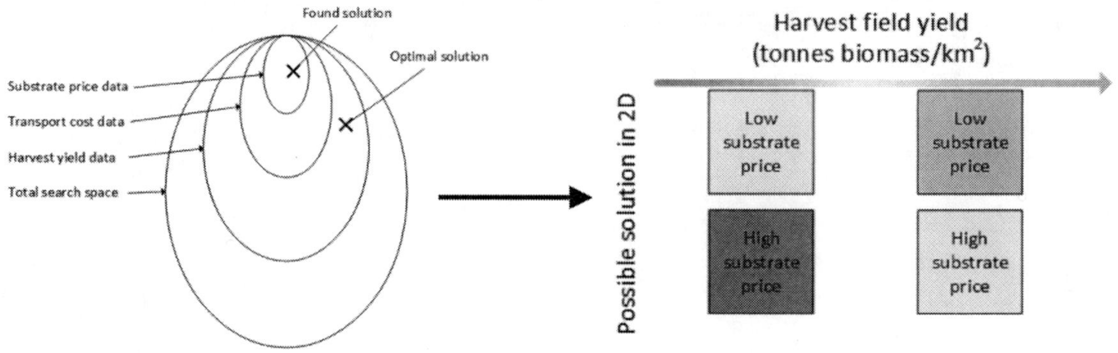


DTU Chemical Engineering
Department of Chemical and Biochemical Engineering

Danmarks Tekniske Universitet DTU

# Biomass Selection

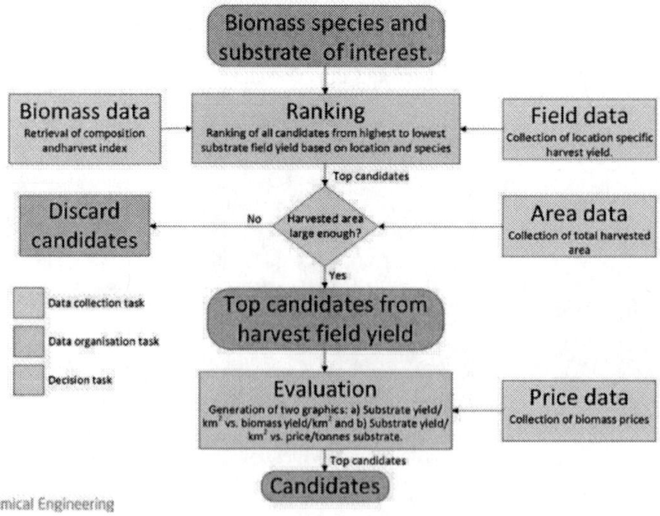


DTU Chemical Engineering
Department of Chemical and Biochemical Engineering

# Case study – Step 1

- 750.000 tonnes biomass/year
- Biomass specie and location: Selected with one parameter screening
- Intermediate: glucose
- Biomass to glucose: Milling, steam explosion and enzymatic hydrolysis
- Products: 1,3-propanediol, 2,3-butanediol, succinic acid, itaconic acid
- Ideal reactions based on redox potential
- Transport and storage cost of biomass not included



Danmarks Tekniske Universitet DTU

# Case study – Step 1

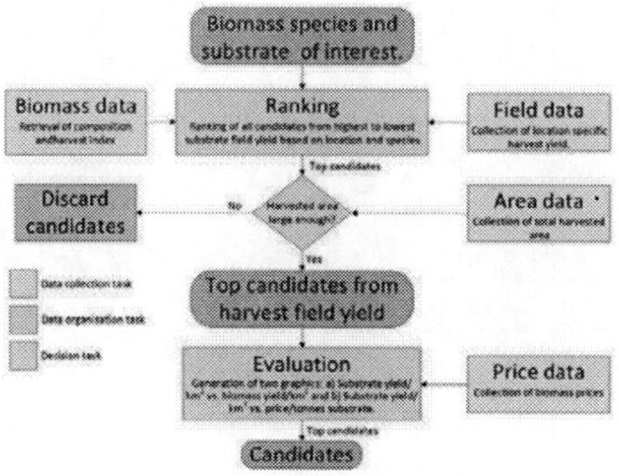


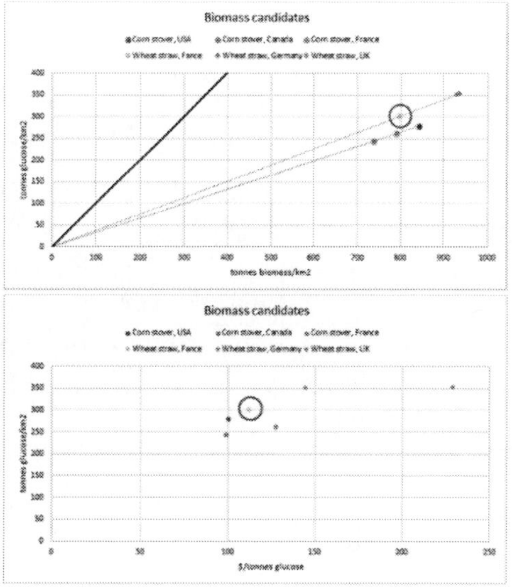


DTU Chemical Engineering
Department of Chemical and Biochemical Engineering

# Case study Step 2 & 3

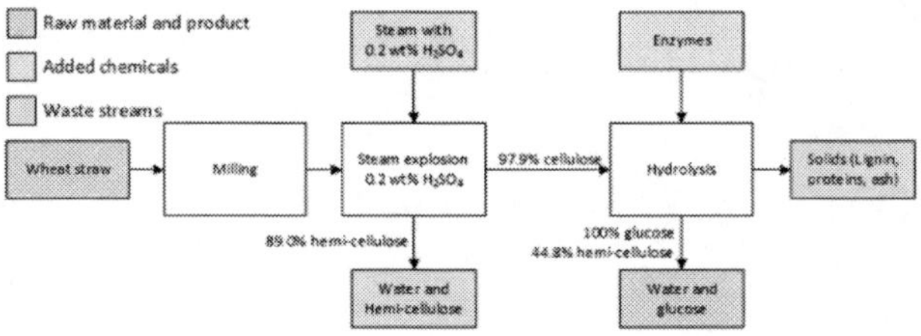


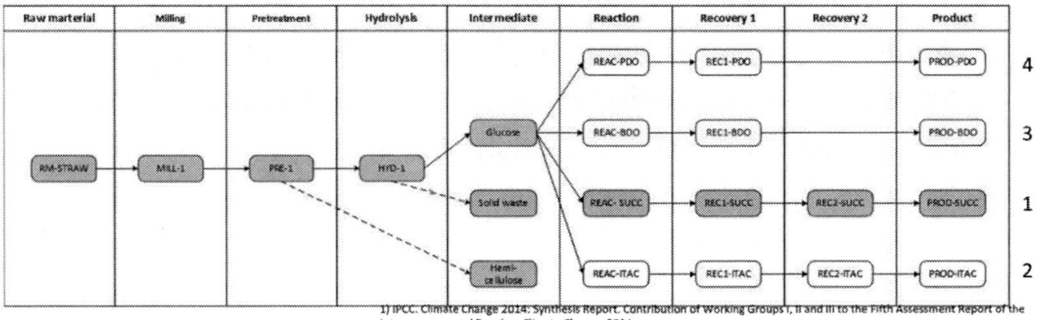


1) IPCC. Climate Change 2014: Synthesis Report. Contribution of Working Groups I, II and III to the Fifth Assessment Report of the intergovernmental Panel on Climate Change. 2014.
2) Deenesh K. Babi, Johannes Holtbruegge, Philip Lutze, Andrzej Gorak, John M. Woodley, and Raqul Gani. Sustainable process synthesis-intensication. Computers and Chemical Engineering, 81:218{244, 2015.





# Results

| Succinic acid | | Price | Amount | Value |
|---|---|---|---|---|
| Raw materials | Wheat straw | 0.045$/kg | 85616.4 | 3852.7 $/hr |
| Added chemicals | Cellulase | 15.67 $/kg | 684,6 | 10728.1 $/hr |
| Products | Succinic acid | 3 $/kg | 29560.7 | 88682.1 $/hr |
| Utilities | Steam | 0.00983 $/MJ | 3680214.4 MJ/hr | 36176.5 $/hr |
| | Cold water | 0.00035 $/MJ | 45830.5 MJ/hr | 16.0 $/hr |
| | Electricity | 0.0295 $/MJ | 3532.6 MJ/hr | 104.1 $/hr |
| | Total | - | 3729577.5 MJ/hr | 36296.7 $/hr |
| Operating profit | | | | 37804.6$/hr |
| Capital investment | | | | 59000635.0 $ |
| Payback time | | | | 1561 hr = 65 days |
| Hourly profit over 20 years | | | | 37467.8 $/hr |

# Results – Comparison

| | Product | Theo. yield | Prac. yield | Operating profit | Capital investment | Payback time |
|---|---|---|---|---|---|---|
| Scenario 1 | 1,3-propanediol | 0.84 g/g | 0.57 g/g | 12186.8 $/hr | 38.6m $ | 132 days |
| | 2,3-butanediol | 0.75 g/g | 0.48 g/g | 26856.5 $/hr | 35.5m $ | 55 days |
| | Itaconic acid | 0.96 g/g | 0.81 g/g | 29053.1 $/hr | 47.0m $ | 67 days |
| | Succinic acid | 1.12 g/g | 0.89 g/g | 37804.6 $/hr | 59.0m $ | 65 days |

# Sensitivity Analysis

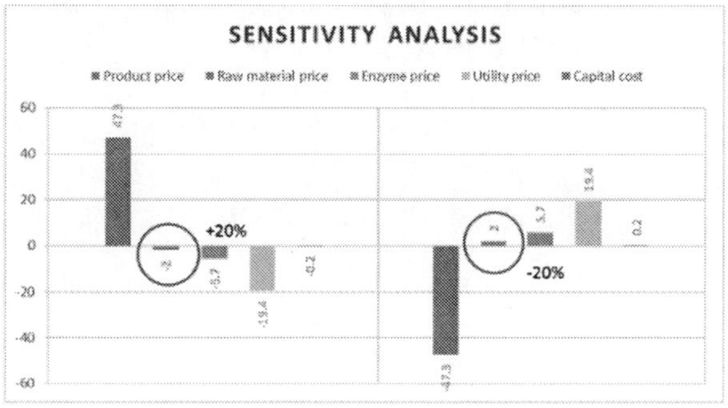






# Biomass Selection in Three Parameters

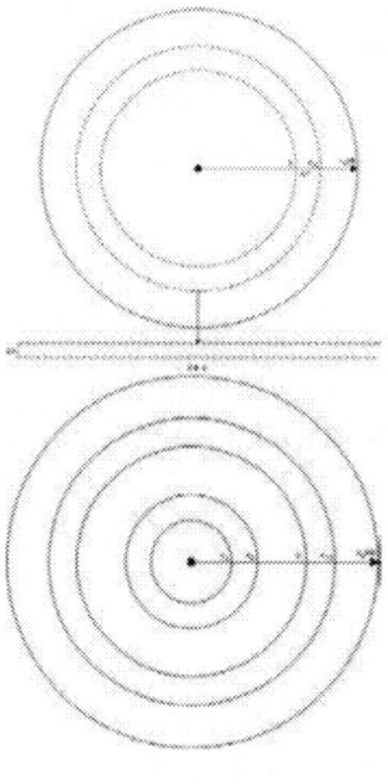

Radius as a function of capacity and yield:

$$R = \sqrt{\frac{C}{\pi Y}}$$

Transportation cost normalised by capacity as a function of radius, truck load and fuel cost:

$$\frac{T}{C} = \frac{\frac{4R^3\pi Y}{3L}F}{C} = \frac{4R^3\pi Y}{3LR^2\pi Y}F = \frac{4R}{3L}F$$

Transportation normalised by capacity as a function of capacity, yield, truck load and fuel cost

$$\frac{T}{C} = \frac{4\sqrt{\frac{C}{\pi Y}}}{3L}F = \frac{4\sqrt{C}}{3L\sqrt{Y\pi}}F$$

| Symbol | Description | Unit |
|---|---|---|
| $Y$ | Biomass harvest field yield | kg biomass/km$^2$ |
| $L$ | Truck load | kg biomass |
| $R$ | Outer diameter | km |
| $F$ | Fuel cost | $/km |
| $C$ | Biomass plant capacity | kg biomass |
| $T$ | Total transport cost | $ |
| $T_i$ | Transport cost of annulus | $ |
| $B$ | At-field biomass price | $/kg biomass |
| $r$ | Radius (axis) | km |
| $\Delta r$ | Differential radius piece | km |

# Biomass Selection in Three Parameters

- Transportation estimate
- France vs. Brazil
  - 796.58 t/km$^2$ vs. 239.27 t/km$^2$
  - Estimation gives: 42.3 $/t vs. 39.35 $/t
  - 941.5 km$^2$ vs. 3134.3 km$^2$

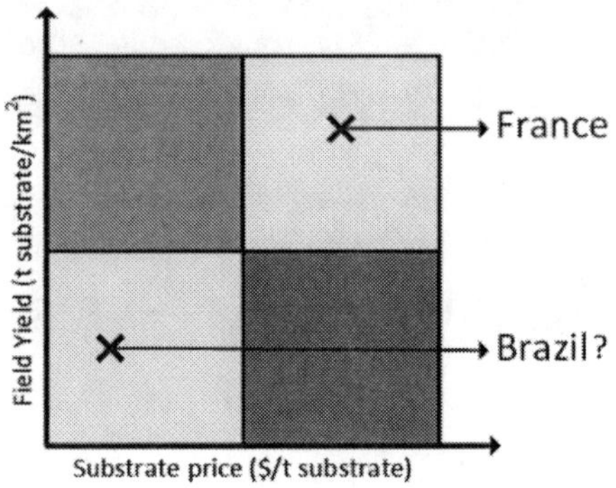


$$\frac{4\sqrt{C}}{3L\sqrt{Y_1\pi}}F + B_1 = \frac{4\sqrt{C}}{3L\sqrt{Y_2\pi}}F + B_2$$





# Conclusion

- Location and specie dependent biomass variables can be reduced to three parameters: Harvest field yield, at-field cost and transportation cost, all based on desired substrate.
- A one-parameter based on harvest field yield has been developed.
- In a case-study a biorefinery network superstructure was developed with feedstock found from the screening.
- An operational profit of 37804.6 $/hr was found for succininc acid. The payback time was 65.
- A sensitivity analysis (+/- 20%) showed that the feedstock price only impacted the operating profit by +/- 2%.
- Expanding the potential selection of biomass to the original three parameters showed that inclusion of transport cost is small compared to at-field price.

Danmarks Tekniske Universitet DTU

# Thank you for listening
# Questions?

**DTU Chemical Engineering**
Department of Chemical and Biochemical Engineering

# AUTHOR INDEX

# AUTHOR INDEX

# AUTHOR INDEX

# AUTHOR INDEX

**AIChE**
120 Wall Street, FL 23
New York, NY 10005-4020

ISBN 978-1-5108-5781-0

CURRAN ASSOCIATES INC.
proceedings.com

# Computational Molecular Science and Engineering Forum

Held at the 2024 AIChE Annual Meeting

San Diego, California, USA
27-31 October 2024

AIChE
The Global Home of Chemical Engineers